T0292956

Advanced Problem Solving Using Maple™

Applied Mathematics, Operations Research, Business Analytics, and Decision Analysis

Textbooks in Mathematics

Series editors:
Al Boggess and Ken Rosen

ORDINARY DIFFERENTIAL EQUATIONS: AN INTRODUCTION TO THE FUNDAMENTALS, SECOND EDITION

Kenneth B. Howell

SPHERICAL GEOMETRY AND ITS APPLICATIONS

Marshall A. Whittlesey

COMPUTATIONAL PARTIAL DIFFERENTIAL PARTIAL EQUATIONS USING MATLAB®, SECOND EDITION

Jichun Li and Yi-Tung Chen

AN INTRODUCTION TO MATHEMATICAL PROOFS

Nicholas A. Loehr

DIFFERENTIAL GEOMETRY OF MANIFOLDS, SECOND EDITION

Stephen T. Lovett

MATHEMATICAL MODELING WITH EXCEL

Brian Albright and William P. Fox

THE SHAPE OF SPACE, THIRD EDITION

Jeffrey R. Weeks

CHROMATIC GRAPH THEORY, SECOND EDITION

Gary Chartrand and Ping Zhang

PARTIAL DIFFERENTIAL EQUATIONS: ANALYTICAL METHODS AND APPLICATIONS

Victor Henner, Tatyana Belozerova, and Alexander Nepomnyashchy

ADVANCED PROBLEM SOLVING USING MAPLE™: APPLIED MATHEMATICS, OPERATIONS RESEARCH, BUSINESS ANALYTICS, AND DECISION ANALYSIS

William P. Fox and William C. Bauldry

https://www.crcpress.com/Textbooks-in-Mathematics/book-series/CANDHTEXBOOMTH

Advanced Problem Solving Using Maple™

Applied Mathematics, Operations Research, Business Analytics, and Decision Analysis

William P. Fox
William C. Bauldry

CRC Press
Taylor & Francis Group
Boca Raton London New York

CRC Press is an imprint of the
Taylor & Francis Group, an **informa** business
A CHAPMAN & HALL BOOK

CRC Press
Taylor & Francis Group
6000 Broken Sound Parkway NW, Suite 300
Boca Raton, FL 33487-2742

Printed on acid-free paper

International Standard Book Number-13: 978-1-138-60187-1 (Hardback)
978-0-429-46962-6 (Ebook)

Visit the Taylor & Francis Web site at
http://www.taylorandfrancis.com

and the CRC Press Web site at
http://www.crcpress.com

To our wives:
Hamilton Dix-Fox and Sue Bauldry

Contents

Preface

From Volume I: *Advanced Problem Solving with Maple™: A First Course*

The study of problem solving is essential for anyone who desires to use applied mathematics to solve real-world problems. We present problem-solving topics using the computer algebra system Maple™ for solving mathematical equations, creating models and simulations, as well as obtaining plots that help us perform our analyses. We present cogent applications of applied mathematics, demonstrate an effective use of a computational tool to assist in doing the mathematics, provide discussions of the results obtained using Maple, and stimulate thought and analysis of additional applications. This book serves as either an introductory course or capstone course, depending on the chapters covered, to start to prepare students to apply the mathematical modeling process by formulating, building, solving, analyzing, and criticizing mathematical models. It is intended for a first course at the sophomore or junior level for applied mathematics or operations research majors that introduces these students to mathematical topics that they will revisit within their major. This text can also be used for a beginning graduate-level course or as a capstone experience for mathematics majors as well as mathematics education majors, as modeling has a much bigger role in secondary education with the newest National Council of Teachers of Mathematics (NCTM) standards. We also introduce many additional mathematics topics that students may study more in depth later in their majors.

Although calculus (either engineering or business) is the prerequisite material, many sections and chapters, especially in Volume II, require multivariable calculus. In addition, the use of linear algebra is required in some chapters. For students without the necessary background, these chapters can be omitted for a specific course. We realize that there are more chapters in this text than could ever be covered within one semester. The increased number of topics and chapters provides flexibility for designing a course appropriate to the background of your students.

Goals and Orientation

This course bridges the study of mathematics topics and the applications of mathematics to various fields of mathematics, science, and engineering. This text affords the student an early opportunity to see how assumptions drive the models, as well as an opportunity to put the mathematical modeling process together. The student investigates real-world problems from a variety of disciplines such as mathematics, operations research, engineering, computer science, business, management, biology, physics, and chemistry. This book provides introductory material to the entire modeling process. Students will find themselves applying the process and enhancing their problem-solving capabilities, becoming competent, confident problem solvers for the twenty-first century. Students are introduced to the following facets of problem solving:

- Creative Problem Solving. Students learn the problem-solving process by identifying the problem to be solved, making assumptions and collecting data, proposing a model (or building a model), testing their assumptions, refining the model as necessary, fitting the model to the data if appropriate, and analyzing the mathematical structure of the model to appraise the sensitivity of the results when the assumptions are not strictly met.
- Problem Analysis. Given a model, students will learn to work backward to uncover the assumptions, assess how well those assumptions fit the scenario, and estimate the sensitivity of the results when the assumptions are not strictly met.
- Problem Research. The students investigate a specific area to gain understanding of behavior and to learn how to apply or extend what has already been created or developed to new scenarios.

Course Content

We introduce problem solving early. Volume I, *Advanced Problem Solving with Maple™: A First Course*, is a typical applied mathematics or introduction to operations research course with topics including ODES, mathematical programming, data fitting with regression, probabilistic problem solving, and simulation. This text, Volume II, contains discrete dynamical systems, both constrained and unconstrained optimization, linear systems, advanced regression, game theory, and multi-attribute decision making.

Organization of Text

Volume I covers introductory topics. Chapter 1 of Volume I is repeated in Volume II, introducing Maple and its basic command structure as well as introducing the problem-solving process. Because the book uses Maple as the tool in mathematical modeling, the chapter provides the foundation or cornerstone of using technology in the modeling process. In Volume I, Chapter 2 introduces ordinary differential equations, and Chapter 3 covers systems of ordinary differential equations. Chapter 4 covers linear, integer, and mixed integer programming as well as the Simplex method. Chapter 5 covers model fitting, concentrating on regression methods to fit data. Chapter 6 covers statistical and probabilistic problem solving; Chapter 7 extends these ideas into Monte Carlo simulations. Scenarios are developed within the scope of the problem-solving process. Student thought and exploration are required.

Volume II covers more advanced topics. Chapter 1, the introduction to problem solving and to Maple, is repeated. Chapter 2 covers discrete dynamical systems, and is complementary to Chapters 2 and 3 of Volume I. Chapters 3 and 4 cover optimization, both constrained and unconstrained, in single-variable and multivariable topics. Chapter 5 deals with solving problems from engineering, economics, and chemistry with linear systems. Chapter 6 continues with regression but introduces more advanced topics such as non-linear regression, logistic regression, and Poisson regression. Chapter 7 covers game theory and relies heavily on linear and non-linear programming methods. Chapter 8 completes Volume II by discussing multi-attribute or multi-criteria decision making with methods such as Data Envelopment Analysis (DEA), Simple Additive Weighting (SAW), Analytic Hierarchy Process (AHP), and Technique of Order Preference by Similarity to the Ideal Solution (TOPSIS), and finishes with investigating methods of choosing weightings.

Student Projects

Student projects form the backbone of this course. In each project, students apply the mathematical modeling process and the mathematical tools they have learned. Each chapter and many sections have collections of student projects. We have seen significant student growth over the course of a semester in project reports from their first project to the final submission. Student projects take time to apply the modeling process, so we typically do not assign more than one project to a student. Most of these projects are designed to be group projects with two or three students working together, although they can be done as individual projects with strong students. COMAP's Mathematical Contest in Modeling[1] provides a rich source of further significant projects.

Technology

Technology is fundamental to serious mathematical modeling. We chose the computer algebra system Maple as our platform for this text; any technology could be used—from graphing and symbolic calculators to spreadsheets.

Emphasis on Numerical Approximations

Numerical solutions techniques are used for solving dynamical systems, for explicative modeling with some numerical analysis approaches, and in optimization search procedures. These are the methods most easily employed in iterative and recursive formulas. Early on, the student is exposed to numerical techniques. We present numerical procedures as iterative and algorithmic.

Focus on Algorithms

All algorithms are provided with step-by-step formats to aid students in learning to do mathematical modeling with these methods and Maple. Examples follow the summary to illustrate an algorithm's use and application. Throughout, we emphasize the process and interpretation, not the rote use of formulas.

Problem Solving and Applications

Each chapter includes examples of models, real-world applications, and problems and projects. Problems are modeled, formulated, and solved within the scenario of the application. These models and applications play an important role in student growth in working in today's complex world.

[1]See https://www.comap.com/undergraduate/contests/matrix/index.html

Exercises

Exercises are provided at the end of each section and chapter so the student can practice the solution techniques and work with the mathematical concepts discussed. Review problems are given at the end of each chapter, some of which combine elements from several chapter sections. Projects are also provided at the end of each chapter to enhance student understanding of the concepts and their application to real-world problems.

Computer Usage

Maple is the computer algebra system used for this text. Tutorial labs are widely available for learning Maple's syntax and command structure. Our emphasis is on providing the student with the ability to use Maple to assist with mathematical modeling. We illustrate graphing in both two and three dimensions. We provide Maple packages *PSM* and *PSMv2* containing programs, functions, and data to use with each volume. The packages are freely available from the Maple Cloud (https://maple.cloud).

Acknowledgments

We need to begin by thanking Frank R. Giordano for being a mentor and choosing to involve each of us in mathematical modeling.

William Fox. I am indebted to my colleagues who taught with me over the years and who were involved with problem and project development: Jack Pollin, David Cameron, Rickey Kolb, Steve Maddox, Dan Hogan, Paul Grimm, Rich West, Chris Fowler, Mike Jaye, Mike Huber, Jack Piccuito, Jeff Appleget, Steve Horton, and Gary Krahn. A special thanks to William Hank Richardson, from Francis Marion University, who assisted in developing and refining many of our Maple programs used within this text.

William Bauldry. I am also indebted to my colleagues who have shared their expertise and wisdom over the years and shaped my approach to modeling, to teaching, and to technology: Wade Ellis, Joe Fiedler, Rich West, Jeff Hirst, Greg Rhoads, and most recently Mike Bossé. The folks at Maplesoft have been incredibly supportive and helpful from the beginnings when I used Maple running as tool in the *Macintosh Programmer's Workshop* in 1987 to today's Maple 2019.

We also wish to thank the great folks at CRC/Taylor & Francis for their help and support with this project.

William P. Fox
William C. Bauldry

1

Introduction to Problem Solving and Maple

Objectives:

(1) Understand the nature of problem solving.

(2) Understand the use of Maple commands.

(3) Understand the Maple Applications Center and its uses.

1.1 Problem Solving

What do we mean by *problem solving*? We interpret this as having a real problem whose understanding and solution requires quantitative analysis and one or more solution techniques using mathematics. To put into context, we say we need a well-defined problem. After we have a well-defined problem, we must brainstorm variables and assumptions that might impact the problem. We build or select a known model and choose a solution technique or combinations of techniques to obtain an answer. We solve and perform sensitivity analysis. We interpret all results, implement, and if necessary, refine the entire process.

In many ways, this process is very similar to the mathematical modeling processes described in other texts: Albright [A2011], Giordano *et al.* [GFH2013], and Meerscheart [M2007] to name a few. Readers may want to examine these texts for a more detailed approach. As a co-author of the Giordano text, my approach is most similar to the approach we describe in that text.

There are four- and five-step processes for problem solving. We present and describe a simple five-step method.

Step 1. Define and understand the problem.

Step 2. Develop strategies to solve the problem. This includes a problem-solving formulation including a methodology to obtain a solution. If data is available, examine the data, plot it, and look for patterns.

Step 3. Solve the problem formulated in Step 2.

Step 4. Perform a self-reflection of your process. You want to make sure the solution answers the problem from Step 1. You also want to ensure the results pass the "commonsense" test. If not go back to Step 2 and reformulate the strategy.

Step 5. If necessary, extend the problem.

We will not concentrate on the modeling portion but on the selection of the model and the solution technique processes including the use of technology in the solution process.

One key point is that our results must pass the "commonsense" test. For example, we were conducting spring-mass experiments in a classroom on the 3rd floor of our mathematics and science building. The simple purpose was to investigate Hooke's Law. The springs were small and the weights varied from a fraction to about 50 grams. After the experiments, we asked the students to calculate the stretch of their spring if it were attached to a seat, and they sat on the seat. Every student found an answer, but none said the spring would most likely break long before it stretched that far.

Let's preview a problem we will see in Volumes I and II. We have data for time (t) and an index (y) from $[0, 100]$. Our plot shows a negative linear trend. We compute the correlation which is -0.94, and is interpreted as a strong negative linear relationship. We use linear regression to build a regression equation which has some very good diagnostics, but one questionable diagnostic from the residual plot. The main goal is predicting the future, which is why the problem is being solved in the first place. The answer for y comes out negative which is not a possible answer for y. So we continue our problem solving and correct the residual plot issue by adding a quadratic term to the regression equation. Again, our diagnostics are all excellent this time. We attempt to use the model to predict, but our answer does not pass the commonsense test as it is too large. A simple plot shows that for the time value in the future we are on the increasing past of the quadratic polynomial. If we cannot use our regression equation then our work is useless. Now, we continue on the nonlinear regression and use an exponential function to fit our data. Finally, not only are all the diagnostics excellent, but our use of the new regression equation passes the commonsense test.

We also believe that in the twenty-first century, technology is a key element in all problem solving. Technology does not tell you what to do, but its use provides insights and the ability to check out possibilities. In this book our technology of choice is the computer algebra system, Maple™.[1]

1.2 Introduction to Maple

Maple is a symbolic computation system or computer algebra system (CAS) that manipulates information in a symbolic or algebraic manner. You can use these symbolic capabilities to obtain exact, analytical solutions to many mathematical problems. Maple also provides numerical estimates to whatever precision is desired when exact solutions do not exist.

Maple 2019 and higher is different from previous versions of Maple. With Maple 2019 you can create profession quality documents, presentations, and custom computational tools. You can access the power of the Maple computational engine through a variety of interfaces: standard worksheet, classical worksheet, command line version, graphing calculator (Windows only), or Maple applications. Although you type in the commands in a very similar manner as in previous versions of Maple, the statement appears in a "pretty print" format on the screen; that is, the statement appears more like typeset mathematics.

Standard Worksheet

This is a full-featured graphical user interface offering features that help to create documents that show all assumptions, the calculations, and any margin of error in your results. You can even hide the computations to focus on problem setup and final results.

Classic Worksheet

The basic worksheet environment works best for older computers with limited memory.

Command-Line Version

Command-line interface, without graphical user interface features, is used for solving very large, complex problems or batch processing.

[1]Maple is a trademark of Waterloo Maple, Inc.

Maplesoft™ Graphing Calculator

The graphical interface to the Maple computational engine allows you to perform simple computations and create customizable, zoomable graphs. The Graphing Calculator is only available in the Windows version.

Maple Applications

The graphical user interface containing windows, textbox regions, and other visual features gives you point-and-click access to the power of Maple. It allows you to perform calculations and plot functions without using the worksheet or command-line interfaces.

Maple's extensive mathematical functionality is most easily accessed through all these interfaces. Previous older versions relied on its advanced worksheet-based graphical interface. A worksheet is a flexible document for exploring mathematical ideas or mathematical alternatives and even creating technical reports.

Experimental mathematical modeling, a natural stepping stone to statistical analysis, has an obvious coupling with computers, which can quickly solve equations, and plot and display data to assist in model test and evaluation. The software computer algebra system, Maple, is a powerful tool to assist in this process. When dealing with real-world problems, data requirements can be immense. When evaluating immigration trends, for example, and the political, social, and economic effects of these trends, thousands of data points are used; in some cases, millions of data points. Such problems cannot be analyzed by hand, effectively or efficiently. The manipulation required to plot, curve fit, and statistically analyze with goodness-of-fit techniques, cannot feasibly be done without the assistance of a computer software system.

The Maple system is easy to learn and can be applied in many mathematical applications. While these demonstrate its versatility, Maple is also an extremely powerful software package. Maple provides over 5000 built-in definitions and mathematical functions, covering every mathematical interest: calculus, differential equations, linear algebra, statistics, and group theory to mention only a few. The statistical package reduces many standard time-consuming statistical questions into one-step solutions, including mean, median, percentile, kurtosis, moments, variance, standard deviation, and so forth. There are many references for Maple, and a short list would include:

- Maple Quick Reference
- Maple Flight Manual
- Maple Language Reference Manual
- Maple Library Reference Manual

- Maple "What's New" Release Notes
- Maple "Getting Started" tutorials and worksheets

This chapter presents a quick review of some basic Maple commands. It is not intended to be a self-contained tutorial for Maple, but will however provide a quick review of the basics prior to more sophisticated commands in the modeling chapters. There are many good references for those new to Maple. Additionally, there is a collection of self-contained tutorials accessed by clicking the "Getting Started" button on Maple's opening screen. See Figure 1.1.

1.3 The Structure of Maple

Maple is an example of a computer algebra system (CAS). It is composed of thousands of commands, to execute operations in algebra, calculus, differential equations, discrete mathematics, geometry, linear algebra, numerical analysis, linear programming, statistics, and graphing. It has been logically designed to minimize storage allocation, while remaining user friendly. Maple allows a user to solve and evaluate complicated equations and calculations, analytically or numerically, such as optimization problems, least square solutions to equations, and solving equations that involve special functions.

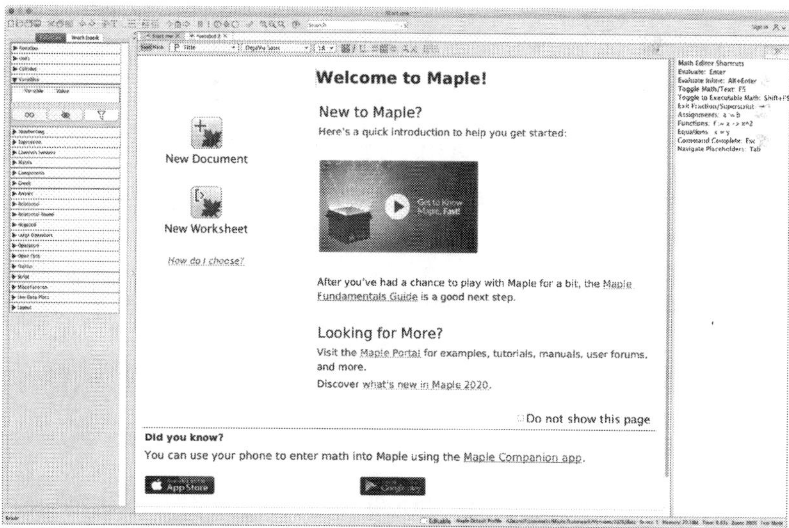

FIGURE 1.1: Maple 2020 Opening Screen

The "New to Maple?" section in the opening screen will lead to videos, tutorials, and links to more information as seen in Figure 1.2.

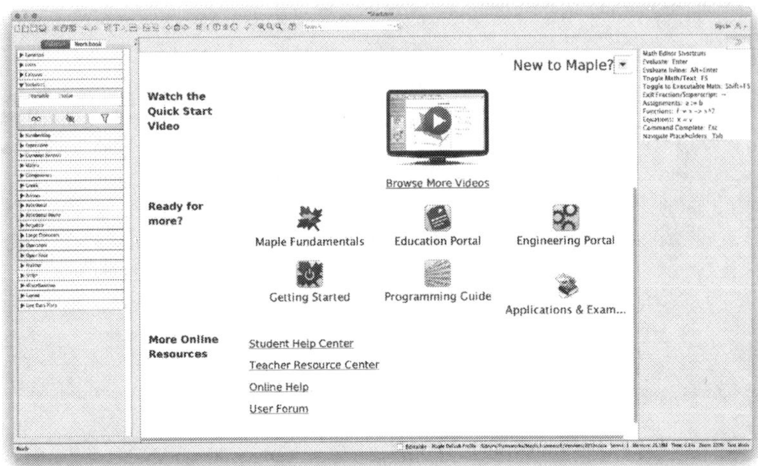

FIGURE 1.2: "New to Maple" Section of Maple 2019 Expanded

Commands

To begin using Maple from the opening screen, we suggest clicking on the "New Worksheet" button shown in Figure 1.1. Enter commands at Maple's *command prompt*: >. See Figure 1.3. A *Worksheet* shows command prompts, while a *Document* does not.

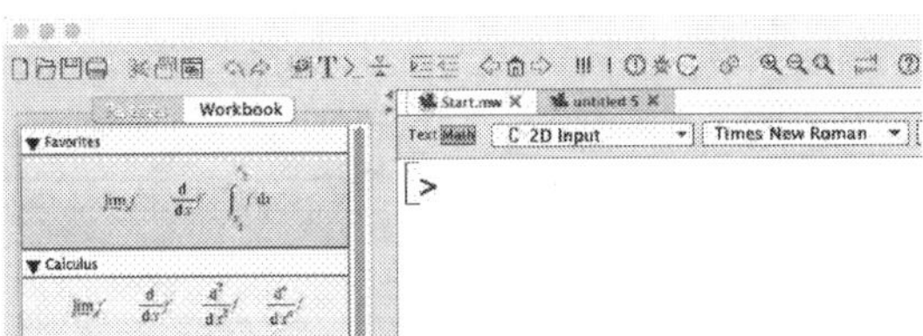

FIGURE 1.3: Maple's Command Prompt

Notation and Conventions

Throughout this book, different types of fonts and styles are used to distinguish between Maple commands, Maple output, and other information. Maple commands are copied directly from Maple 2019 and the output will immediately follow the Maple commands. In the first example below, the variable a has been assigned an expression as its value. Notice that the symbol := is used to indicate this assignment. In the second line of the example, the value of a, the expression, is differentiated with respect to x.

Type the command in either the Worksheet or the Document mode as

`a := 2*x^3` [→] `- 5*x/6.`

(The [→] indicates pressing the "right arrow" cursor key to exit superscript mode.) You will type the statement this way, but the screen version appears as follows:

$$> a := 2 \cdot x^3 - \frac{5 \cdot x}{6}$$

$$a := 2x^3 - \frac{5}{6}x$$

$$> diff(a, x)$$

$$6x^2 - \frac{5}{6}$$

The screen version shows Maple's mathematical interpretation of what you have typed.

1.4 General Introduction to Maple

Maple has many types of windows: the worksheet or document window, help window, 2-D plot window, 3-D plot window, and the animation window. Many "Maple Assistants" use a "maple" window. In this book, the worksheet and help windows are used most often. We provide a brief explanation of each.

The worksheet is where all interaction between Maple and the user occurs. Within a worksheet, commands (input) and text (remarks for clarification) are entered by the user, and results, numerical or symbolic (output and graphics) are produced by Maple. The user may manipulate these interactions to create a flowing document that can be saved by clicking File, then clicking *SaveAs* followed by naming the document. The name of the document can be any word or group of letters and/or numbers, which has a length of less than nine characters. Once a document has initially been saved, it can be retrieved by clicking File, clicking Open, and entering the document name. Then it can be re-saved after modification by clicking File, followed by clicking Save. We will use Worksheet mode rather than Document mode since Worksheets show a command prompt, but Documents do not.

The input and text regions of the worksheet can be modified to change a document, but the graphic and output regions cannot be modified once Maple inserts them into a document. A command must be edited and re-executed to alter its associated output. The input region is identified by the > prompt which precedes all command entries into Maple. (Note: Maple only recognizes the Maple generated > prompt. If the symbol > itself is typed by the user, Maple does not respond to it as an input prompt, but as the "greater than" relation.) The input commands, output characters and text regions are all of different font size and color to assist the user in distinguishing between them. Text regions assist in documentation and explanation of the input/output regions of the document and they may be placed anywhere in a document. Graphic regions, once they are generated by input commands, can be copied and pasted into a worksheet or into another document. Once the graphic is pasted into the worksheet, it can no longer be edited or manipulated. The output regions are generated by the user's input commands and cannot be manipulated once they appear in a document, although the user is allowed to delete these results.

The Maple menu bar is located at the top of the screen on a Macintosh, and immediately below the Maple title in Windows. The menu bar provides easy access and *collocation* (collocation is defined as a sequence of words or terms which co-occur more often than would be expected by chance) of many commonly used options. The menu bar includes File, Edit, View, Insert, Format, Plot, Tools, Window, and Help, much like any software. Enter *"? worksheet,reference,standard Menubar"* for detailed information on all menus. Immediately below the Maple window's title bar (Macintosh) or menu bar (Windows) is the Maple tool bar. The tool bar provides accelerated access to the most commonly used options; see Figure 1.4. Enter *"? Worksheet-Toolbar"* for detailed information.

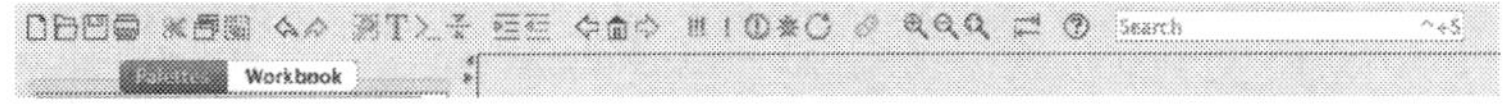

FIGURE 1.4: Maple's Tool Bar

Maple syntax uses either a semicolon or a colon to end a statement. A single command ends with a semicolon implicitly. Multiple statements may be on the same line, but each must have its own colon or semicolon. The colon suppresses output of the command, while the semicolon signals that the results are to be printed to the screen immediately after the enter key is pressed. There is a large set of commands either readily available in Maple memory or stored separately in Maple packages, which assist in more efficient memory storage. Standard commands such as addition and multiplication are *built-in*, not contained in packages. Enter *"? inifcns"* to see the complete list of functions always available.

The Calling Sequence

A "package" is a collection of related definitions and functions that can be brought into a Maple session using the *with* command. The syntax for *with* is

```
> with(<package_name>);
```

When using commands stored in packages, such as graphing commands in the *plots* package, the command "*with(plots)*:" is issued prior to using any of the commands in that package. The *with* command is required only once. Several of the Maple packages will be used in this text. Specifically, *plots* will be used extensively for creating graphs, *LinearAlgebra* for linear algebra operations, and *Statistics*, for statistical and linear regression commands. Other important packages include *DETools* for differential equations and *MultivariableCalculus*, *Optimization*, and *simplex* for optimization.

The Help Command

The Maple help database can provide all the information found in the Maple Library Reference Manual. See Figure 1.5. However, help can be obtained immediately to assist the user in solving problems without leaving the document. Help can be acquired by a number of methods: click on Help in the menu bar, type "help" at the > prompt, or type a "?" at the > prompt. By using the ? or the *help* at the > prompt, the user must type the keyword for the help search. If the specific syntax of a command is in question, for example, the syntax for differentiation, type *? differentiate* at the > prompt. This procedure is perhaps the most convenient one for help on syntax. "Quick Help," the last item on the tool bar, gives quick feedback.

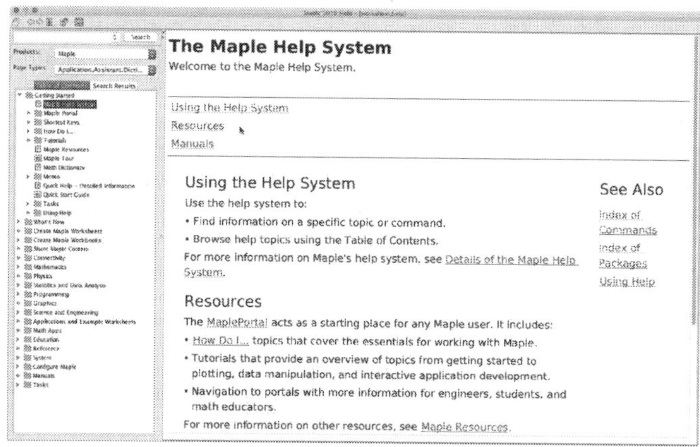

FIGURE 1.5: The Maple Help System

Begin typing a command, say *dif*, then press esc, the escape key. A pop-up menu appears with a list of commands beginning with or related to *dif* appears. Use the mouse to select the desired choice. This technique also works with variables or names that have been defined. Enter and execute *myVariable* := 10. Now type *myVar*, pause, and press esc. Maple fills in the rest of the name. If more than one possibility exists, a pop-up menu will appear with the choices.

Maple's help database also includes a mathematical encyclopedia. Try entering

```
> ? Definition, integral
```

Data Entry

Commonly, a set of data will describe a process. Entering the data is the first step to analyzing the data. Maple provides a convenient method for manually entering data via a list. A list is a group of data to which Maple's many operations are applied. Suppose a group of five college students' weights are known to be 185, 202, 225, 195, and 145. We demonstrate the command required to enter the data into Maple.

```
> weights := [185, 202, 225, 195, 145];
```

$$weights := [185, 202, 225, 195, 145]$$

The use of brackets in the command indicates a list. The brackets will maintain the initial ordering of the data and allows for duplicate values, and may be manipulated by the methods described later in this book. Commas are required between the list elements. Any alphanumeric element may be included in a list; integers, rational numbers, decimals, strings, variable names, and expressions are all allowed. If any data should be missing, a place holder can be entered in its place, such as the letter x. If there was a sixth student in the group with an unknown weight, a symbol could be used to represent the sixth student's weight.

```
> weights := [185, 202, 225, 195, 145, x];
```

$$weights := [185, 202, 225, 195, 145, x]$$

Data Entry and Verification

The Maple lines below illustrate entering, verifying, and naming data pertaining to the length and weight of bass caught during a fishing derby. In subsequent chapters, several models for predicting the weight of a snook fish as a function of length of the fish is suggested. We enter the data in rows. The data is printed to verify correct entry.

```
> length_inches := [12, 14, 12.5, 16, 21.5];
```

$$length\_inches := [12, 14, 12.5, 16, 21.5]$$

```
> weight_oz := [15, 21, 10, 33, 41];
```

$$weight\_oz := [15, 21, 10, 33, 41]$$

Correcting Erroneous Entries

In the above illustration, Maple displayed the snook fish data immediately after its input. This is done to help the user verify that all elements have been entered correctly. Reentry of each command is typically the best method for correcting an erroneous entry with many errors.

Transformation and Functions

As described, the symbol := is used to assign the value on the right-hand side of the statement to the name on the left-hand side of the statement. An example, *x:=3*; assigns the value 3 to x, to unassign x, use the command x='x'. Functions can be defined using the mapping arrow symbol $\rightarrow$. The function can be evaluated numerically or symbolically, we provide a short example.

> $f := x \rightarrow x^2 - 3 \cdot x + 4.5;$

$$f := x \mapsto x^2 - 3x + 4.5$$

> $f(5);$

$$14.5$$

> $f(x-2);$

$$(x-2)^2 - 3x + 10.5$$

A problem solution may suggest transforming a variable. Perhaps we want to transform both x and y by taking natural logarithms (ln) of each. In the two-dimensional case, the model suggests a functional relationship between a dependent and an independent variable. Given values of an independent variable, a function can transform the given data to yield predicted values for the dependent variable. Transforming data requires the understanding of algebraic operations and functions that are used in Maple. Table 1.1 presents the regular arithmetic operations that Maple recognizes.

TABLE 1.1: Maple Arithmetic Functions

Symbol	Operation
+	Addition
−	Subtraction
*	Multiplication
/	Division
^	Exponentiation

Many other functions are also recognized by Maple. Table 1.2 has an abbreviated listing.

TABLE 1.2: Maple Algebraic Functions

Command	Operation
abs	absolute value of real or complex argument
arg	argument of a complex number
ceil	least integer $\geq x$
conjugate	conjugate of a complex number
exp	the exponential function: $e^x = \exp(x) = \sum_{k=0}^{\infty} x^k/k!$
factorial, !	the factorial function, factorial$(n) = n!$
floor	greatest integer $\leq x$
ln	natural logarithm (with base $e = 2.718...$)
log[b]	logarithm to arbitrary base b
log 10	log to the base 10
max, min	maximum/minimum of a list of real numbers
RootOf	function for expressing roots of algebraic equations

We illustrate by taking the natural logarithm of our length and weight data.

$> ln\_length := map(x \to evalf(\ln(x)), length\_inches);$

$$ln\_length := [2.484906650, 2.639057330, 2.525728644, 2.772588722, 3.068052935]$$

$> ln\_weight := map(x \to evalf(\ln(x)), weight\_oz);$

$$ln\_weight := [2.708050201, 3.044522438, 2.302585093, 3.496507561, 3.713572067]$$

Some other functions that are available will be discussed in a later chapter when considering statistical operations that may be performed on columns of data: sums, mean, standard deviation, and so forth. Table 1.3 presents a few examples of data transformations, with the Maple commands.

TABLE 1.3: Examples of Maple Commands in the Worksheet

Expression	Typed Command
x^2	$x\hat{}2$
$2x^2 + 2.5 + 9$	$2 * x\hat{}2 + 2.5 * x + 9$
$(x^2 + 2)^{0.5}$	$(x\hat{}2 + 2)\hat{}0.5$

Column Operations

In this section, the operations that can be performed directly on worksheet columns are present. The operations require the *LinearAlgebra* package to be loaded prior to use. The following demonstrates these commands with six examples using two columns of data; $c1 := [1, 2, 3, 4]$ and $c2 := [5, 6, 7, 8]$.

$> with(LinearAlgebra) :$

$> c1 := \langle 1|2|3|4 \rangle;$

$$c1 := [1, 2, 3, 4]$$

$> c2 := \langle 5|6|7|8 \rangle;$

$$c2 := [5, 6, 7, 8]$$

Summing the vectors $c1$ and $c2$

$> VectorAdd(c1, c2);$

$$[6, 8, 10, 12]$$

$> c1 + c2;$

$$[6, 8, 10, 12]$$

To add a constant to the entries of $c1$, use '+~'

$> 10 +\sim c1;$

$$[11, 12, 13, 14]$$

To multiply $c2$ by constant, use either

$> VectorScalarMultiply(c2, 0.5);$

$$\begin{bmatrix} 2.50000000000000000 & 3.0 & 3.50000000000000000 & 4.0 \end{bmatrix}$$

$> 0.5 \cdot c2;$

$$\begin{bmatrix} 2.5 & 3.0 & 3.5 & 4.0 \end{bmatrix}$$

To apply a function to the elements of a vector use *map* or '~'.

$> map(\ln, c1);$
$evalf(\%);$

$$\begin{bmatrix} 0 & \ln(2) & \ln(3) & 2\ln(2) \end{bmatrix}$$

$$\begin{bmatrix} 0.0 & 0.6931471806 & 1.098612289 & 1.386294361 \end{bmatrix}$$

$> \ln\sim(c2);$

$$\begin{bmatrix} \ln(5) & \ln(6) & \ln(7) & 3\ln(2) \end{bmatrix}$$

$> map(x \to x^2, c2);$

$$\begin{bmatrix} 25 & 36 & 49 & 64 \end{bmatrix}$$

Arrays and Matrices

Arrays and matrices are structured devices used to store and manipulate data. An array is a specialization of a table; a matrix is a two-dimensional array. Both *array* and *matrix* are part of the linear algebra package, and require the *with(LinearAlgebra)*: command prior to use. Table 1.4 presents a few examples of the use of the *Array* and *Matrix* commands.

TABLE 1.4: Array and Matrix Commands

Command	Output
$with(LinearAlgebra)$:	(load the *LinearAlgebra* package)
$a := Array([1, 2, 2, 3, 4]);$	$a := [\ 1\ 2\ 2\ 3\ 4\ ]$
$b := Array([[1, 2], [3, 5]]);$	$b := \begin{bmatrix} 1,2 \\ 3,5 \end{bmatrix}$
$c := Vector[row]([9, 3, 1, 8, 3]);$	$c := [\ 9\ 3\ 1\ 8\ 3\ ]$
$d := Matrix([[7, 3], [2, 3]]);$	$d := \begin{bmatrix} 7,3 \\ 2,3 \end{bmatrix}$

Saving and Printing a Worksheet

To start Maple from Windows or MacOS, launch Maple by double-clicking on the Maple icon. Once Maple has been started, it will automatically open an empty worksheet, with a flashing cursor to the right of a character prompt >. To save the worksheet, click on File and then click on *SaveAs* and then specify a name for the worksheet. Once the worksheet has been named and saved, click on File and then click on Save to re-save the worksheet. To open a previously saved worksheet, click on File and then click on Open, then specify the name of the worksheet.

After re-opening a document, the document will contain the commands and display the results, but not have any values defined. After a command is executed, the result is stored in memory. If the document is closed and then reopened, Maple recovers the commands, but does not recall the results.

After the completion of a document, involving commands and results, the document can be printed by choosing File ▶ Print (which calls up the standard printing dialog).

To quit Maple, choose File ▶ Exit (Windows) or Maple 2019 ▶ Quit (Macintosh); that is, choose Exit from the File menu (Windows) or Quit from the Maple 2019 menu (Macintosh). Saving your work is prompted when quitting

Maple. (In the Command-line version type "quit," "done," or "stop" at the Maple prompt. CAUTION: these are the only commands that do not require a trailing semicolon in the Command-line version; there is no opportunity to save your work when using these commands.)

Procedures

The **proc** command is very useful. The following comes directly from the Maple Help page in Maple 2019 and explains and provides an example for a procedure.

Procedures

Calling Sequence	Evaluation Rules
Parameters	Notes
Description	Examples
Implicit Local Variables	Details
The Operands of a Procedure	

Calling Sequence

proc (parameterSequence) :: returnType; local localSequence; global globalSequence; option optionSequence; description descriptionSequence; uses usesSequence; statementSequence end proc;

Parameters

parameterSequence	- formal parameter declarations
returnType	- (optional) assertion on the type of the returned value
localSequence	- (optional) names of local variables
globalSequence	- (optional) names of global variables used in the procedure
optionSequence	- (optional) names of procedure options
descriptionSequence	- (optional) sequence of strings describing the procedure
usesSequence	- (optional) names of modules or packages the procedure uses
statementSequence	- statements comprising the body of the procedure

Description

- A procedure definition is a valid expression that can be assigned to a name. That name may then be used to refer to the procedure in order to invoke it in a function call.
- The parenthesized **parameterSequence**, which may be empty, specifies the names and optionally the types and/or default values of the procedure's parameters. In its simplest form, the **parameterSequence** is just a comma-separated list of symbols by which arguments may be referred to within the procedure.
- More complex parameter declarations are possible in the **parameterSequence**, including the ability to declare the type that each argument must have, default values for each parameter, evaluation rules for arguments, dependencies between parameters, and a limit on the number of arguments that may be passed. See Procedure Parameters for more details on these capabilities.
- The closing parenthesis of the **parameterSequence** may optionally be followed by **::**, a **returnType**, and a **;**. This is *not* a type declaration, but rather an assertion . If **kernelopts(assertlevel)** is set to **2**, the type of the returned value is checked as the procedure returns. If the type violates the assertion, then an exception is raised.
- Each of the clauses **local localSequence;**, **global globalSequence;**, **option optionSequence;**, **description descriptionSequence;**, and **uses usesSequence;** is optional. If present, they specify respectively, the local variables reserved for use by the procedure, the global variables used or modified by the procedure, any procedure options, a description of the procedure, and any modules or packages used by the procedure. These clauses may appear in any order.
- Local variables that appear in the **local localSequence;** clause may optionally be followed by **::** and a **type**. As in the case of the optional **returnType**, this is not a type declaration, but rather an assertion. If **kernelopts(assertlevel)** is set to **2**, any assignment to a variable with a type assertion is checked before the assignment is carried out. If the assignment violates the assertion, then an exception is raised.
- A global variable declaration in the **global globalSequence** clause cannot have a type specification.
- Several options that affect a procedure's behavior can be specified in the **option optionSequence;** clause. These are described in detail on their own page.
- The **description descriptionSequence;** clause specifies one or more lines of description about the procedure. When the procedure is printed,

this description information is also printed. Even library procedures, whose body is generally elided when printing, have their description (if any) printed. The **descriptionSequence** is also used when information about the procedure is printed by the Describe command.

- The optional **uses usesSequence;** clause is equivalent to wrapping the **statementSequence** with a use statement. In other words,
  proc ... uses LinearAlgebra; ... end proc
  is equivalent to:
  proc ... use LinearAlgebra in ... end use; end proc
- The **statementSequence** consists of one or more Maple language statements, separated by semicolons (;), implementing the algorithm of the procedure.
- A procedure assigned to a name, **f**, is invoked by using **f(argumentSeq)**. See Argument Processing for an explanation of argument passing.
- The value of a procedure invocation is the value of the *last* statement executed, or the value specified in a return statement.
- In both 1-D and 2-D math notation, statements entered between **proc** and **end proc** must be terminated with a colon (:) or semicolon (;).

Implicit Local Variables

- For any variable used within a procedure without being explicitly mentioned in a **local localSequence;** or **global globalSequence;** the following rules are used to determine whether it is local or global:
  The variable is searched for amongst the locals and globals (explicit or implicit) in surrounding procedures, starting with the innermost. If the name is encountered as a parameter, local variable, or global variable of such a surrounding procedure, that is what it refers to.
  Otherwise, any variable to which an assignment is made, or which appears as the controlling variable in a 'for' loop, is automatically made local.
  Any remaining variables are considered to be global.
- **Note:** Any name beginning with **_Env** is considered to be an environment variable, and is not subject to the rules above.

The Operands of a Procedure

- A Maple procedure is a valid expression like any other (e.g., integers, sums, inequalities, lists, etc.). As such, it has sub-parts that can be extracted using the op function. A procedure has eight such operands:

  op 1 is the **parameterSequence**,

  op 2 is the **localSequence**,

  op 3 is the **optionSequence**,

op 4 is the remember table,

op 5 is the **descriptionSequence**,

op 6 is the **globalSequence**,

op 7 is the lexical table (see note below), and

op 8 is the **returnType** (if present).

- Any of these operands will be NULL if the corresponding sub-part of the procedure is not present.
- **Note:** The lexical table is an internal structure used to record the correspondence between undeclared variables and locals, globals, or parameters of surrounding procedures. It does not correspond to any part of the procedure as written.

Evaluation Rules

- Procedures have special evaluation rules (like tables) so that if the name **f** has been assigned a procedure, then:
  f evaluates to just the name **f**,
  eval(f) yields the actual procedure, and
  op(eval(f)) yields the sequence of eight operands mentioned above (any or all of which may be **NULL**).
- Within a procedure, during the execution of its **statementSequence**, local variables have *single level evaluation.* This means that using a variable in an expression will yield the current value of that variable, rather than first evaluating that value. This is in contrast to how variables are evaluated outside of a procedure, but is similar to how variables work in other programming languages.

Notes

- Remember tables (option remember) should not be used for procedures that are intended to accept mutable objects (e.g., rtables or tables) as input, because Maple does not detect that such an object has changed when retrieving values from remember tables.

Examples

```
> lc := proc( s, u, t, v )
    description "form a linear combination of the arguments";
    s · u + t · v;
    end proc;
    lc := proc(s, u, t, v)
          description "form a linear combination of the arguments";
          s * u + t * v
      end proc
```

```
> lc(π, x − I, y)
                         π x − I y
> Describe(lc)
# form a linear combination of the arguments
lc( s, u, t, v )
> lc
                            lc
> eval(lc)
  proc (s, u, t, v)
        description "form a linear combination of the arguments";
        s * u + t * v
   end proc
> op(1, eval(lc))
                          s, u, tv
> addList := proc(a::list, b::integer)::integer;
      local x,i,s;
      description "add a list of numbers and multiply by a constant";
      x:=b;
      s:=0;
      for i in a do
         s:=s+a[i];
      end do;
      s:=s*x;
    end proc;
  addList := proc(a :: list, b :: integer) :: integer;
  localx, i, s;
  description)"add a list of numbers and multiply by a constant";
  x := b; s := 0; for iin ado s := s + a[i]end do; s := s · x;
  end proc
> sumList := addList([1, 2, 3, 4, 5], 2)
                       sumList := 30
```

Details

For details on defining, modifying, and handling parameters, see Procedure Parameters.

See Also

_nresults, assertions, Function Calls, kernelopts, Last-Name Evaluation, Procedure Options, ProcessOptions, procname, Reading and Saving Procedures, remember, return, separator, Special Evaluation Rules, use

We add an example procedure for using Newton's Method to find the roots of a differentiable function. First, recall the iterating formula for Newton's root-finding algorithm:

$$x_{new} = x_{old} - \frac{f(x_{old})}{f'(x_{old})}$$

We iterate the formula, finding new values of x_{new}, until the absolute difference $|x_{new} - x_{old}| < tolerance$ or $|f(x_{new})| < tolerance$.

The Maple procedure: Newton's Method for Finding Roots of a polynomial.

Assumptions:

1. The function must be differentiable.
2. You have a fair guess at a starting point x_0 for the method. We suggest plotting the function and estimating a value near the desired root.

Algorithm:

1. Pick a tolerance, *tol* (*tol* is small), and a maximum number of iterations, $MaxN$, allowed.
2. Pick an initial value *xold*.
3. Iterate $xnew := xold - f(xold)/f'(xold)$.
4. Stop when either $|xnew - xold| < t$ or $|f(x(new))|$ approximately equals 0.

Maple Procedure:

```
> Newton := proc(f, x0, tol, MaxN)
    local df, xold, x_new, i;
    df := D(f);
    xold := x0;
    print(xold);
    for i from 1 to MaxN do
      xnew := xold - f(xold)/f'(xold);
      if |f(xnew)| < tol or |xnew - xold| < tol then
        return(xnew);
        end if;
      xold := xnew;
      print(xold);
      end do;
    return(cat(MaxN, " iterate "), xold);
    end proc :
```

Let's consider an example.

$> f := x \to (x-5)^2 - 3;$

$$f := x \mapsto (x-5)^2 - 3;$$

We plot the function to be able to estimate the roots for the procedure.

$> plot(f(x), x = -1..10, thickness = 3, color = black);$

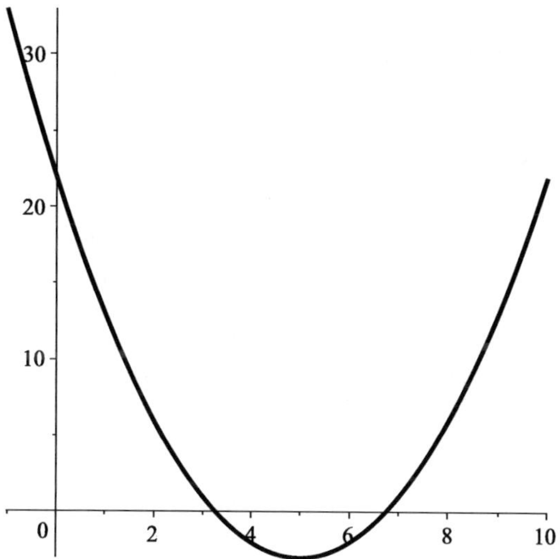

$> Newton(f, 2.0, 10^{-9}, 30);$

2.0
3.000000000
3.250000000
3.267857143
3.267949190
3.267949192
3.267949192

$> f(3.267949192);$

$$1.\ 10^{-9}$$

$> Newton(f, 7.0, 10^{-9}, 30);$

7.0
6.750000000
6.732142857
6.732050810
6.732050808
6.732050808

$> f(6.732050808);$

$$1.\ 10^{-9}$$

We found approximations to the two roots as $x = 3.2679449192$ and $x = 6.732050808$.

We will often create **proc** s in both volumes of this text to solve problems using specialized techniques.

A Quick Review of Key Commands

Assignments and Basic Mathematics

For example, let's compute $(1.2^{4.1} + 4.3(9.8))/34$.

> $(1.2^{4.1} + 4.3 \cdot (9.8))/34;$

$$1.301522146$$

All Maple statements are entered after the > prompt. Again note that Maple commands end with a semicolon; in the document interface, semicolons are optional. Maple output, printed in blue, is centered in the page.

To assign a label or name to a number or an expression, we use :=. For example, let's use two Maple statements to assign a the value 11.5 and b the value 9.

> $a := 11.5;$

$$a := 11.5$$

> $b := 9;$

$$b := 9$$

We can enter multiple commands on the same line or in the same input cell. If we separated the two commands with a colon instead of a semicolon, only the second command would be displayed in the output even though both commands were executed. A colon suppresses output.

> $a := 11.5 : b := 9;$

$$b := 9$$

Once the assignments have been made we can perform arithmetic operations. For example, let's compute $a^2 + b^3$, $a^2 \cdot b^3$, and $\sqrt{a^2 + b^3}$.

> $a^2 + b^3;$
> $a^2 \cdot b^3;$
> $sqrt(a^2 + b^3);$

$$861.25$$
$$96410.25$$
$$29.34706118$$

A very useful command is *evalf*. This command produces the decimal equivalent of a given expression. Its mnemonic is *eval*uate as *f*loating point (decimal).

> $c := evalf(a^2/b^3);$

$$0.1814128944$$

For expressions, we assign y using the same assignment operator :=.

> $y := x^2 + 21.6 \cdot x - 1$

$$y := x^2 + 21.6x - 1$$

To evaluate this type of expression for specific values of x, we use the *subs* command (substitution) or the *eval* command (evaluate). For example, we want to substitute 3 for x.

> $subs(x = 3, y);$

$$72.8$$

> $eval(y, x = 3);$

$$72.8$$

To use functional notation, such as $f(3)$, we must start with a different form of assignment. To create a function assignment f we use the *arrow operator* as follows. (Type '–', then '>' for the arrow operator; the image changes to '→'.)

> $f := x \to x^2 + 21.6 \cdot x - 1;$

$$f := x \mapsto x^2 + 21.6x - 1$$

> $f(3);$

$$72.8$$

We can even substitute variables such as $(x + h)$ for x;

> $f(x + h);$

$$(x + h)^2 + 21.6x + 21.6h - 1$$

The two forms, expressions (objects) and functions (operations), are very important in both programming and plotting as we shall see later. Creating a function from an expression is easy using the *unapply* command.

> $y := x^3 + 3 \cdot \cos(x) - 4;$

$$x^3 + 3\cos(x) - 4$$

> $f := unapply(y, x);$

$$f := x \mapsto x^3 + 3\cos(x) - 4$$

> $f(x);$

$$x^3 + 3\cos(x) - 4$$

> $f(2);$

$$4 + 3\cos(2)$$

Maple can easily handle functions of more than one variable. For example, consider a surface area defined by $\pi(x^2y^3 + 3)$. Suppose we want to evaluate the surface area at the point $(2, 5)$.

> $s := (x, y) \rightarrow \text{Pi} \cdot (x^2 \cdot y^3 + 3)'$

$$s := (x, y) \mapsto \pi(x^2 \cdot y^3 + 3)'$$

> $s(2, 5);$

$$503\,\pi$$

> *evalf*(%)

$$1580.221105$$

The expression *evalf*(%) contains the symbol '%' that means insert the result of the last expression evaluated. *(This may not be the result just above in your Maple document if you've re-executed another statement.)*

Algebra and Calculus

Let's return to our expression $y = x^2 + 21.6x - 1$.

> $y = x^2 + 21.6x - 1;$;

$$y = x^2 + 21.6x - 1$$

Let's factor y. We can use the *factor* command or the *solve* command. The *solve* command has more utility.

> $factor(y);$

$$(x + 21.64619749)(x - 0.04619749036)$$

> $solve(y = 0);$

$$0.04619749036, -21.64619749$$

Let's consider the function $q(x) = ax^2 + bx + c$. We use the *solve* command and obtain the result:

> *restart*;

> $q := x \rightarrow ax^2 + bx + c$

$$q := x \mapsto ax^2 + bx + c$$

> $solve(q(x) = 0, x);$

$$\frac{-b + \sqrt{-4ac + b^2}}{2a}, \frac{-b + \sqrt{-4ac + b^2}}{2a}$$

Note the results are the quadratic formula. We also used the command *restart*. A *restart* forgets all previous assignments to f, a, b, c, and anything else we defined.

In calculus, we can differentiate and integrate in one and many variables. The commands for differentiation and integration are:

diff & *Diff*: differentiation or partial differentiation and 'indefinite' differentiation

int & *Int*: definite and indefinite integration

For example, let's differentiate an expression $y = 2x^2 + 24.1x - 1$, and then find the area under the curve in general and from $x = 1$ to $x = 4$.

> $y := 2x^2 + 24.1 \cdot x - 1$

$$y := 2x^2 + 24.1x - 1$$

> $Diff(y, x)$

$$\frac{d}{dx}\left(2x^2 + 24.1x - 1\right)$$

> $diff(y, x)$

$$4x + 24.1$$

> $Int(y, x)$

$$\int \left(2x^2 + 24.1x - 1\right) dx$$

> $int(y, x)$

$$0.6666666667x^3 + 12.05000000x^2 - x$$

> $Int(y, x = 1..4)$

$$\int_1^4 \left(2x^2 + 24.1x - 1\right) dx$$

> $int(y, x = 1..4)$

$$219.7500000$$

Note the difference between *diff* and *Diff*. The capital letters indicate the "inert form" of the command.

Often, we want to find critical points of the first derivative (where $y' = 0$). We can use the *solve* command as follows,

> $solve(diff(y, x) = 0, x);$

$$-6.025000000$$

Plotting and Graphs

Maple has an extremely detailed and developed *plot* command, which provides graphs in both two and three dimensions. For modeling purposes, 2-D plots will typically be used. We suggest loading the *plots* package, via *with(plots)*, prior to plotting. The syntax for *plot* is *plot(y, hr, vr, options)*, where y is the expression to be plotted, *hr* is the horizontal range and *vr* is the vertical range. Additionally, many other options can be added after the vertical range to control a variety of items. Table 1.5 presents a short list of options (defaults values are in the third column). See the Maple Help topic "plot/options" for the full list.

TABLE 1.5: Plot Command Options

Option	Description	Default
scaling =	*constrained* or *unconstrained*	*unconstrained*
style =	*point*, *line*, *patch* or *patchnogrid*	*line*
title =	"*a title*"	no title
thickness =	0, 1, 2, or 3	1
axes =	*framed*, *boxed*, *normal* or *none*	*normal*
view =	[*xmin..xmax*, *ymin..ymax*]	entire graph

Continuous Plots

The command, *plot(y):* will generate a 2-D plot of y with a default horizontal range of $-10..10$ and a vertical range that shows the curve. No other command information is required; however, the axes ranges and options may be specified to generate a specific plot. The default range can be specified as a finite range or an infinite range. By constraining the scale, equal units occur in both the x and y directions. However, a plot is generally easier to see when the scale is unconstrained, although it would be distorted (e.g., a circle would appear as an ellipse). Maple automatically scales the axes to spread the data over as large a space as possible, but this procedure does not imply that the area of interest will be plotted most effectively. As a result, the *view* option must be employed to ensure that the correct portion of the plot is best displayed. To demonstrate the plot command with a variety of options, a few examples are provided of $\sin(x)$ below, the first with the default ranges, the second with an infinite range.

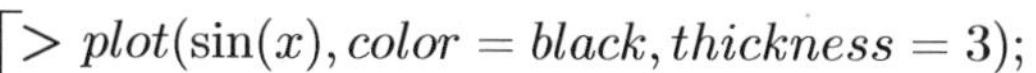

> $plot(\sin(x), color = black, thickness = 3);$

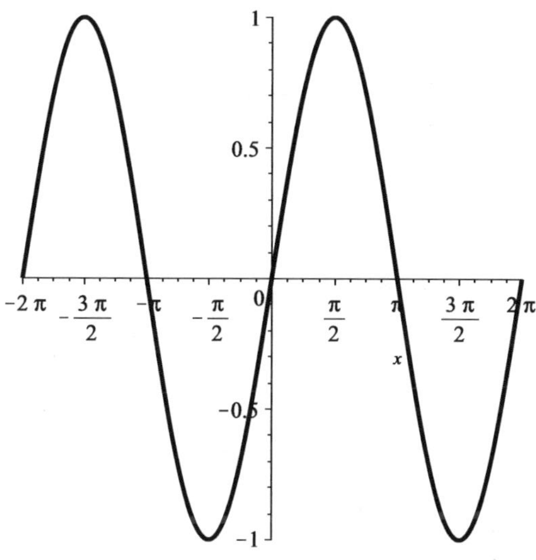

> $plot(\sin(x), x = 0..\text{infinity}, color = black, thickness = 3);$

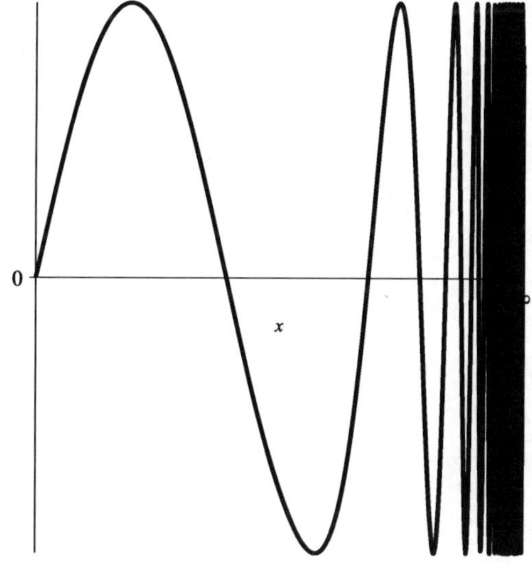

Scatterplots

The previous plots demonstrate functions with continuous x-values; either defaulted to $-10..10$ or selected by the user. However, Maple can also plot discrete sets of data. Using the data provided in Table 1.6, we present an example of plotting the ages of five people versus their respective weights.

TABLE 1.6: Age-Weight Data

Age (years)	1	5	13	17	24
Weight (lbs)	15	40	90	160	180

> $age := [1, 5, 13, 17, 24]:$
> $wt := [15, 40, 90, 160, 180]:$

> $agewt := \{seq([age[k], wt[k]], k = 1..5)\}$

$$agewt := \{[1, 15], [5, 40], [13, 90], [17, 160], [24, 180]\}$$

> $plot(agewt, style = point, symbol = diamond, symbolsize = 14)$

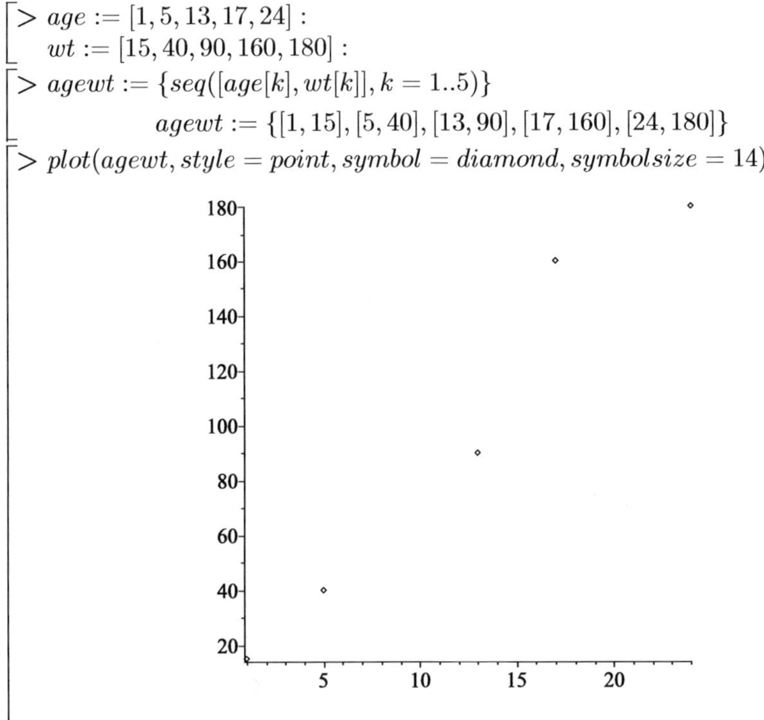

We could also have used

> $with(plots):$
> $pointplot(agewt, symbol = diamond, symbolsize = 14);$

Multiple Plots

It is also an option to plot multiple functions on one set of axes. One method requires that both functions have the same domain. The second method uses the *display* command which requires the *plots* package be loaded via *with(plots):*. This method does not restrict the domain, as presented below.

> $myOptions := (scaling = constrained, axes = framed, thickness = 2):$
> $plot([\sin(x), x^{0.5}], x = 0..5, myOptions);$

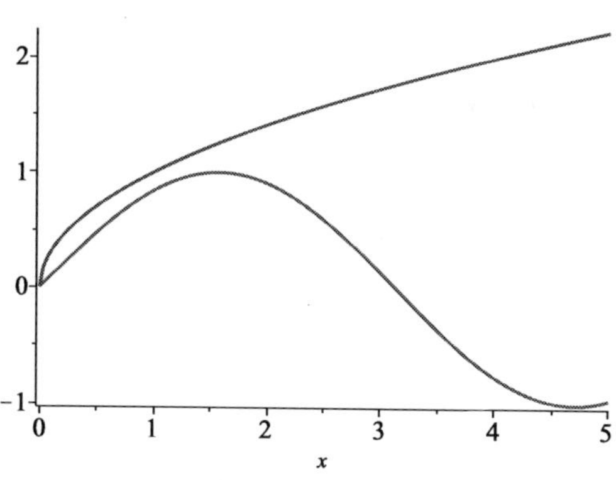

> $with(plots):$

> $curve := plot(30 \cdot \sin(x), x = 0..30):$

> $points := pointplot(agewt, symbol = diamond, symbolsize = 14):$

> $display(curve, points);$

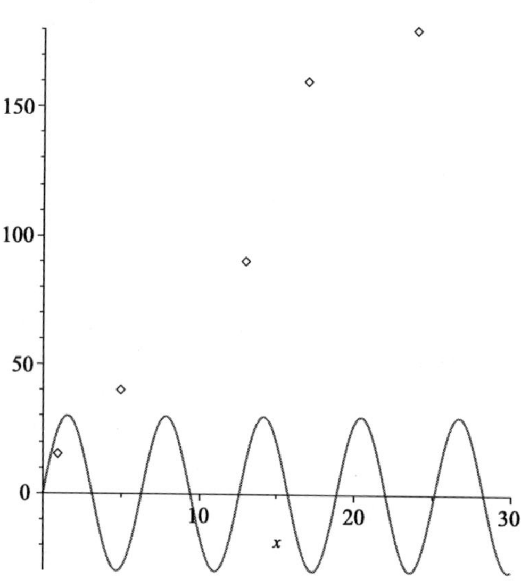

An intriguing method of adding curves to an existing graph is to select the output (blue font on your screen) expression desired, then "drag and drop" that selection onto an existing plot image. Try it!

1.5 Maple Training

A Maple training video of the basic features is provided at the website:
https://www.maplesoft.com/support/training/quickstart.aspx.
The website has a self-contained video tutorial "Maple Quick Start." Maple's basic features covered include:

- Numeric Computations
- Symbolic Computations
- Programming Basic Maple Procedures
- Visualization
- Calculus
- Linear Algebra
- Student Packages
- Other Maple Packages

This training serves as a basic refresher of Maple commands. The website also has a "QuickStart" PDF reference.

1.6 Maple Applications Center

The *Maple Applications Center* found at
https://www.maplesoft.com/applications/
contains a large collection of worksheets available to students and practitioners alike. The authors have several worksheets available on the site.

Exercises

Perform the following operations in Maple:

1. $\sqrt{2.3^5(4.5)}$
2. $11.3^3 + 5.1^2$
3. $\sqrt[3]{21.6}$
4. Let $a = 8$, $b = 7$, then compute $\left(a^2 - b^2\right)$.
5. $2(11.5) + 6.2^2(0.7)$

Enter the following functions. Obtain a graph. Find roots and solve for the intercepts.

6. $f(x) = -x^2 + 3x + 3$
7. $f(x) = x^2 - 3x - 1$
8. $f(x) = -0.1213x^4 + 3.462x^3 - 29.22x^2 + 64.68x + 97.69$
9. $g(x) = x^3 - 2x^2 - 5x + 6$
10. $s(x) = 2x^3 - 3x^2 - 11x + 7$

Enter the following date sets into Maple and obtain a scatterplot for each:

11.

x	1	3	8	10
y	0.7	5	15.2	36

12.

t	7	14	21	28	35	42
P	8	41	133	250	280	297

13.

x	29	48	72.7	92	118	140	165	199
y	0.49	0.82	1.23	1.54	1.97	2.34	2.74	3.30

Perform the required function in Maple.

14. $\dfrac{d}{dx}\left(1.104x - 0.542x^2\right)$
15. $\displaystyle\int 1.104x - 0.542x^2\,dx$
16. $\displaystyle\int_1^5 1.104x - 0.542x^2\,dx$

References

Modeling References

[A2011] Brian Albright, *Mathematical Modeling with Excel,* Jones & Bartlett Publishers, 2011.

[GFH2013] Frank Giordano, William P. Fox, and Steven Horton, *A First Course in Mathematical Modeling*, Nelson Education, 2013.

[M2007] Mark Meerscheart, *Mathematical Modeling*, Third Edition. 2007.

Maple References and Help

[T2017] Ian Thompson, *Understanding Maple*, 2017.

[MUM2018] Maplesoft, *Maple User Manual*, 2018.
https://www.maplesoft.com/support/help/category.aspx?CID=2317

[MPG2018] Maplesoft, *Maple Programming Guide*, 2018.
https://www.maplesoft.com/support/help/category.aspx?CID=2318

[MPG2012] L. Bernardin, P. Chin, P. DeMarco, K. O. Geddes, D. E. G. Hare, K. M. Heal, G. Labahn, J. P. May, J. McCarron, M. B. Monagan, D. Ohashi, and S. M. Vorkoetter, *Maple Programming Guide*, 2012.

[EM2007] Richard H. Enns and George McGuire, *Computer Algebra Recipes: An Advanced Guide to Scientific Modeling*, 2007.

2

Discrete Dynamical Models

Objectives:

(1) Define and build a discrete dynamical system for a real problem.

(2) Iterate a solution.

(3) Graph a solution.

(4) Understand equilibrium values and stability.

(5) Solve both linear and nonlinear systems of discrete dynamical systems.

(6) Solve systems of discrete dynamical systems.

2.1 Introduction

Consider a disease that is spreading throughout the Unites States such as a new influenza or flu. The U.S. Centers for Disease Control and Prevention (CDC) is interested in knowing and experimenting with a model for this new disease prior to it actually becoming a "real" epidemic.[1] Let us consider the population being divided into three categories: susceptible (able to catch this new flu), infected (currently has the flu and is contagious), and removed (already had the flu and will not get it again, or has died from the flu). We make the following assumptions for our model:

- No one enters or leaves the community.
- There is no contact outside the community.
- Each person is either susceptible S, infected I, or removed R.
- Initially, every person is either an S or I.
- Once someone gets the flu this year, they cannot get again ('developed immunity').

[1]The CDC website "FluView" monitors the current situation in the U.S.

- The average length of the disease is 2 weeks over which the person is deemed infected and can spread the disease.
- The time step for the model will be one week.

Can we build a model to determine useful information to the CDC?

We will revisit this model later in the chapter.

2.2 Modeling with Dynamical Systems[2]

Buying a new car is a major purchase. Once you have looked at the makes and models to determine what type of car you like, it's time to consider the "costs" and finance packages available. Payments are made typically at the end of each month. The amount owed is predetermined depending on the total price of the car, fees and charges, and the interest rate of a loan. This process can be modeled as a dynamical system.

Another name for a discrete dynamical system (DDS) is a "difference equation." We will use DDS extensively in this chapter. Let's begin with basic definitions:

A **sequence** is a *function* whose domain is the set of all nonnegative integers and whose range is a subset of the real numbers. A **dynamical system** is a relationship among the terms in a sequence; a DDS describes the evolution of an item over time. A **numerical solution** is a table of values satisfying the dynamical system.

Let's start with a brief review of the concepts of 'relation' and 'function' since these are foundational in the definition above. A **relation** is a set of ordered pairs, often written as (*input*, *output*). As an ordered pair, a relation can indicate how one variable depends on another. The **domain** of a relation is the set of all first coordinates of the ordered pairs, the **range** is the set of all second coordinates of the ordered pairs. A function is a special type of relation. A **function** is a relation for which each element in the domain has exactly one related element in the range. The domain of a function is the set of all of the *independent variables* (allowed inputs) and the range is the set of all possible *dependent values* (outputs).

The concepts of function, domain, and range can be seen clearly in a dynamical system model. Let's define a recurrence relation, a relation where the next term depends on the previous values, as an equation of the form

$$a(n+1) = f(a(n), a(n-1), \ldots, a(0), n) \qquad \text{or} \qquad a_{n+1} = f(a_n, a_{n-1}, \ldots, a_0, n).$$

Using subscripts for arguments makes the notation much more compact and easy to read.

[2]Modified from USMA Math 103 Study Guide.

Recurrence relations, also called recursive formulas, occur in many branches of applied mathematics. A **discrete dynamical system** (DDS or *difference equation*) is a sequence defined by a recurrence relation. A DDS is a "changing system," where the change of the system at each discrete iteration depends on its previous states. Suppose we have a function $y = f(x)$. A *first-order* discrete dynamical system is given by the sequence of numbers $a(n)$ for $n = 0, 1, \ldots$ such that each number after the first is related to the previous number by

$$a(n+1) = f(a(n)).$$

(Many books and articles refer to the relationship $a(n+1) - a(n) = g(a(n))$ as a *first-order difference equation.*)

The **order** of a dynamical system is the difference between the largest and the smallest arguments (or variable of the function) appearing in the formula. For example, $A(n+2) = 0.5A(n+1) + 2A(n)$ is a second order DDS because $n+2$, the largest argument, minus n, the smallest, gives $(n+2) - n = 2$.

Example 2.1. A Simple First-Order Discrete Dynamical System. Define the dynamical system A to be the amount of antibiotic in a patient's blood stream after n time periods. The domain is the nonnegative integers representing the time periods from 0, 1, 2, $\ldots$, n that will be the inputs to the function. Since the domain is discrete, we have a DDS. The range is the set of values of $A(n)$ determined for each value of the domain. Thus, $A(n)$ also represents the dependent variable. For each input value of the domain from 0, 1, 2, $\ldots$, n, the result is one and only one amount $A(n)$, thus A is a function.

There are three components to a dynamical system:

- a formula for the sequence representing $A(n)$,
- the time period n is well defined, and
- at least one starting value $(n_0, A(n_0))$. (The number of required starting values corresponds to the order of the DDS.)

The starting value set is called the **initial condition(s)**. In the example above, if we start with no antibiotic in our system then $A(0) = 0$ mg is our initial condition. However, if we started after we took an initial 200 mg tablet, then $A(0) = 200$ mg would be our initial condition. An example of a discrete dynamical system with its initial condition would be:

$$A(n+1) = 0.5A(n)$$
$$A(0) = 200$$

We are mainly concerned with one of several aspects of a DDS: Does the DDS have a stable equilibrium? What is the value of the system after period n for a specified n. What is the long-term behavior for the DDS?

Next, let's look at Maple commands that can assist us with investigating a DDS.

Maple Commands for Discrete Dynamical Systems

We will use familiar commands and libraries from Maple such as *plots*, and we will add new commands to our Maple tool box.

> **rsolve** – *recurrence equation solver*
>
> Calling Sequence
> *rsolve*(*eqns*, *fcns*)
> *rsolve*(*eqns*, *fcns*, 'genfunc'(z))
> *rsolve*(*eqns*, *fcns*, 'makeproc')
> *rsolve*(*eqns*, *fcns*, 'series')
>
> Parameters
> eqns – single equation or a set of equations
> fcns – function name or set of function names
> z – name, the generating function variable

> **seq** – *create a sequence*
>
> Calling Sequence
> $seq(f, m..n)$
> $seq(f, i = m..n)$
> $seq(f, i = m..n, \text{step})$
> $seq(f, i = x)$
>
> Parameters
> f – any expression
> i – name
> m, n – numerical values
> x – expression
> step – (optional) numerical value

Look at Maple's help pages for *rsolve* and *seq* for detailed information and to see many examples.

Several of the models that we will solve have closed-form solutions; we can use *rsolve* to obtain a formula, then we can use *seq* to obtain numerical values of the solution.

Many dynamical systems do not have closed-form analytical solutions, so we cannot use *rsolve* and *seq* to obtain solutions. When this occurs, we will write a small program using **proc** to obtain the numerical solutions through iteration. To graph the solution of the dynamical systems, we will use plot commands to see sequential data pairs $(k, A(k))$. We will illustrate Maple use in our examples in the next section.

2.3 Linear Systems

We are interested in modeling discrete *change*. Modeling with discrete dynamical systems employs a method to explain certain discrete behaviors or make long-term predictions. A powerful paradigm that we use to model with discrete dynamical systems is:

$$\textit{future value} = \textit{present value} + \textit{change}$$

The dynamical systems that we will study with this paradigm will differ markedly in appearance and composition, but we will be able to solve a large class of these "seemingly" different dynamical systems with similar methods. In this chapter, we will use iteration and graphical methods to answer questions about discrete dynamical systems.

We will use flow diagrams to help us see how the dependent variable changes. Flow diagrams help to see the paradigm and to put the problem into mathematical terms. Let's consider financing a new Ford Mustang. The total cost is \$25,000. We can put down \$2,000, so we need to finance \$23,000. The dealership offers us 2% APR financing over 72 months. Consider the flow diagram for financing the car below in Figure 2.1 that depicts this situation.

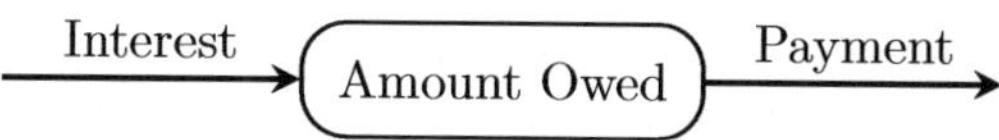

FIGURE 2.1: Car Financing Flow Diagram

We'll use the flow diagram to help build the discrete dynamical model. Let $A(n)$ = the amount owed after n months. Notice that the arrow pointing into the *Amount Owed* is the interest which increases the unpaid balance. The arrow pointing out of the oval is the monthly payment which decreases the debt.

$$\begin{aligned} A(n+1) &= \text{amount owed next month } (\textit{future}) \\ A(n) &= \text{amount currently owed } (\textit{present}) \\ i &= \text{monthly interest rate} \\ P &= \text{monthly payment} \end{aligned}$$

then our paradigm *future* = *present* + *change* gives

$$A(n+1) = A(n) + (i \cdot A(n) - P)$$

We will model dynamical systems that have *constant coefficients*. For example, a third-order discrete dynamical system with constant coefficients that is homogeneous may be written in the form

$$a(n+3) = b_2 a(n+2) + b_1 a(n+1) + b_0 a(n)$$

where b_0, b_1, and b_2 are arbitrary constants. If we added a term not involving $a(n)$s to the right side, this DDS would be nonhomogeneous.

The "Tower of Hanoi" puzzle, that we will illustrate shortly, involves moving a tower of disks from one pole to another. The number of discrete moves $H(n)$ that it takes to move n disks depends on the number of discrete moves it took to move $n-1$ disks. For the drug dosage model of Example 2.1, the model shows the amount of drug in the bloodstream after n hours depends on the amount of drug in the bloodstream after $n-1$ hours. For financial matters, such as our Mustang purchase, the amount we still owe after n months depends upon the amount we owed after $n-1$ months. We will also find this process based on transitioning states useful when we discuss *Markov chains* as an example of a discrete dynamical system.

Recall that a *first-order* DDS is a system where the next iteration $a(n)$, the state of the system after n iterations, is related only to the previous iteration $a(n-1)$ by the relation

$$a(n) = f\left(a(n-1)\right);$$

i.e., $a(n)$ is a function of only $a(n-1)$ for $n = 2, 3, \ldots$. For example, $a(n) = 2a(n-1)$ is a first-order DDS. If the function f of $a(n)$ is just a constant multiple of $a(n)$, a constant coefficient, we will say that the discrete dynamical system is **linear**. If the function f involves powers (like $a(n)^2$), or a functional relationship (like $a(n)/a(n-1)$ or $\sin(a(n-1))$), we will say that the discrete dynamical system is **nonlinear**.

Example 2.2. Iteration with the Tower of Hanoi.
The Tower of Hanoi puzzle was invented in 1883 by the French mathematician Édouard Lucas (1842-1891) under the pseudonym "Professor N. Claus (of Siam) from the Mandarin of the College of Li-Sou-Stian."[3] The puzzle consists of a board with three upright pegs and disks (now usually 6 to 10) with successively smaller outside diameters. The disks begin on the first peg with the largest disk on the bottom, topped by the next largest disk, and so on up to the smallest disk on top forming a tower or pyramid as we see in Figure 2.2.

The object of the game is to transfer the disks, one at a time, using the smallest possible number of moves, to the third peg to form an identical pyramid. During each transfer step, a larger disk may not be placed on top of a smaller disk—this is why a second and third pegs are needed. Lucas' original rules were the following.

[3]The puzzle is also called the "Tower of Brahma" or the "Tower of Lucas." There are several intriguing legends about the game. See Paul Stockmeyer's webpage http://www.cs.wm.edu/~pkstoc/toh.html for the original game box and instructions.

FIGURE 2.2: Tower of Hanoi Puzzle[4]

> *The game consists of moving this, by threading the disks on another peg, and by moving only one disk at a time, obeying the following rules:*
>
> *I. After each move, the disks will all be stacked on one, two, or three pegs, in decreasing order from the base to the top.*
>
> *II. The top disk may be lifted from one of the three stacks of disks, and placed on a peg that is empty.*
>
> *III. The top disk may be lifted from one of the three stacks and placed on top of another stack, provided that the top disk on that stack is larger.*

Lucas' original *La Tour d'Hanoï* game had eight disks. The puzzle was a model to illustrate moving the legendary "Sacred Tower of Brahma," which actually had "sixty-four levels in fine gold, trimmed with diamonds." This tower was attended by monks of the Temple of Bernares who moved one disk a minute according to the long-established ritual that no larger disk could be placed on a smaller disk. The monks believed, so Lucas's story went, that as soon as all sixty-four disks were transferred, the earth would collapse in a cloud of dust.

This legend is somewhat alarming, so let's use induction to find a formula for the number of moves required to transfer n disks from the first peg to the third peg. Suppose that we have $n+1$ disks. Then we can move n smaller disks from the first peg to the second peg in $a(n)$ moves, the number of moves for n disks. Then we move the largest disk from the first peg to the third peg. Finally we move the n smaller disks from the second peg to the third peg in $a(n)$ moves. Therefore, $a(n{+}1) = (a(n){+}1){+}a(n) = 2a(n){+}1$. So the recursion relation that gives the number of moves required is $a(n+1) = 2a(n) + 1$; the initial value is $a(1) = 1$, meaning that we start with the case where there is only one disk to move. Figure 2.3 shows a plausible flow diagram for the Tower of Hanoi.

[4]Photo source: Bjarmason (2005). Creative Commons License.

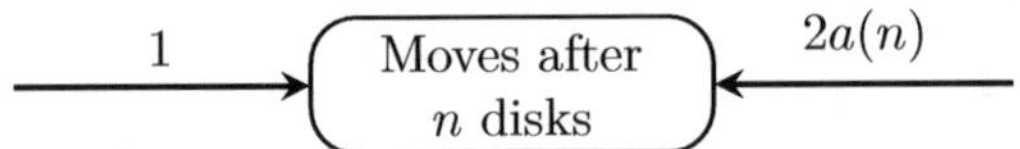

FIGURE 2.3: Tower of Hanoi Flow Diagram

There is a formula for the number of moves required to move n disks from the first peg to the third peg. We will consider methods of solving this recursion relation to find this formula. The first method is *iteration*. Later we will learn a better method by explicitly solving the DDS.

Let's build Table 2.1 by iterating the recursion relation

$$a(n+1) = 2a(n) + 1$$
$$a(1) = 1.$$

TABLE 2.1: Tower of Hanoi Recursion

Number of Disks	Recursion Relationship	Number of Moves
1	$a(1) = 1$	1
2	$a(2) = 2a(1) + 1 = 2(1) + 1$	3
3	$a(3) = 2a(2) + 1 = 2(3) + 1$	7
4	$a(4) = 2a(3) + 1 = 2(7) + 1$	15
5	$a(5) = 2a(4) + 1 = 2(15) + 1$	31
6	$a(6) = 2a(5) + 1 = 2(31) + 1$	63
7	$a(7) = 2a(6) + 1 = 2(63) + 1$	127
$\vdots$	$\vdots$	$\vdots$
n	$a(n) = 2a(n-1) + 1$	$2a(n-1) + 1$

Using iteration, we repeatedly calculate the recursion relation, each time for the next value of n. To find out how many minutes it would take to move sixty-four disks, we would have to iterate the given recursion relation 64 times. We'll use Maple to help us with the computations for tabulating Table 2.2.

TABLE 2.2: Tower of Hanoi with Sixty-Four Disks

Disks	Moves Required
1	1
5	31
10	1023
15	32767
20	1048575
25	33554431
30	1073741823
35	34359738367
40	1099511627775
45	35184372088831
50	1125899906842623
55	36028797018963967
60	1152921504606846975
64	18446744073709551615

The table above shows that $18446744073709551615 \approx 1.84467 \times 10^{19}$ minutes are needed. Let's convert to units that are equivalent. There are 1440 minutes in a day and 525,600 minutes in a year. At one move per minute, $a(64)$ is equivalent to 3.5135×10^{13} years. Imagine how long that amount of time really is. (*The estimated age of the universe is* 1.37×10^{10} *years.*)

The power of discrete dynamical systems is that they can always be solved by iteration. Between the table of iteration values and a graph of those values, we are able to analyze difficult modeling problems.

In Maple, to obtain the iteration values, we can either write a procedure with **proc** or, when the recurrence has a closed-form solution, use the *rsolve* and *seq* commands. We illustrate both below to obtain the values seen in Tables 2.1 and 2.2. First, *rsolve* and *seq*.

$$> rsolve(\{H(n+1) = 2 \cdot H(n) + 1, H(1) = 1\}, H(n));$$

$$-1 + 2^n$$

We have a closed-form analytical solution $H(n) = 2^n - 1$. If there had not been a formula, *rsolve* would just have echoed the input. Using *rsolve* is a quick way to see if a DDS has a closed-form solution.

Now we can use *seq* with the formula from *rsolve* to generate values. Also, since there is a formula, we don't have to iterate through all n up to the number we want.

> $seq([n, 2^n - 1], n = 1..64, 7);$ #*Note the* step *size of* 7

$[1, 1], [8, 255], [15, 32767], [22, 4194303], [29, 536870911], [36, 68719476735],$
$[43, 8796093022207], [50, 1125899906842623], [57, 144115188075855871],$
$[64, 18446744073709551615]$

When a dynamical system does not have a closed-form analytical solution, the *rsolve* command echoes its input, so we cannot use a formula in the (**seq**) command to generate values. We'll need to use **proc** to create a program to iterate the discrete dynamical system. Call the program *Tower* and use inputs $n0$, $t0$, and n where $n0$ is the first nonnegative integer in our domain, $t0$ is our first range value (that is, $(n0, t0)$ is the initial condition), and n is the nonnegative integer value from our domain that we need. We use '**option** *remember*' so the program stores and can recall the previously iterated values. Without remembering values, a simple recursion can take huge numbers of calculations to compute a value. For example, the Fibonacci numbers given by $F(n) = F(n-1) + F(n-2)$ take on the order of 2^n calculations to compute $F(n)$ when previous values aren't remembered.

```
> Tower := proc(n0, t0, n)
    option remember;
    local T;
    if n > n0 then
      T := 2 · Tower(n0, t0, n − 1) + 1;
    else
      T := t0;
    end if;
    return T;
    end proc :
```

Now we can use *seq* with the procedure *Tower* to generate values. (*Tower* is in the book's *PSMv2* Maple package.) Also, since there is a formula, we don't have to iterate through all n up to the number we want.

> $seq([n, Tower(1, 1, n)], n = 1..64, 7);$ #*Note the* step *size of* 7

$[1, 1], [8, 255], [15, 32767], [22, 4194303], [29, 536870911], [36, 68719476735],$
$[43, 8796093022207], [50, 1125899906842623], [57, 144115188075855871],$
$[64, 18446744073709551615]$

Are the values the same as with the formula?
Let's plot the recursion.

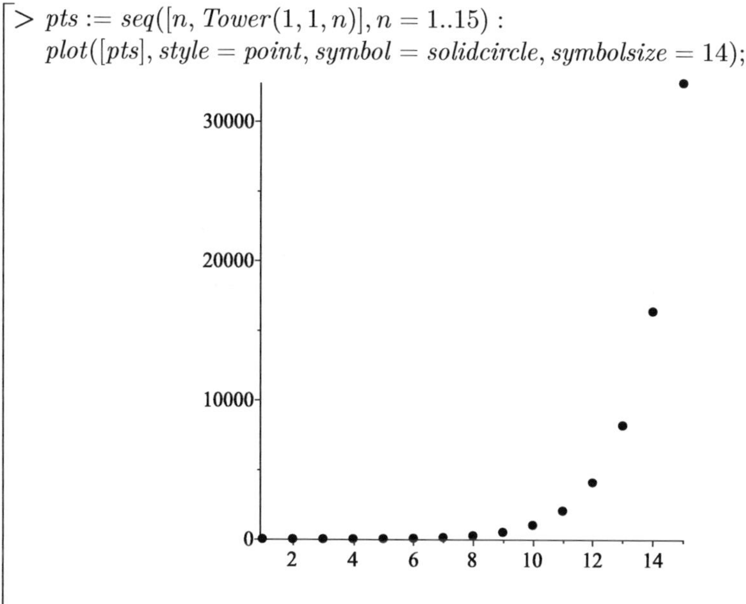

It's easy to see how quickly this recursion grows.

Example 2.3. A Drug Dosage Problem.
A doctor prescribes an oral dose of 100 mg of a certain drug every hour for a patient. Assume that the drug is immediately absorbed into the bloodstream once taken. Also, assume that every hour the patient's body eliminates 25% of the drug that is in the bloodstream. Suppose that the patient's bloodstream had 0 mg of the drug prior to taking the first pill. How much of the drug will be in the patient's bloodstream after 72 hours?

General Problem Statement.

Determine the relationship between the amount of drug in the bloodstream and time.

Assumptions.

- The problem can be modeled by a discrete dynamical system.
- The patient is of normal size and health.
- There are no other drugs being taken that will affect the prescribed drug.
- There are no internal or external factors that will affect the drug absorption or elimination rates.
- The patient always takes the prescribed dosage at the correct time.

Variables.

Define $a(n)$ = amount of drug that is in the bloodstream after a period of $n = 0, 1, 2, \ldots$ hours.

Flow Diagram:

We diagram the flow in Figure 2.4.

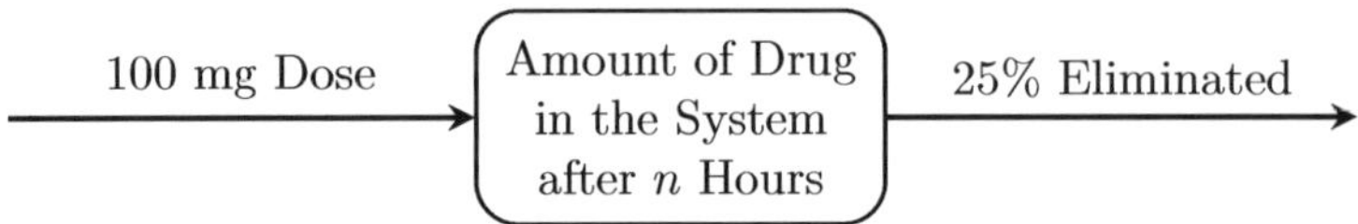

FIGURE 2.4: Amount of Drug in the Bloodstream Flow Diagram

Model Construction.

Let $a(n)$ represent the amount of drug in the system after time period n. We calculate the change in the drug's amount as: *change* = *dose* − *system's loss.* The the *future* = *present* + *change* paradigm gives

$$\begin{aligned} a(n+1) &= a(n) - 0.25 \cdot a(n) + 100 \\ &= 0.75 \cdot a(n) + 100 \end{aligned}$$

From the DDS we see that since the body loses 25% of the amount of drug in the bloodstream every hour, there would be 75% of the amount of drug in the bloodstream remaining every hour. After one hour, the body has 75% of the initial amount, 0 mg, plus the dose of 100 mg that is added every hour. The bloodstream has 100 mg of drug after one hour. After two hours the body has 75% of the amount of drug that was in the bloodstream after one hour (100 mg) plus an additional 100 mg of drug added from the new oral dose. There would be 175 mg of drug in the bloodstream after two hours. After three hours the body has 75% of the amount of drug in the bloodstream after two hours plus an additional 100 mg of drug added to the bloodstream. Thus there would be 231.25 mg of drug in the bloodstream after three hours.

The values are tabulated in Table 2.3.

TABLE 2.3: Amount of Drug in the Bloodstream - First Computations

Hour	Amount of Drug (mg)
0	$a(0) = 0$
1	$a(1) = 0.75 \cdot a(0) + 100 = 100$
2	$a(2) = 0.75 \cdot a(1) + 100 = 175$
3	$a(3) = 0.75 \cdot a(2) + 100 = 231.25$

We can see the change that occurs every hour within this system (amount of drug in the bloodstream), and the state of the system after any hour, is dependent on the state of the system after the previous hour. This is a *discrete dynamical system* (DSS).

To find the value of $a(72)$ we can either iterate the recurrence with *seq* or use *rsolve* to obtain a formula. Let's start with iteration. (Remember to either use *restart* or open a new Worksheet to have a fresh Maple environment.)

Since this is a simple recurrence, we'll use a 'short-cut' procedure.

$> a := n \to piecewise(n > 0, 0.75 \cdot a(n-1) + 100, 0);$

$$a := n \mapsto \begin{cases} 0.75\, a(n-1) + 100 & 0 < n \\ 0 & otherwise \end{cases}$$

Asking Maple for $a(72)$ will perform the iterations and display the result.

$> a(72);$

$$399.9999996$$

We could see all the iterates by using $seq([k, a(k)], k = 0..72)$.

Let's attempt to find a formula by using *rsolve*. (Remember to *restart*.)

$> rsolve(\{a(n+1) = 0.75 \cdot a(n) + 100, a(0) = 0\}, a(n));$

$$400 - 400\left(\frac{3}{4}\right)^n$$

Note that, in this case, Maple returned an exact answer even though we entered decimals.

What is the long-term level of drug in the patient's bloodstream?

$> limit(400 - 400 \cdot 0.75^n, n = infinity);$

$$400.$$

Checking iterates would show that a patient's bloodstream would have approximately 400 mg of the drug after 24 hours.

It's time for a plot.

$> pts := [seq([n, 400 - 400 \cdot 0.75^n], n = 0..48)] :$

```
> plot(pts, style = point, symbol = solidcircle, symbolsize = 14);
```

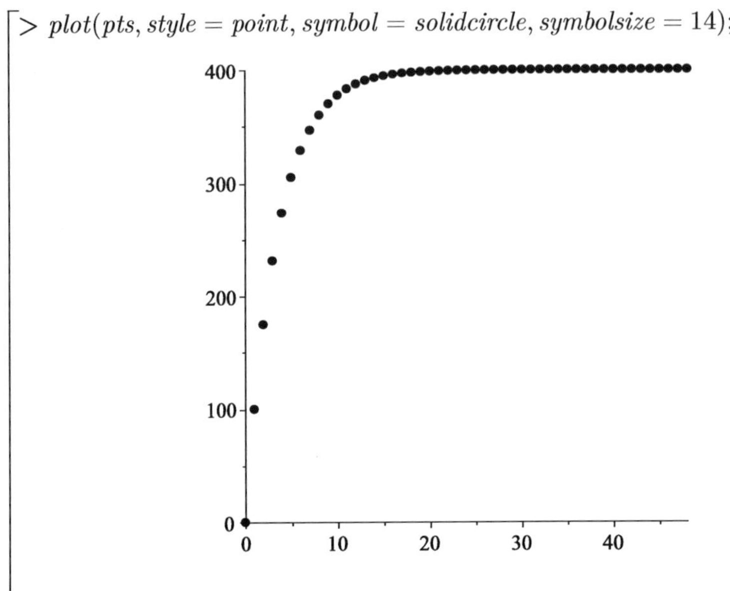

Interpretation of Results.

The DDS shows that the drug reaches a value where change stops. The concentration of the drug in the bloodstream eventually levels at 400 mg. If 400 mg is both a safe and effective drug level, then this dosage schedule is an acceptable treatment. Note that we're looking at discrete points, but the drug concentration is a continuous quantity. What is the shape of the continuous curve modeling drug concentration?

We discuss the concept of change stopping—equilibrium—later in this chapter. We used the Maple command *limit* to quickly determine long-term behavior of the DDS discovering the equilibrium.

Example 2.4. The Time Value of Money.
A bank customer wishes to purchase a \$1000 savings certificate that pays 1.2% interest a year (APR) compounded monthly at 0.1% = 1.2%/12 per month. Why use a discrete model here? At our local financial institution, there is a sign that says that interest is compounded (and paid) at the end of each month based on the average monthly balance. Therefore, we conclude that a discrete model for interest is appropriate.

Remark: Always divide the annual interest rate (APR) by the number of compounding periods per year to compute the actual interest rate being used. Here, the annual rate is 1.2% or 0.012 and interest is calculated and paid monthly (12 times per year). So, use $0.012/12 = 0.001$ as the monthly interest rate.

General Problem Statement.

Find a relationship between the amount of money invested and the time over which it is invested.

Assumptions.

The interest rate is constant over the entire time period. No additional money is added or withdrawn other than the interest.

Variables.

Let $n = 1, 2, 3, \dots$ be number of months passed.

Let $a(n)$ = the amount of money in the certificate after month n.

Flow Diagram:

We diagram the flow in Figure 2.6.

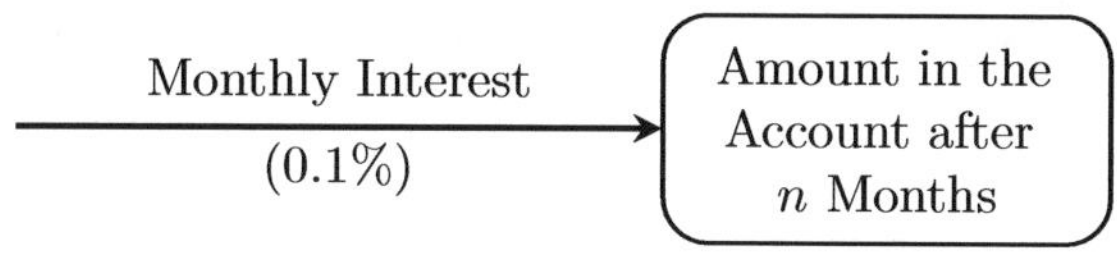

FIGURE 2.5: Flow Diagram for Money in a Certificate

Model Construction.

Since

$$\begin{aligned} a(n+1) &= \text{ the future—account balance at the end of month } n+1 \\ a(n) &= \text{ the present—account balance at the end of month } n \\ 0.001 \cdot a(n) &= \text{ the change—interest compounded at the end of month } n \end{aligned}$$

Using our paradigm *future* = *present* + *change*, the worth of the certificate with interest accumulated each month is

$$\begin{aligned} a(n+1) &= a(n) + 0.001a(n) \quad \text{for } n = 1, 2, \dots \\ &= 1.001a(n) \quad \text{for } n = 1, 2, \dots \end{aligned}$$

The initial deposit of \$1000 gives

$$a(0) = 1000.$$

Use the discrete time periods $(1, 2, 3, \dots)$ to iterate as follows:

$$\begin{aligned} a(0) &= 1000 \\ a(1) &= 1.001 \cdot a(0) = 1001 \\ a(2) &= 1.001 \cdot a(1) = 1.001^2 \cdot a(0) = 1002.001 \\ a(3) &= 1.001 \cdot a(2) = 1.001^3 \cdot a(0) = 1003.003 \end{aligned}$$

Calling $1.001 = r = (1 + i)$, where i is the monthly interest rate, gives

$$a(1) = r \cdot a(0),\ a(2) = r^2 \cdot a(0),\ a(3) = r^3 \cdot a(0),\ a(4) = r^4 \cdot a(0),\ \dots,$$

suggesting that

$$a(n) = r^n \cdot a(0) \quad \text{for } n = 0, 1, 2, \ldots$$

Let's check with Maple.

> $rsolve(\{a(n+1) = 1.001 \cdot a(n), a(0) = 1000\}, a(n)) :$
> $Money := unapply(\%, n);$

$$Money \mapsto 1000 \left(\frac{1001}{1000}\right)^n$$

Note that Maple converts to rational numbers when using *rsolve*. Recall that *unapply* turns an expression into a function.

What is the overall trend?

> $Limit(Money(n), n = infinity) = limit(Money(n), n = infinity);$

$$\lim_{n \to \infty} 1000 \left(\frac{1001}{1000}\right)^n = \infty$$

> $with(plots) :$

> $pointplot([seq([i, Money(i)], i = 0..12 \cdot 4)], title =$ "Money over Time");

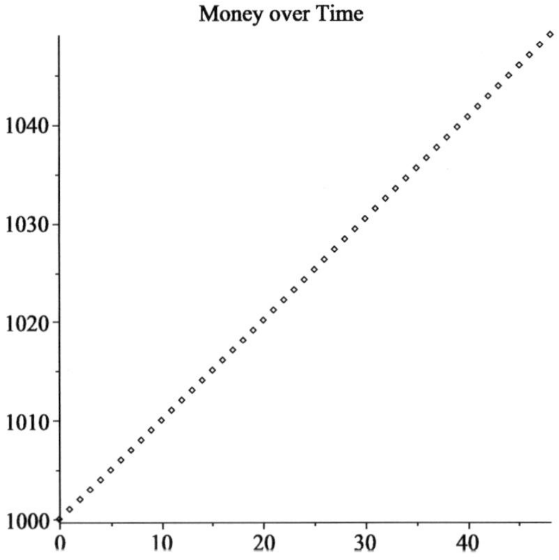

Money over Time

The graph looks like a line even though this function is obviously nonlinear. How many years does it take for the plot to show some curvature?

Let's look at the sequence of 4 years of values to see how *Money* grows.

$> Money\_table := seq(evalf(Money(i)), i = 0..12 \cdot 4);$

$$Money\_table := 1000., 1001., 1002.001000, 1003.003001, 1004.006004, 1005.010010, 1006.015020, 1007.021035, 1008.028056, 1009.036084, 1010.045120, 1011.055165, 1012.066220, 1013.078287, 1014.091365, 1015.105456, 1016.120562, 1017.136682, 1018.153819, 1019.171973, 1020.191145, 1021.211336, 1022.232547, 1023.254780, 1024.278035, 1025.302313, 1026.327615, 1027.353943, 1028.381297, 1029.409678, 1030.439088, 1031.469527, 1032.500996, 1033.533497, 1034.567031, 1035.601598, 1036.637199, 1037.673836, 1038.711510, 1039.750222, 1040.789972, 1041.830762, 1042.872593, 1043.915465, 1044.959381, 1046.004340, 1047.050345, 1048.097395, 1049.145492$$

Can the account ever reach \$10,000? We can determine that it will take about 192 years to reach \$10,000.

$> months := fsolve(10000 = Money(n), n);$

$$months := 2303.736194$$

$> years := \dfrac{months}{12};$

$$years := 191.9780162$$

Since $Money(n)$ goes to infinity as n grows, the account continues to grow as long as there are no withdrawals.

Example 2.5. A Simple Mortgage.
Five years ago, your parents purchased a home by financing \$150,000 over 20 years with an interest rate of 4% APR. Their monthly payments are \$908.97. They have made 60 payments and wish to know what they actually owe on the house at this time. They can use this information to decide whether or not to refinance their house at a lower interest rate of 3.25% APR for the next 15 years or 3.5% APR for 20 years.

The change in the amount owed each period increases by the amount of the interest and decreases by the amount of the payment.

General Problem.
Build and use a model that relates the time to the amount owed on a mortgage for a home.

Assumptions.

- Initial interest was 4% APR.
- The monthly interest rate is $4\%/12 = 0.3\bar{3}\%$.

- Payments are made on time each month.
- The current rates for refinancing are 3.25% for 15 years and 3.50% for 20 years.

Variables.
Let $b(n)$ = amount owed on the home after n months.

Flow Diagram:
We diagram the flow in Figure 2.6.

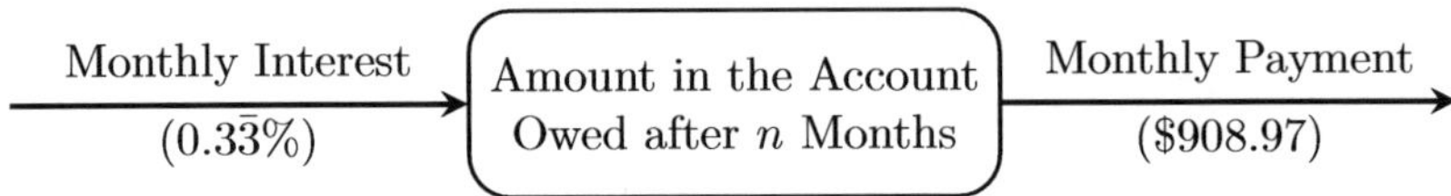

FIGURE 2.6: Flow Diagram for Amount Owed on Mortgage

Model Construction.

$$\mathit{future} = \mathit{present} + \mathit{change}$$

$$\begin{aligned} b(n+1) &= b(n) + (0.04/12 \cdot b(n) - 908.97) \quad \text{and} \quad b(0) = 150000 \\ &= (1 + 0.04/12) \cdot b(n) - 908.97 \quad \text{and} \quad b(0) = 150000 \end{aligned}$$

Model Solution:
Use *rsolve* to find a formula for $b(n)$. Since Maple converts to rational numbers for *rsolve*, we'll embed an *evalf* to return to decimals.

$$> \mathit{rsolve}\left(\left\{b(n+1) = \left(1 + \frac{0.04}{12}\right) \cdot b(n)\right\}, b(n)\right) :$$
$$b := \mathit{unapply}(\mathit{evalf}(\%), n);$$
$$b := n \mapsto -122691.0273 \cdot 1.003333333^n + 272691.0273$$

First, a reasonableness check.

$$> b(12 * 20);$$
$$0.1695$$

The last payment leaves 0.17¢ to pay; this is due to round-off error in the computations as 0.04/12 is a repeating decimal value.

Let's plot the DDS over the entire 20 years (240 months). Note the graph shows that the balance essentially reaches 0 in 240 months.

```
> with(plots) :
  |MCpts := [seq([i, b(i)], i = 0..240)] :
  pointplot(pts, title = "Mortgage Balance");
```

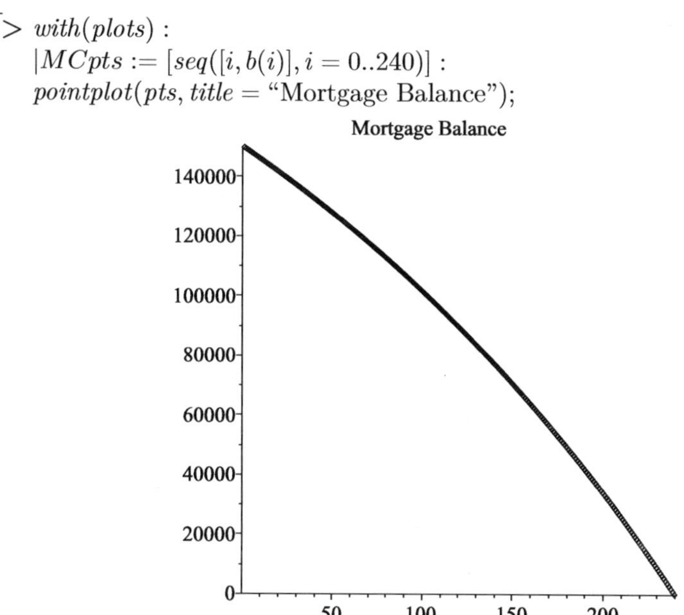

Now let's make a table of the mortgage balance up to month 60.

```
> Mortgage_table := seq(b(i), i = 0..60);
```

Mortgage_table := 150000.0000, 149591.0299, 149180.6967, 148768.9957, 148355.9222, 147941.4719, 147525.6401, 147108.4222, 146689.8135, 146269.8095, 145848.4056, 145425.5968, 145001.3787, 144575.7466, 144148.6957, 143720.2214, 143290.3187, 142858.9831, 142426.2096, 141991.9936, 141556.3303, 141119.2146, 140680.6419, 140240.6074, 139799.1061, 139356.1329, 138911.6833, 138465.7522, 138018.3347, 137569.4258, 137119.0206, 136667.1138, 136213.7009, 135758.7764, 135302.3357, 134844.3734, 134384.8846, 133923.8641, 133461.3071, 132997.2080, 132531.5619, 132064.3638, 131595.6083, 131125.2903, 130653.4045, 130179.9458, 129704.9090, 129228.2886, 128750.0796, 128270.2764, 127788.8740, 127305.8668, 126821.2496, 126335.0171, 125847.1639, 125357.6843, 124866.5733, 124373.8251, 123879.4345, 123383.3959, 122885.7038

From the table, we see the balance on the mortgage is $b(60) = 122885.7038$. After paying for 60 months, your parents still owe \$122,885.70 of the original \$150,000. They have paid a total of \$54,538.20, but only \$27,114.30 went towards the principal, the rest, \$27,423.90, was interest. If the family continues with this loan, they will make 240 payments of \$908.97 for a total of \$218,152.80. They would pay a total of \$68,152.80 in interest. They've already paid \$27,423.90 in interest, so they would pay an additional \$41,038.50 in interest. Should they refinance?

The alternatives for refinancing were 3.25% for 15 years or 3.50% for 20 years. Assuming no additional costs for refinancing, only securing a new mortgage for what they currently owe, \$122,885.70, what is the best choice?

You will be asked to solve this problem in the exercise set.

Exercises

Iterate and graph the following DDSs. Explain their long-term behavior. For each DDS, find a realistic scenario that it might explain/model.

1. $a(n+1) = 0.5\,a(n) + 0.1, \quad a(0) = 0.1$
2. $a(n+1) = 0.5\,a(n) + 0.1, \quad a(0) = 0.2$
3. $a(n+1) = 0.5\,a(n) + 0.1, \quad a(0) = 0.3$
4. $a(n+1) = 1.01\,a(n) - 1000, \quad a(0) = 90000$
5. $a(n+1) = 1.01\,a(n) - 1000, \quad a(0) = 100000$
6. $a(n+1) = 1.01\,a(n) - 1000, \quad a(0) = 110000$
7. $a(n+1) = -1.3\,a(n) + 20, \quad a(0) = 9$
8. $a(n+1) = -4\,a(n) + 50, \quad a(0) = 9$
9. $a(n+1) = 0.1\,a(n) + 1000, \quad a(0) = 9$
10. $a(n+1) = 0.9987\,a(n) + 0.335, \quad a(0) = 72$
11. Consider the mortgage problem of Example 2.5 (pg. 49). Determine what the cost of each refinancing alternative is compared to the current mortgage. Make a recommendation to your parents.

2.4 Equilibrium Values and Long-Term Behavior

Equilibrium Values

Recall our original paradigm

$$\textit{Future} = \textit{Present} + \textit{Change}.$$

When the DDS stops changing, *change* equals zero, and so *future* equals *present*, and remains there. A value for which *change* equals 0, if any exist, is the *equilibrium value.* Change having stopped gives us a context for the concept of the equilibrium value.

We will define the *equilibrium value*, also called *fixed point*, as the value where change stops; i.e., where *change* equals zero. The value A is an equilibrium value for the DDS

$$a(n+1) = f(a(n), a(n-1), \ldots, a(0)),$$

if whenever for some N, we have $a(N) = A$ and all future values $a(n) = A$ for $n > N$.

Formally, we define the equilibrium value as follows:

Definition 2.1. Equilibrium Value.
The number ev is called an **equilibrium value** or a **fixed point** for a discrete dynamical system $a(n)$ if $a(k) = ev$ for all values of k when the initial value $a(0)$ is set to ev. That is, $a(k) = ev$ is a constant solution to the recurrence equation for the dynamical system.

Another way of characterizing such values is to note that a number ev is an equilibrium value for a dynamical system $a(n+1) = f(a(n), a(n-1), \ldots, n)$ if and only if ev satisfies the equation $ev = f(ev, ev, \ldots, ev)$. This characterization is the genesis of the term *fixed point.*

Definition 2.1 shows that a linear homogeneous dynamical system of order 1 only has the value 0 as an equilibrium value.

In general, dynamical systems may have no equilibrium values, a single equilibrium value, or multiple equilibrium values. Linear systems have unique equilibrium values. The more nonlinear a dynamical system is, the more equilibrium values it may have.

Not all DDSs have equilibrium values, many DDSs have equilibrium values that the system will never achieve *unless* $a(0)$ *equals that value.*

The DDS from the Tower of Hanoi $a(n+1) = 2a(n)+1$ has an equilibrium value $ev = -1$. If we begin with $a(0) = -1$ and iterate, we get

$$\begin{aligned}
a(0) &= -1,\\
a(1) &= 2a(0) + 1 = -1,\\
a(2) &= 2a(1) + 1 = -1,\\
a(3) &= 2a(2) + 1 = -1,
\end{aligned}$$

etc. (Note that the physical Tower of Hanoi puzzle cannot have a negative number of disks, so for the puzzle, -1 is an unreachable equilibrium value.)

We can use the observation that $ev = f(ev, ev, \ldots, ev)$ to find equilibrium values, and to determine whether or not a given DDS has an equilibrium value. Look again at the DDS $a(n+1) = 2a(n) + 1$. Substituting $a(n+1) = ev$ and $a(n) = ev$ into our DDS yields

$$ev = 2ev + 1.$$

Solving for ev gives the equilibrium value for the DDS as $ev = -1$.

Now, let's consider the DDS $a(n+1) = a(n) + 1$. Using our observation, we write

$$ev = ev + 1.$$

This equation has no solution, so the DDS $a(n+1) = 2a(n) + 1$ *does not have* any equilibrium values.

Example 2.6. Finding Equilibrium Values.
Consider the following four DDSs. Find their equilibrium value(s), if any exist.

1. $a(n+1) = 0.3\,a(n) - 10$
2. $a(n+1) = 1.3\,a(n) + 20$
3. $a(n+1) = 0.5\,a(n)$
4. $a(n+1) = -0.1\,a(n) + 11$

Solutions.

1. $ev = 0.3ev - 10 \implies ev = -10/0.7 \approx -14.286$
2. $ev = 1.3ev + 20 \implies ev = -20/0.3 \approx -66.667$
3. $ev = 0.5ev \implies ev = 0$
4. $ev = -0.1ev + 11 \implies ev = 11$

Above, we observed that DDSs that have equilibrium values **may never attain** their equilibrium value, given some specific initial condition $a(0)$. In DDS 3. from Example 2.6 above, choose any initial value A not equal to 0, the ev. Then, by iteration, $a(k) = 2^{-k}A \neq 0$ for any k. This DDS can never achieve its equilibrium when starting from a nonzero value even though $\lim_{n\to\infty} a(n) = 0$, the equilibrium value.

We found the equilibrium value was -1 for the DDS from the Tower of Hanoi puzzle. Suppose that we begin iterating the Tower's DDS with an initial value $a(0) = 0$. Then

$$\begin{aligned} a(1) &= 2\,a(0) + 1 = 3, \\ a(2) &= 2\,a(1) + 1 = 7, \\ a(3) &= 2\,a(2) + 1 = 15, \\ a(4) &= 2\,a(3) + 1 = 31, \end{aligned}$$

etc. The values continue to get larger and will never reach the value of -1. Since $a(n)$ represents the number of moves of the disks, the value of -1 makes no real sense in the context of the number of disks to move.

We will study equilibrium values in many of the applications of discrete dynamical systems. In general, a linear nonhomogeneous discrete dynamical system, where the nonhomogeneous part is a constant, will have an equilibrium value. *Can you find any exceptions?* A linear homogeneous discrete dynamical system will always have an equilibrium value of zero. *Why?*

2.5 A Graphical Approach to Equilibrium Values

We can examine a graph of the DDS's iterations using Maple. If the values reach a specific value and remain constant, the graph levels, then that value is an equilibrium value (change has stopped).

Dynamical systems often represent real-world behavior that we are trying to understand. We model reality to be able to predict future behavior and gain deeper insights into the behavior and how to influence or alter that behavior. Thus, we have great interest in the predictions of the model. Understanding how the model changes in the future is our goal.

Models of the Form: $a(n+1) = r \cdot a(n) + b$ with Constant r and b

Consider a savings account with 12% APR interest compounded monthly where we deposit \$1000 per month. Repeating our earlier analysis, but adding the monthly deposit. (Remember to start with a fresh Maple worksheet, then load the *plots* package.)

$$> DDS := a(n+1) = \left(1 + \frac{0.12}{12}\right) \cdot a(n) + 1000 :$$

$$> rsolve(\{DDS, a(0) = 0\}, a(n)) :$$
$$a := unapply(\%, n);$$

$$a \mapsto 101000 \left(\frac{101}{100}\right)^n - 100000$$

```
> pts := [seq([k, a(k)], k = 0..24)] :
pointplot(pts, title = "Savings Account with Monthly Deposit");
```

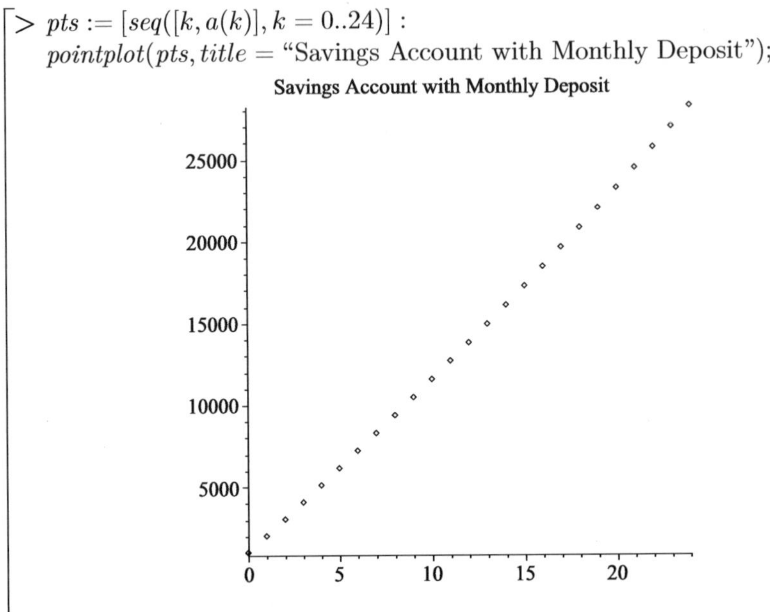

The DDS $a(n+1) = 1.01\,a(n) + 1000$, with $r = 1.01 > 1$, grows without bound. The graph suggests that there is no equilibrium value—a value where the graph levels at a constant value. Analytically, we have $ev = 1.01ev + 1000 \implies ev = -100000$. There is an equilibrium value $ev = -100000$, but that value will never be reached in our savings account problem. Without withdrawals, the savings account can never be $100,000 in the hole!

What happens if r is less than 0?

Proceed as before, but with a negative r. The new DDS is

$$a(n+1) = -1.01\,a(n) + 1000$$

with $a(0) = 0$. Analytically solve for the equilibrium value.

$$ev = -1.01\,ev + 1000$$
$$ev = 497.5124378$$

The definition of equilibrium value implies that if we start at $a(0) = 497.5124378$, we stay there forever. From the plot of the DDS in Figure 2.7, we note the oscillations between positive and negative numbers, each growing without bound as the oscillations fan out. Although there is an equilibrium value, the solution to our example does not tend toward this fixed point.

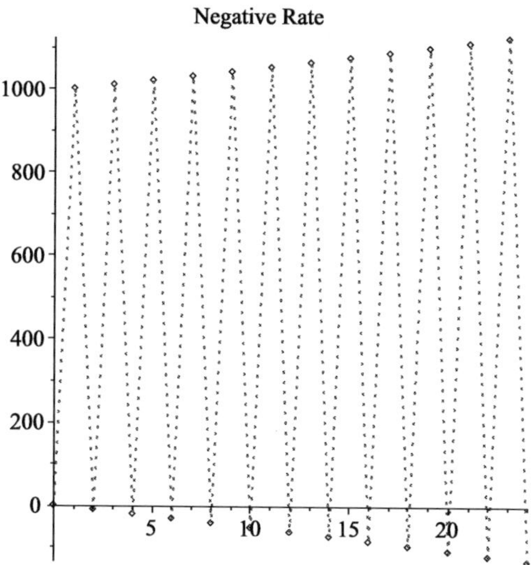

FIGURE 2.7: A Linear DDS with Negative r

Let's examine an example with a value of r between 0 and 1, that is, $0 < r < 1$.

A drug concentration model with a constant dosage of 16 mg each time period (4 hours) has the DDS

$$a(n+1) = 0.5a(n) + 16$$

with an initial dosage of $a(0)$ applied prior to beginning the regime. Figure 2.8 shows the DDS with 4 different initial values, $a(0) = 10$, 20, 32, and 50 mg.

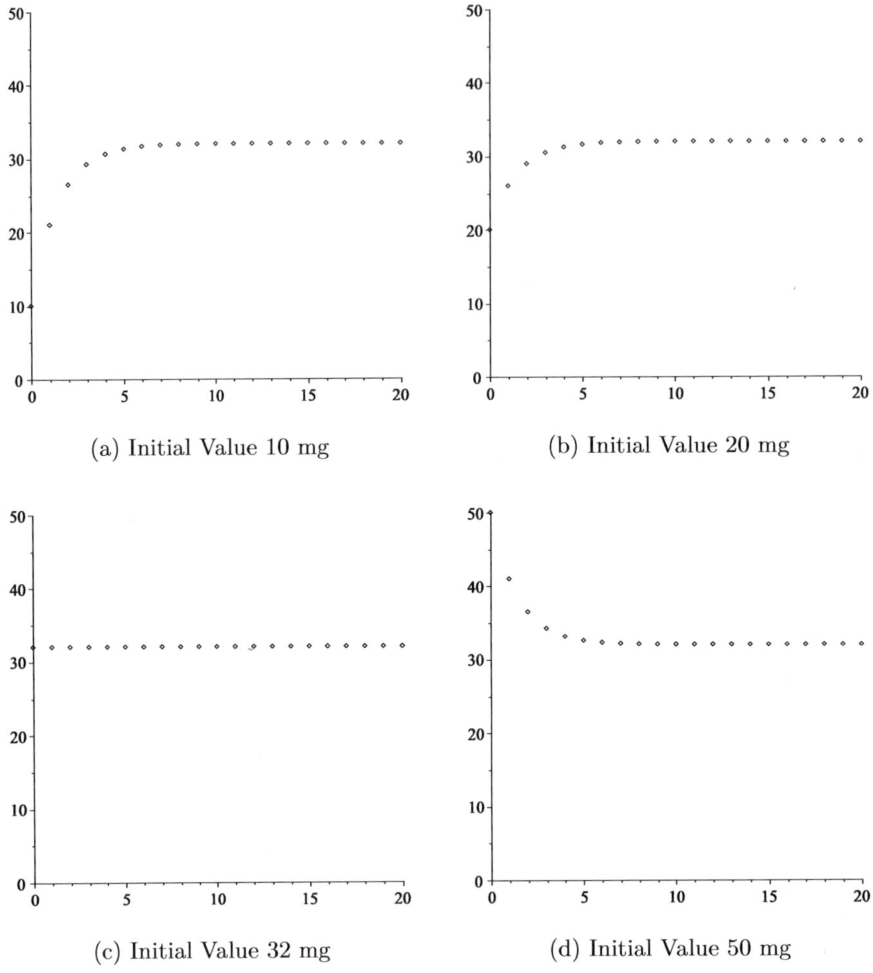

(a) Initial Value 10 mg

(b) Initial Value 20 mg

(c) Initial Value 32 mg

(d) Initial Value 50 mg

FIGURE 2.8: Drug Concentration DDS with Four Different Initial Values

Regardless of the starting value, each graph shows the future terms of $a(n)$ approach 32. Thus, 32 is the equilibrium value. We could have easily solved for the equilibrium value algebraically as well.

$$a(n+1) = 0.5\,a(n) + 16 \quad \longrightarrow \quad ev = 0.5\,ev + 16 \quad \longrightarrow \quad ev = 32$$

In general, finding the equilibrium value for this type of DDS requires solving the equation $ev = r \cdot ev + b$ for ev. We find

$$a = \frac{b}{1-r}, \quad \text{when } r \neq 1.$$

Applying this formula to the drug concentration DDS aboves calculates the equilibrium value as

$$ev = \frac{16}{1 - 0.5} = 32,$$

the value we observed from the graphs.

Stability and Long-Term Behavior

For a dynamical system $a(n)$ with a specific initial condition $a(0) = A_0$, we have shown that we can compute $a(1)$, $a(2)$, and so forth, by iteration. Often the particular values are not as important as the long-term behavior. By long-term behavior, we refer to what will eventually happen to $a(n)$ for larger and larger values of n. There are many types of long-term behavior that can occur with DDSs; we will only discuss a few here.

If the values of $a(n)$ for a DDS eventually get close[5] to the equilibrium value ev, for all initial conditions in a range, then the equilibrium is called a *stable equilibrium* or an *attracting fixed point.*

Example 2.7. A Stable Equilibrium or Attracting Fixed Point. Consider the DDS $a(n+1) = 0.5\,a(n) + 64$ with initial conditions either $a(0) = 0$, 50, 100, 150, or 200. In each case, the ev is 128; therefore, the $ev = 128$ is a stable equilibrium. Examine Figure 2.9.

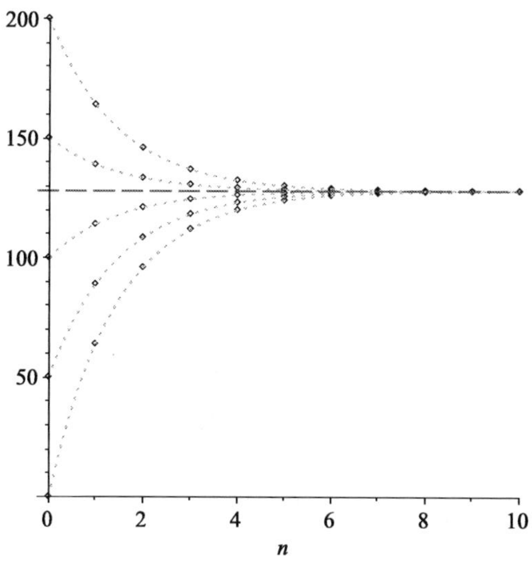

FIGURE 2.9: Drug Concentration with Varying Initial Values

[5]"Eventually get close to ev" can be defined formally as: For any positive tolerance ϵ, there is an integer $N(\epsilon)$, such that whenever $n > N$, then $|a(n) - ev| < \epsilon$, i.e., $a(n)$ is within ϵ of ev whenever $n > N$. See, e.g., Bauldry *Intro. to Real Analysis* [Bauldry2009].

Let's use *rsolve* to find a general formula, and then generate a table of values for the DDS.

$> rsolve(\{a(n+1) = 0.5 \cdot a(n) + 64, a(0) = A\}, a(n));$
$a := unapply(\%, [A, n]):$

$$A\left(\frac{1}{2}\right)^n - 128\left(\frac{1}{2}\right)^n + 128$$

$> inits := [0, 50, 100, 150, 200]:$

$> gen := (i, j) \to evalf_4(a(inits_j, i)));$
$DrugConcTable := Matrix(10, 5, gen);$

$$DrugConcTable := \begin{bmatrix} 64.0 & 89.0 & 114.0 & 139.0 & 164.0 \\ 96.0 & 108.5 & 121.0 & 133.5 & 146.0 \\ 112.0 & 118.2 & 124.5 & 130.8 & 137.0 \\ 120.0 & 123.1 & 126.2 & 129.4 & 132.5 \\ 124.0 & 125.6 & 127.1 & 128.7 & 130.2 \\ 126.0 & 126.8 & 127.6 & 128.3 & 129.1 \\ 127.0 & 127.4 & 127.8 & 128.2 & 128.6 \\ 127.5 & 127.7 & 127.9 & 128.1 & 128.3 \\ 127.8 & 127.8 & 127.9 & 128.0 & 128.1 \\ 127.9 & 127.9 & 128.0 & 128.0 & 128.1 \end{bmatrix}$$

Notice that all the sequences are converging to 128, the *attracting fixed point.* (To see more than 10 rows, first execute *interface*(*verboseproc*=100).)

Example 2.8. A Financial Model with an Unstable Equilibrium. Consider a financial model where \$100 is deposited each month in an account that pays 12% APR compounded monthly. The DDS is

$$B(n+1) = \left(1 + \frac{0.12}{12}\right) B(n) + 100, \quad B(0) = 100.$$

Let's first determine a formula for $B(n)$

$> rsolve(\{B(n+1) = \left(1 + \frac{0.12}{12}\right) \cdot B(n) + 100, B(0) = 100\}, B(n)):$
$B := unapply(\%, n);$

$$B := n \mapsto 10100 \cdot \left(\frac{101}{100}\right)^n - 10000$$

The algebraic technique easily gives us the equilibrium value.

$> ev := solve(ev = 1.01 \cdot ev + 100, ev);$

$$ev := -10000.$$

The *ev* value is −10000. If the DDS ever achieves an input of −10000, then the system stays at −10000 forever. Since \$100 is deposited each month and there are no withdrawals, the account must stay positive—we can't reach the equilibrium value! We say this equilibrium value is unstable or a repelling fixed point.

Figure 2.10 shows a graph of our financial model and *ev*.

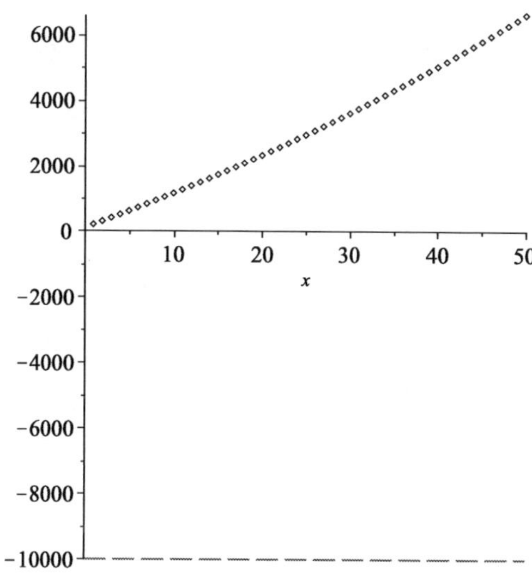

FIGURE 2.10: Unstable Equilibrium Value

Generate a table of values.

$> \langle\langle seq(i, i = 0..9)\rangle \mid \langle seq(evalf_6(B(i)), i = 0..9)\rangle\rangle;$

$$\begin{bmatrix} 0 & 100.0 \\ 1 & 201.0 \\ 2 & 303.010 \\ 3 & 406.040 \\ 4 & 510.101 \\ 5 & 615.202 \\ 6 & 721.354 \\ 7 & 828.567 \\ 8 & 936.853 \\ 9 & 1046.22 \end{bmatrix}$$

The values increase over time and never move toward -10000. Therefore, the *ev* is unstable or repelling.

Try this model with different initial values and different deposit values. Can the equilibrium become attracting?

Often, we characterize the long-term behavior of the system in terms of its stability. If a DDS has an equilibrium value, and if the solution tends to the equilibrium value from starting values near the equilibrium value, then the DDS is said to be stable. We summarize the results for the dynamical system $a(n+1) = r \cdot a(n) + b$, where $b \neq 0$ in Table 2.4.

TABLE 2.4: Linear Nonhomogeneous Discrete Dynamical Systems

Value of r	Equilibrium	Solution Stability	Long-Term Behavior
$r < -1$	$\frac{b}{1-r}$	Unstable	Unbounded Oscillations
$r = -1$	$\frac{b}{2}$	Unstable	Oscillates between $a(0)$ and $b - a(0)$
$-1 < r < 0$	$\frac{b}{1-r}$	Stable	Oscillates about equilibrium
$r = 0$	b	Stable	Constant
$0 < r < 1$	$\frac{b}{1-r}$	Stable	Goes to equilibrium from $a(0)$
$r = 1$	*None*	Unstable	Unbounded (Goes to $\text{sgn}(b) \cdot \infty$)
$1 < r$	$\frac{b}{1-r}$	Unstable	Unbounded

From the table we see that the DDS $a(n+1) = r \cdot a(n) + b$ has an equilibrium $b/(1-r)$ whenever $r \neq 1$ and has no equilibrium when $r = 1$. Further, we see that the equilibrium is stable only when $|r| < 1$.

Relationship of the Equilibrium to Analytical Solutions

If a discrete dynamical system has an equilibrium *ev*, we can use the value to find the analytical solution.

Consider the DDS for a 6.5% APR mortgage having monthly payments of \$639.34:

$$B(n+1) = 1.00541667B(n) - 639.34, \quad B(0) = 73{,}395$$

The *ev* value is analytically found to be 118031.9927. Substituting the *ev* and r values into $B(k) = r^k C + ev$ and then setting $k = 0$ gives

$$B(0) = 1.00541667^0 C + 118031.9927 = 73395$$

which yields

$$C = -44636.99270$$

The solution to our DDS is

$$B(n) = -44636.99270 \cdot 1.005416667^n + 118031.9927$$

(Check this with *rsolve*!)

We can also use this method to determine the mortgage payment. The DDS is now

$$B(n+1) = 1.00541667\,B(n) - P, \quad B(0) = 73395$$

with solution

$$B(n) = 1.00541667^n\,c + ev.$$

Build a system of two equations and two unknowns.

$$73395 = c + d \qquad \textit{(from } B(0) = 73395\textit{)}$$
$$0 = 1.00541667^{180}\,c + d \qquad \textit{(from } B(180) = 0\textit{)}$$

Solve the system.

```
> solve({73395 = c + d, 0 = 1.00541667^180 c + d}, {c, d});
          {c = -44638.66498, d = 118033.6650}
```

Thus

$$B(n) = -44638.66498 \cdot 1.005416667^n + 118033.6650.$$

(Note: there is a little round-off error creeping into the calculations.)

To find the payment P, we know that

$$ev = 1.005416667\,ev - P.$$

Substitute $ev = 118033.6650$ and solve to find

$$P = 639.3494.$$

This value agrees with our original payment other than some round-off error. Round-off in financial calculations can be very serious[6]—always be careful!

Equilibrium Values and the Limit

Studying the behavior of equilibrium values relies fundamentally on the concept of *limit*. Elementary calculus is based on limit and includes a careful study of the concept. Advanced calculus or real analysis presents a precise, carefully crafted definition, but using that would overly complicate our investigation of equilibrium values. The informal definition of *limit of a sequence*

$$\lim_{k\to\infty} a(k) = L \text{ if and only if for } n \text{ large enough } a(n) \text{ must be close to } L.^7$$

will serve our needs and keep us focused on the behavior of the discrete dynamical system in question.

[6]See Whiteside's *Computer capers: Tales of electronic thievery, embezzlement, and fraud,* Crowell, New York, c1978.

[7]A more formal definition that specifies *large enough* and *close to* precisely is: $\lim_k a(k) = L$ iff given any tolerance $\varepsilon > 0$, there is a positive integer N^* such that whenever $n > N^*$, then $|a(n) - L| < \varepsilon$. See, e.g., Bauldry *Intro. to Real Analysis* [Bauldry2009].

We are interested in what happens to $a(k)$ as k gets larger and larger without bound. (In other words, when we iterate many times.) It may happen that for large values of k, $a(k)$ is close or even equal to some number L. If increasing k (doing more iterations) causes $a(k)$ to get even closer to L until for very large values of k, $a(k)$ is essentially equal to L, then L is the limit of $a(k)$. If $a(k)$ has a limit, then we also say $a(k)$ *converges to* L or simply $a(k)$ *converges*. It is important to understand that if $a(k)$ converges to L, then further increases in k will never cause $a(k)$ to move "too far" away from L. Also remember that we are looking at large values of k. How large is large depends upon the DDS. Compare how large k must be for the two sequences DDS_1: $a(k) = e^{-k}$ and DDS_2: $b(k) = 1/(\ln(k) + 1)$.

If increasing k causes $a(k)$ to continue to increase or to oscillate, and $a(k)$ does not begin to converge to a number L, we say $a(k)$ is *diverging* and the limit does not exist.

Example 2.9. Convergent Difference Equations.
Consider

1. the difference equation $a(n+1) = 0.9\,a(n) + 2$ with $a(0) = 1$. Iterating with an accuracy of 6 decimal places produces the following:

$$
\begin{aligned}
a(1) &= 2.9 \\
&\vdots \\
a(150) &= 19.999997 \\
a(151) &= 19.999998 \\
&\vdots \\
a(156) &= 19.999999 \\
&\vdots \\
a(166) &= 20.000000
\end{aligned}
$$

   and for all $k > 166$, $a(k) = 20$. Thus, for this difference equation, the limit of $a(k)$ is 20.

2. the difference equation $b(n+1) = -0.1\,b(n) + 11$ with $b(0) = 1$. Iterating with an accuracy of 6 decimal places, we see the following:

$$
\begin{aligned}
b(4) &= 9.999100 \\
b(5) &= 10.000090 \\
b(6) &= 9.999991 \\
b(7) &= 10.000001 \\
b(8) &= 9.999999 \\
b(9) &= 10.000000
\end{aligned}
$$

and for all $k > 9$, $b(k) = 10$. Thus, for this difference equation, the limit of $b(k)$ is 10.

Even though $b(4)$ is less than 10 and $b(5)$ is greater than 10, $b(5)$ is closer to 10 than $b(4)$ is. Increasing k caused $a(k)$ to get closer to 10. Thus, the limit as $k \to \infty$ of $b(k)$ is 10. *Graph it!*

If $a(k)$ does not converge to L, then the sequence *diverges*. There are several ways in which a sequence may diverge. If $a(k)$ gets infinitely large as k gets infinitely large, then $a(k)$ diverges. If $a(k)$ gets infinitely large in the negative direction as k gets infinitely large, then $a(k)$ also diverges. If $a(k)$ oscillates between large positive and large negative values, always getting further from 0 as k gets infinitely large, then $a(k)$ diverges. If $a(k)$ oscillates in a pattern between two or more fixed values as k gets infinitely large, $a(k)$ diverges. If $a(k)$ shows absolutely no pattern of behavior as k gets infinitely large, then $a(k)$ diverges. Given a random sequence $a(k)$, it will likely diverge.

Example 2.10. Divergent Difference Equations.
Consider

1. the difference equation $a(n+1) = 4\,a(n) + 2$ with $a(0) = 1$. Iteration of this difference equation produces

$$
\begin{aligned}
a(10) &= 1747626 \\
a(11) &= 6990506 \\
a(12) &= 27962026 \\
a(13) &= 111848106
\end{aligned}
$$

   Further increase in k causes increase in $a(k)$. Thus, $a(k)$ diverges.

2. the difference equation $a(n+1) = -4\,a(n) + 2$ with $a(0) = 1$. In this case, iteration produces

$$
\begin{aligned}
a(10) &= 629146 \\
a(11) &= -2516582 \\
a(12) &= 1006630 \\
a(13) &= -40265318
\end{aligned}
$$

   As k gets larger and larger, $a(k)$ gets further from 0, always oscillating between positive and negative values. Thus, $a(k)$ diverges.

We close this section with a very useful theorem.

Theorem. Limits are Stable Equilibria.
If a dynamical system has a limit, then that limit is a stable equilibrium value; i.e., is an attracting fixed point.

Exercises

1. For each of the following DDSs, find the equilibrium value(s), if any exist. Classify the DDS as *stable* or *unstable*.

   (a) $a(n+1) = 1.23a(n)$

   (b) $a(n+1) = 0.99a(n)$

   (c) $a(n+1) = -0.8a(n)$

   (d) $a(n+1) = a(n) + 1$

   (e) $a(n+1) = 0.75a(n) + 21$

   (f) $a(n+1) = 0.80a(n) + 100$

   (g) $a(n+1) = 0.80a(n) - 100$

   (h) $a(n+1) = -0.80a(n) + 100$

2. Build a numerical table for the following DDSs. Observe the patterns and provide information on equilibrium values and stability.

   (a) $a(n+1) = 1.1a(n) + 50$ with $a(0) = 1010$

   (b) $a(n+1) = 0.85a(n) + 100$ with $a(0) = 10$

   (c) $a(n+1) = 0.75a(n) - 100$ with $a(0) = -25$

   (d) $a(n+1) = a(n) + 100$ with $a(0) = 500$

2.6 Modeling Nonlinear Discrete Dynamical Systems

In this section, we build nonlinear DDS to describe the change in behavior of the quantities we study. We also will study systems of DDS to describe the changes in various quantities that act together in some way or influence each other in deterministic ways. We define a nonlinear DDS by

> If the DDS $a(n+1) = f(a(n), \ldots, n)$ involves powers of $a(k)$ [such as $a^2(k)$], or a functional relationship [such as $a(k)/a(k-1)$ or $\sin(a(k))$], we say the discrete dynamical system is **nonlinear**.

We will restrict our investigations to numerical and graphical solutions of nonlinear models. Analytical solutions of nonlinear models are studied in more advanced mathematics courses.

We often model population growth by assuming that the change in population is directly proportional to the current size of the given population.

This produces a simple, first-order DDS similar to those seen earlier. It might appear reasonable at first examination, but the long-term behavior of growth without bound is disturbing. Why would growth without bound of a yeast culture in a jar (or confined space) be alarming?[8]

There are certain factors that affect population growth. These factors include resources: food, oxygen, space, etc. The available resources can support some maximum population. As this number is approached, the change, or growth rate, should decrease and the population should never exceed its resource supported amount. If the population does exceed the maximum supported amount, the growth should become negative.

Example 2.11. Growth of a Cancer Culture.
Problem.

To determine whether our treatments are successful, we need to predict the growth of cancer cells in a controlled environment as a function of the resources available and the current cell 'population.'

Assumptions and Variables.

We assume that the population size is best described by the weight of the biomass of the culture. Define $y(n)$ as the population size of the cell culture after period n. There exists a maximum carrying capacity M that is sustainable by the resources available. The culture is growing under the conditions established.

Model.

$$\begin{aligned} n &: \text{ the time period measured in hours} \\ y(n) &: \text{ the population size after period } n \\ M &: \text{ the } \textit{carrying capacity} \text{ of our system} \\ y(n+1) &= y(n) + k\,y(n)\,(M - y(n)) \end{aligned}$$

where k is the constant of proportionality (a function of *growth rate*).

Using the data collected in our experiment, we first plot $y(n)$ versus n and find a stable *ev* of approximately 670. Next, we plot $y(n+1) - y(n)$ versus $y(n)(670 - y(n))$ to find the slope, which gives k, is approximately 0.00090. With $k = 0.00090$ and carrying capacity in biomass is 670. Our specific model[9] is then

$$y(n+1) = y(n) + 0.0009\,y(n)(670 - y(n))$$

Again, this is nonlinear because of the $y^2(n)$ term (after expanding the right side of the model). There is no closed-form analytical solution for this model. The numerical solution iterated in Maple from an initial condition of $\textit{biomass} = 30.0$ follows.

[8]See Rev. Malthus' 1798 book *An Essay on the Principle of Population* [Malthus1798].
[9]For an approach to fitting this type of model to data, see Bauldry, "Fitting Logistics to the U.S. Population," MapleTech, 4(3), 73-77.

```
> k, M, init := 0.0009, 670, 30.0 :
```

```
> biomass := proc(n :: integer)
    option remember;
    piecewise(n > 0,
      biomass(n − 1) + k · biomass(n − 1) · (M − biomass(n − 1)),
      init);
    end proc :
```

We use **option** *remember* in the procedure to 'remember' (store) values of *biomass* as they are computed in order to dramatically reduce the number of calculations needed. *Try it without the option!*

```
> pts := [seq([n, biomass(n)], n = 0..30)] :
```

```
> pointplot(pts, view = [0..30, 0..700], title = "Biomass");
```

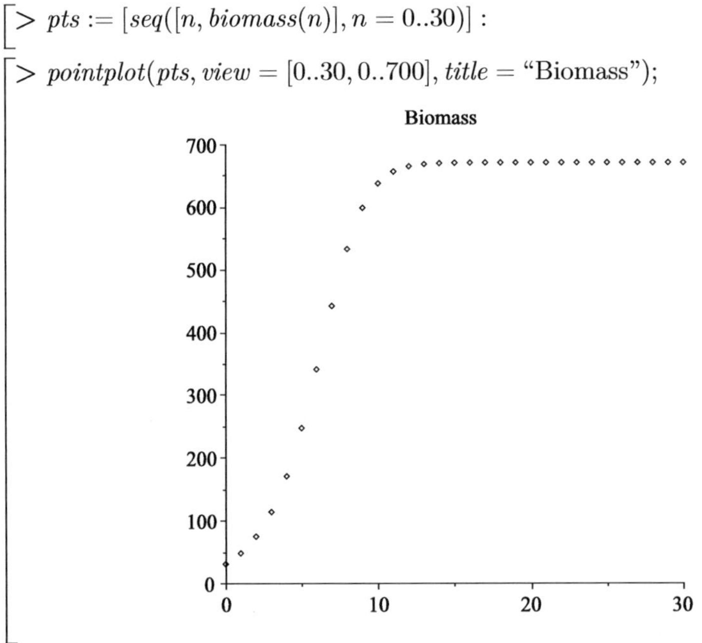

The model shows stability in that the population (biomass) of the cell culture quickly approaches 670 as n gets large. Thus, the population is eventually stable at approximately 670 units. We would then proceed with treatments, collect new data, and evaluate the new trend of the biomass to determine whether the treatment is successful or not.

Example 2.12. Spread of a Contagious Disease.
There are 1000 students in a college dormitory. Several students have returned from a Spring Break trip where they were exposed to mumps, a highly contagious disease. The Health Center wants to build a model to determine how fast the disease will spread in order to determine effective vaccination and quarantine procedures.

Problem.

Predict the number of students affected with mumps as a function of time.

Assumptions and Variables.

We assume all students are susceptible to the disease. The possible contacts of infected and susceptible students are proportional to their product (as an *interaction term*). Let $m(n)$ be the number of students affected with mumps after n days.

Model.

$$m(n+1) = m(n) + k \cdot m(n) \cdot (1000 - m(n))$$

Two students have come down with mumps. The rate of spreading per day is characterized by $k = 0.0005$. A vaccine can be delivered and students vaccinated within 1-2 weeks. Hence

$$m(n+1) = m(n) + 0.0005m(n) \cdot (1000 - m(n))$$

This model matches the cancer model above. We modify its Maple code accordingly.

```
> k, M, init := 0.0005, 1000, 2 :
> Mumps := proc(n :: integer)
    option remember;
    piecewise(n > 0,
      Mumps(n - 1) + k · Mumps(n - 1) · (M - Mumps(n - 1)), init);
    end proc :
> pts := [seq([n, round(Mumps(n))], n = 0..30)] :
```

We *rounded* the value of *Mumps* as it represents an integer: the number of infected students.

Graph the points to see the trend of the disease.

> *pointplot*(*pts*, *title* = "Mumps Infections");

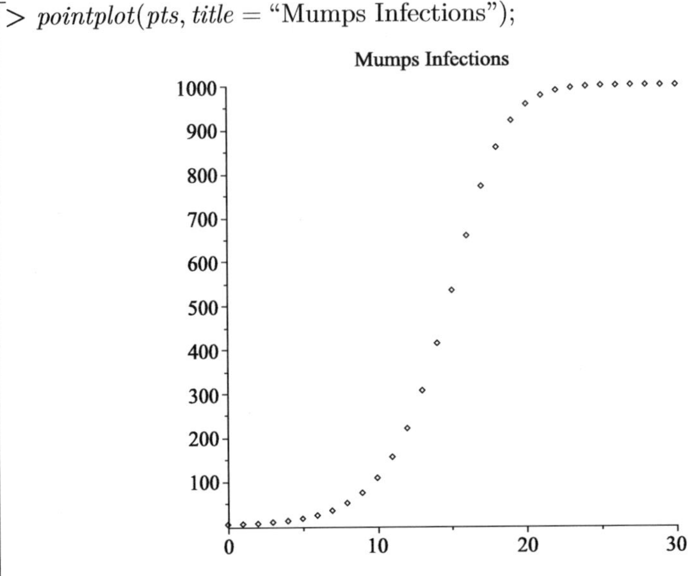

By 21 days, essentially all the students are infected. Let's look at the weekly counts.

> $seq(round(Mumps(k)), k$ **in** $[7, 14, 21, 28])$

$$33, 413, 977, 1000$$

Interpretation.

The results show that essentially all the students will be infected by 3 weeks. Since only about 3% will be infected within one week, every effort must be made to get the vaccine to the school and to vaccinate the students within one week.

We can model the effect of the vaccine on the spread of mumps by reducing k in the model and using $Mumps(7)$ as the initial value of the new curve.

Exercises

Iterate and graph the following DDSs. Explain their long-term behavior. For each DDS, find a realistic scenario that it might explain/model.

1. Consider the model $a(n+1) = r \cdot a(n)(1 - a(n))$. Let $a(0) = 0.2$. Determine the numerical and graphical solutions for the following values of r. Find the pattern in the solution.

   (a) $r = 2$

   (b) $r = 3$

(c) $r = 3.6$
(d) $r = 3.7$

2. Find the equilibrium value of the given DDS by iteration. Determine if the equilibrium value is stable or unstable.

   (a) $a(n+1) = 1.7a(n) - 0.14a(n)^2$
   (b) $a(n+1) = 0.8a(n) + 0.1a(n)^2$
   (c) $a(n+1) = 0.2a(n) - 0.2a(n)^3$
   (d) $a(n+1) = 0.1a(n)^2 + 0.9a(n) - 0.2$

3. A rumor spreads through a company of 1000 employees all working in the same building. We assume that the spread of a rumor is similar to the spread of a contagious disease in that the number of people hearing the rumor each day is proportional to the product of the number hearing the rumor and the number who have not yet heard the rumor. This model is given by

$$r(n+1) = r(n) + 1000\,k \cdot r(n) - k \cdot r(n)^2$$

   where k is the parameter that depends on how fast the rumor spreads. Assume $k = 0.001$. Further assume that 4 people initially know the rumor. How soon will everyone know the rumor?

Projects

Project 2.1. Consider the highly contagious and deadly Ebola virus, which in 2018 appeared to be spreading again. Determine how deadly this virus actually is. (Visit https://www.cdc.gov/vhf/ebola/index.html.) Consider an animal research laboratory in Restin, VA (pop. $\sim$ 58,000), a suburb of Washington, DC, (pop. $\sim$ 602,000). A monkey carrying the Ebola virus has escaped its cage and infected one employee (unknown at the time) during its escape from the research laboratory. This employee reports to University hospital later with Ebola symptoms. The Centers for Disease Control and Prevention (CDC) in Atlanta gets a call and begins to model the spread of the disease. Build a model for the CDC with the following growth rates to determine the number infected after 2 weeks:

(a) $k = 0.00025$
(b) $k = 0.000025$
(c) $k = 0.00005$
(d) $k = 0.000009$

List some possible ways of controlling the spread of the virus.

Project 2.2. A rumor concerning termination spreads among 10,000 employees of a major company. Assume that the spreading of a rumor is similar to the spread of contagious disease in that the number hearing the rumor each day is proportional to the product of those who have heard the rumor and those who have not yet heard the rumor. Build a model for the company with the following rumor growth rates to determine the number having heard the rumor after 1 week:

(a) $k = 0.25$

(b) $k = 0.025$

(c) $k = 0.0025$

(d) $k = 0.00025$

List some possible ways of controlling the spread of the rumor.

2.7 Systems of Discrete Dynamical Systems

In this section, we examine models of systems of DDS. For selected initial conditions, we'll build numerical solutions to get a sense of long-term behavior of the system. We'll find the equilibrium values of the systems we study. We'll then explore starting values near the equilibrium values to see if by starting close to an equilibrium value, the system will:

a. Remain close to the equilibrium value

b. Approach the equilibrium value

c. Move away from the equilibrium value

What happens near the equilibrium values gives great insight into the long-term behavior of the system. We can study the resulting numerical solutions for patterns.

Example 2.13. Location Merchants Choose.
Consider an attempt to revitalize the downtown section of a small city with merchants. There are some merchants downtown, and others in the large city shopping plaza. Suppose historical records showed that 60% of the downtown merchants remain downtown, while 40% move to the shopping plaza. We find that 70% of the plaza merchants want to remain in the plaza, but 30% want to move downtown. Build a model to determine the long-term behavior of these merchants based upon the historical data. See Figure 2.11. There are initially 100 merchants in the plaza and 150 merchants downtown. We seek to find the long-term behavior of this system.

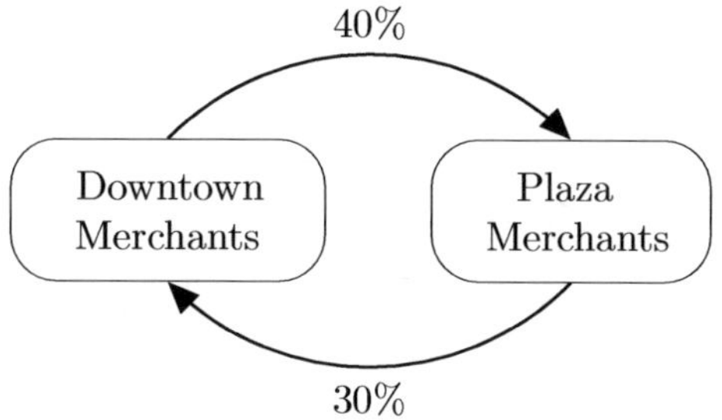

FIGURE 2.11: Diagram of Merchant Movement

Problem.

Determine the behavior of the merchants over time to see whether the downtown will survive.

Assumptions and Variables.

Let n represent the number of months. We define

$$D(n) = \text{ number of merchants downtown at the end of the } n\text{th month}$$
$$P(n) = \text{ number of merchants at the plaza at the end of the } n\text{th month}$$

We assume that no incentives are given to the merchants for either staying or moving.

The Model.

The number of merchants downtown in any time period is equal to the number of downtown merchants that remain downtown plus the number of plaza merchants that relocate downtown. The same is true for the number of plaza merchants: the number is equal to the number that remain in the plaza plus the number of downtown merchants that move to the plaza. Mathematically, we write the model as the system

$$d(n+1) = 0.60d(n) + 0.30p(n)$$
$$p(n+1) = 0.40d(n) + 0.70p(n)$$

with $d(0) = 150$ and $p(0) = 100$ merchants, respectively.

Let's use Maple to explore this system.

```
> d := proc(n)
    option remember;
    if n < 1 then 150 else 0.60 · d(n − 1) + 0.30 · p(n − 1) end if :
    end proc :
  p := proc(n)
    option remember;
    if n < 1 then 100 else 0.40 · d(n − 1) + 0.70 · p(n − 1) end if :
    end proc :
```

Numerically, we have:

```
> Pts := Matrix([seq([k, d(k), p(k)], k = 0..9)]);
```

$$Pts := \begin{bmatrix} 0 & 150 & 100 \\ 1 & 120.0 & 130.0 \\ 2 & 111.0 & 139.0 \\ 3 & 108.300000 & 141.700000 \\ 4 & 107.4900000 & 142.5100000 \\ 5 & 107.2470000 & 142.7530000 \\ 6 & 107.1741000 & 142.8259000 \\ 7 & 107.1522300 & 142.8477700 \\ 8 & 107.1456690 & 142.8543310 \\ 9 & 107.1437007 & 142.8562993 \end{bmatrix}$$

And now, graphically:

```
> g1 := pointplot(Pts[.., [1, 2]], color = red) :
  g2 := pointplot(Pts[.., [1, 3]], color = blue) :
> display(g1, g2, legend = [evaln(d(n)), evaln(p(n))]);
```

Analytically, we can solve for the equilibrium values. Let $X = d(n)$ and $Y = p(n)$. Then, from the DDS, we have

$$X = 0.6X + 0.3Y$$
$$Y = 0.4X + 0.7Y$$

However, both equations reduce to $X = 0.75Y$.

There are 2 unknowns, so we need a second equation. From the initial conditions, we know that $X + Y = 250$. Use the equations

$$X = 0.75Y$$
$$X + Y = 250$$

to find the equilibrium values

$$X = 107.1428571 \text{ and } Y = 142.85714329$$

Iterating from near these values, we find the sequences (quickly) tend toward the equilibrium. We conclude the system has *stable equilibrium values.*

Change the initial conditions and see what behavior follows!

Interpretation.

The long-term behavior shows that eventually (without other influences) of the 250 merchants, about 107 merchants will be in the plaza and about 143 will be downtown. We might want to try to attract new businesses to the community by adding incentives for operating either in the downtown business area or in the shopping plaza.

Competitive hunter models involve species vying for the same resources (such as food or living space) in their habitat. The effect of the presence of a second species diminishes the growth rate of the first and vice versa.[10]

Let's consider a specific example with lake trout and bass in a small lake.

Example 2.14. Competitive Hunter Model[11].

Hugh Ketum owns a small lake in which he stocks fish with the eventual goal of allowing fishing. He decided to stock both bass and lake trout. The Fish and Game Warden tells Hugh that after inspecting his lake for environmental conditions he has a solid base for growth of his fish. In isolation, bass grow at a rate of 20% and trout at a rate of 30% given an abundance of food. The warden tells Hugh that the species' interaction for the food affects the trout more than the bass. They estimate the interaction term affecting bass is $0.0010 \cdot \textit{bass} \cdot \textit{trout}$, and for trout it is $0.0020 \cdot \textit{bass} \cdot \textit{trout}$. Assume no changes in the habitat occur.

[10]See "Bass are bad news for lake trout" from the online version of *nature, International journal of science* [Lawrence1999].

[11]This example comes from a problem developed by Dr. Rich West, Professor Emeritus, Francis Marion University.

Model.

Define the following:

$$\begin{aligned} B(n) &= \text{ the number of bass in the lake after period } n \\ T(n) &= \text{ the number of lake trout in the lake after period } n \\ B(n)\cdot T(n) &= \text{ interaction of the two species} \end{aligned}$$

Then

$$\begin{aligned} B(n+1) &= 1.20B(n) - 0.0010\, B(n)\cdot T(n) \\ T(n+1) &= 1.30T(n) - 0.0020\, B(n)\cdot T(n) \end{aligned}$$

The equilibrium values can be found by substituting $B(n) = x$ and $T(n) = y$, then solving for x and y. We have

$$\begin{aligned} x &= 1.2x - 0.001xy \\ y &= 1.3y - 0.002xy \end{aligned}$$

We rewrite these equations as

$$\begin{matrix} 0.2x - 0.001xy = 0 \\ 0.3y - 0.002xy = 0 \end{matrix} \implies \begin{matrix} x\,(0.2 - 0.001y) = 0 \\ y\,(0.3 - 0.002y) = 0 \end{matrix}$$

The solution is ($x = 0$ or $y = 200$) and ($y = 0$ or $x = 150$) which gives the equilibrium values as $(0, 0)$ and $(150, 200)$.

Our next task is to investigate the long-term behavior of the system and the stability of the equilibrium points.

Hugh initially considered stocking 151 bass and 199 trout in his lake. The solution is left to the student as an exercise. (Don't forget to use **option** *remember* in your programs for bass and trout.) From Hugh's initial conditions, bass will grow without bound and trout will die out ($T(29) = 0$). This is certainly not what Hugh had in mind.

Example 2.15. Fast Food Tendencies.

The Student Union of a university that has 14,000 students plans to have fast food chains available serving burgers, tacos, and pizza. The chains commissioned a survey of students finding the following information concerning lunch: 75% of those who ate burgers will eat burgers again at the next lunch, 5% will eat tacos next, and 20% will eat pizza next. Of those who ate tacos last, 20% will eat burgers next, 60% will stay will tacos, and 35% will switch to pizza. Of those who ate pizza, 40% will eat burgers next, 20% tacos, and 40% will stay with pizza. See Figure 2.12.

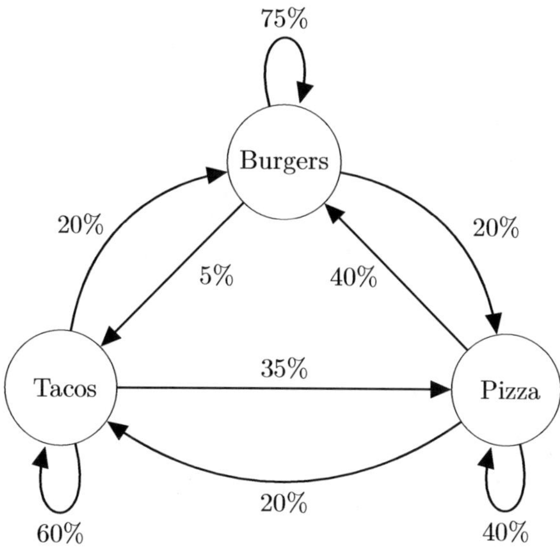

FIGURE 2.12: Diagram of Fast Food Movement

Model.

We formulate the problem as follows. Let n represent the nth day's lunch, and define

$$\begin{aligned} B(n) &= \text{ the number of burger eaters in the } n\text{th lunch period} \\ T(n) &= \text{ the number of taco eaters in the } n\text{th lunch period} \\ P(n) &= \text{ the number of pizza eaters in the } n\text{th lunch period} \end{aligned}$$

Using the values in the problem and the diagram leads us to the discrete dynamical system

$$\begin{aligned} B(n+1) &= 0.75B(n) + 0.20T(n) + 0.40P(n) \\ T(n+1) &= 0.05B(n) + 0.60T(n) + 0.20P(n) \\ P(n+1) &= 0.20B(n) + 0.20T(n) + 0.40P(n) \end{aligned}$$

The same analytic technique that we used in the bass-lake trout example lets us find any equilibria for our fast food problem. Substitute $B = x$, $T = y$, and $P = z$ in the DDS, then solve. Thus

$$\begin{aligned} x &= 0.75x + 0.20y + 0.40z \\ y &= 0.05x + 0.60y + 0.20z \\ z &= 0.20x + 0.20y + 0.40z \end{aligned} \implies \left\{x = \frac{20\,z}{9}, y = \frac{7\,z}{9}, z = z\right\}$$

Since the university has 14,000 students, then $x + y + z = 14000$. Substitute the result above into this equation.

$$\frac{20\,z}{9} + \frac{7\,z}{9} + z = 14000 \implies z = 3500$$

The equilibrium value is then $(B, T, P) = (7777.8, 2722.2, 3500)$.

The campus has 14,000 students who eat lunch. The graphical results in Figure 2.13 also show that an equilibrium value is reached at a value of about 7778 burger eaters, 2722 taco eaters, and 3500 pizza eaters. This information allows the fast food establishments to plan for a projected future. By varying the initial conditions, the initial numbers of who eats where, for 14,000 students we find that these are stable equilibrium values. (*Do this!*)

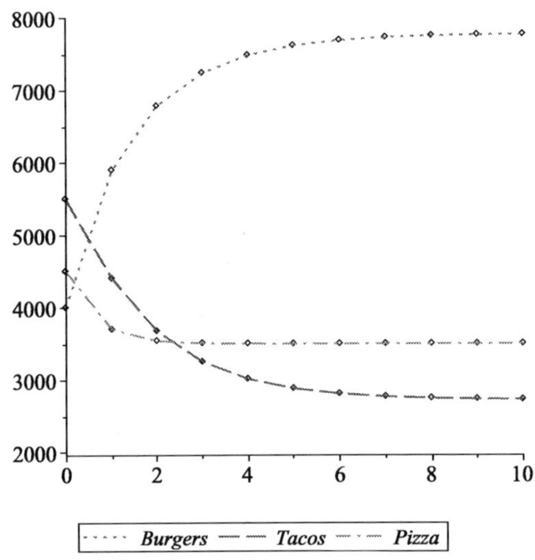

FIGURE 2.13: Graphical Results of Burgers, Tacos, and Pizza

Note the equilibrium value 7778 does not indicate that the *same people* always eat burgers, etc., but rather that the *same number* of people choose burgers, etc.

Exercises

Iterate and graph the following DDSs. Explain their long-term behavior. For each DDS, find a realistic scenario that it might explain/model.

1. What happens to the merchant problem if 200 were initially in the shopping plaza and 50 were in the downtown portion?

2. Determine the equilibrium values of the bass and lake trout. Can these levels ever be achieved and maintained? Explain.
3. Test the fast food models with different starting conditions summing to 14,000 students. What happens? Obtain a graphical output and analyze the graph in terms of long-term behavior.

Projects

Project 2.1. Small Birds and Osprey Hawks.
Problem.
Predict the number of small birds and osprey hawks in the same environment as a function of time. Osprey hawks will eat small birds when fish aren't readily available along the coast. The coefficients given below are assumed to be accurate.
Variables.

$$\begin{aligned} B(n) &= \text{ number of small birds at the end of period } n \\ H(n) &= \text{ number of hawks at the end of period } n \end{aligned}$$

Model.

$$\begin{aligned} B(n+1) &= 1.2\,B(n) - 0.001\,B(n)H(n) \\ H(n+1) &= 1.3\,H(n) - 0.002\,H(n)B(n) \end{aligned}$$

(a) Find the equilibrium values of the system.
(b) Iterate the system from the initial conditions given in Table 2.5 and determine what happens to the hawks and small birds in the long term.

TABLE 2.5: Initial Conditions for Small Birds and Hawks

Small Birds	Hawks
150	200
151	199
149	210
10	10
100	100

Project 2.2. Winning at Racket-Ball.
Rickey and Grant play racket-ball very often and are very competitive. Their racket-ball match consists of two games. When Rickey wins the first game,

he wins the second game 75% of the time. When Grant wins the first game, he wins the second only 55% of the time. Diagram the 'movement.' Model this situation as a DDS and determine the long-term percentages of their racket-ball game wins. What assumptions are necessary?

Project 2.3. Voter Distribution.
It is getting close to election day. The influence of the new Independent Party is of concern to both the Republicans and Democrats. Assume that in the next election that 79% of those who vote Republican vote Republican again, 1% vote Democratic, and 20% vote Independent. Of those that voted Democratic before, 5% vote Republican, 70% vote Democratic again, and 20% vote Independent. Of those who previously voted Independent, 35% will vote Republican, 20% will vote Democratic, and 45% will vote Independent again.

(a) Formulate the discrete dynamical system that models this situation.

(b) Assume that there are 399,998 voters initially in the system. How many will vote Republican, Democratic, and Independent in the long run? (Hint: you can break down the 399,998 voters in any manner that you desire as initial conditions.)

(c) New Scenario. In addition to the above, the community is growing:

18-year-olds+new people moving in−deaths−current people moving out.

Republicans predict a gain of 2,000 voters between elections. Democrats also estimate a gain of 2,000 voters between elections. The Independents estimate a gain of 1,000 voters between elections. If this rate of growth continues, what will be the long-term distribution of the voters?

2.8 Case Studies: Predator–Prey Model, SIR Model, and Military Models

A Predator-Prey Model: Foxes and Rabbits

In the study of the dynamics of a single population, we typically take into consideration such factors as the "natural" growth rate and the "carrying capacity" of the environment. Mathematical ecology often studies populations that interact, thereby affecting each other's growth rates. In this Case Study, we investigate a very special case of interaction, in which there are exactly two species: a predator which eats a prey. Such pairs exist throughout nature:

- lions and gazelles,
- birds and insects,

- pandas and bamboo,
- Venus Flytraps and insects.

Let $x(n)$ be the size of the prey population and $y(n)$ be the size of the predator population at time period n.

To keep our model simple, we will make some assumptions that would be unrealistic in most predator-prey situations. Specifically, we will assume that:

1. the predator species is totally dependent on a single prey species as its only food supply, (e.g., koalas and eucalyptus trees)
2. the prey species has an unlimited food supply, and
3. there are no other threats to the prey, just the specific predator.

We will use the Lotka-Volterra model.[12] If there were no predators, the second assumption would imply that the prey species grows exponentially without bound, then $x(n+1) = a \cdot x(n)$.

But there *are* predators, which must account for a negative component in the prey growth rate. The crucial additional assumptions about predator-prey interactions for developing the model are:

1. The rate at which predators encounter prey is jointly proportional to the sizes of the two populations.
2. A fixed proportion of encounters between predator and prey lead to the death of the prey.

These assumptions lead to the conclusion that the negative component of the prey growth rate is proportional to the product xy of the population sizes. Putting these factors together yields

$$x(n+1) = x(n) + a\,x(n) - b\,x(n) \cdot y(n)$$

Now we consider the predator population. If there were no food supply, the predators would die out at a rate proportional to its size; i.e., we would find $y(n+1) = -c\,y(n)$.

We assume that the "natural growth rate" is a composite of birth and death rates, both presumably proportional to the current population size. In the absence of food, there is no energy supply to support the birth rate. But there is a food supply: the prey. And what's bad for foxes is good for rabbits. That is, the energy to support growth of the predator population is proportional to deaths of prey, so

$$y(n+1) = y(n) - c\,y(n) + p\,x(n) \cdot y(n)$$

[12]Lotka and Volterra independently developed this model. See Lotka's *Elements of physical biology,* Williams & Wilkins Co. 1925, and Volterra's "Variazioni e fluttuazioni del numero d'individui in specie animali conviventi," Mem. R. Accad. Naz. dei Lincei, Ser. VI, vol 2, 1926.

Put all the above together to have the discrete version of the Lotka-Volterra Predator-Prey Model

$$\begin{aligned} x(n+1) &= (1+a)x(n) - b\,x(n)y(n) \\ y(n+1) &= (1-c)y(n) + p\,x(n)y(n) \end{aligned} \qquad \text{for } n = 0, 1, 2, \ldots, \tag{2.1}$$

with $(x(0), y(0)) = (x_0, y_0)$ and where a, b, c, and p are positive constants.

The continuous Lotka-Volterra model, analogous to (2.1), consists of the system of linked differential equations

$$\begin{aligned} x'(t) &= +\alpha\,x(t) - \beta\,x(t)y(t) \\ y'(t) &= -\gamma\,y(t) + \rho\,x(t)y(t) \end{aligned} \qquad \text{for } t \geq 0,$$

that cannot be separated from each other and that cannot be solved in closed form. Nevertheless, the continuous model, just like the discrete, can be solved numerically and graphed in order to obtain insights about the system being studied.

Let's model an isolated population of foxes and hares on a small island with a discrete Lotka-Volterra system. Data collected in the field has yielded the estimates

$$\{a, b, c, p\} = \{0.039, 0.0003, 0.12, 0.0001\}$$

for the parameters of (2.1). Use Maple to investigate the model's behavior.

In order to simplify the Maple code, we will use *global variables*, variables defined outside any program but available inside any program. Remember to load *plots* and to use *option remember* to drastically reduce the amount of computation needed.

```
> PredPrey := proc(n :: integer)
    local u, v, rn, fn;
    global r0, f0, a, b, c, p;
    option remember;
    if n < 1 then
      u := [r0, f0];   # Initial [rabbits, foxes]
    else
      v := PredPrey(n − 1);
      rn := v[1];   # current rabbits
      fn := v[2];   # current foxes
      u := [(1 + a) · rn − b · rn · fn, (1 − c) · fn + p · rn · fn];
    end if;
    return(u);
    end proc :
```

```
> r0, f0 := 900, 110 :
  a, b, c, p := 0.039, 0.0003, 0.12, 0.0001 :
```

> $RF := Matrix([seq(PredPrey(i), i = 0..200)]);$

$$RF := \begin{bmatrix} 900 & 110 \\ 905.4000 & 106.7000 \\ 911.7287460 & 103.5566180 \\ 918.9615035 & 100.5713784 \\ 927.0746346 & 97.74493550 \\ 936.0454902 & 95.07722828 \\ 945.8522811 & 92.56762196 \\ 956.4739312 & 90.21503696 \\ 967.8899152 & 88.01806562 \\ 980.0800826 & 85.97507756 \\ \vdots & \vdots \end{bmatrix}$$

201 × 2 Matrix

(Double click the output (blue) matrix to open the *Matrix Browser* to see all the entries.)

> $Rabbits := [seq([i, RF[i, 1]], i = 1..201)]$:
$pointplot(Rabbits, title =$ "Rabbits");

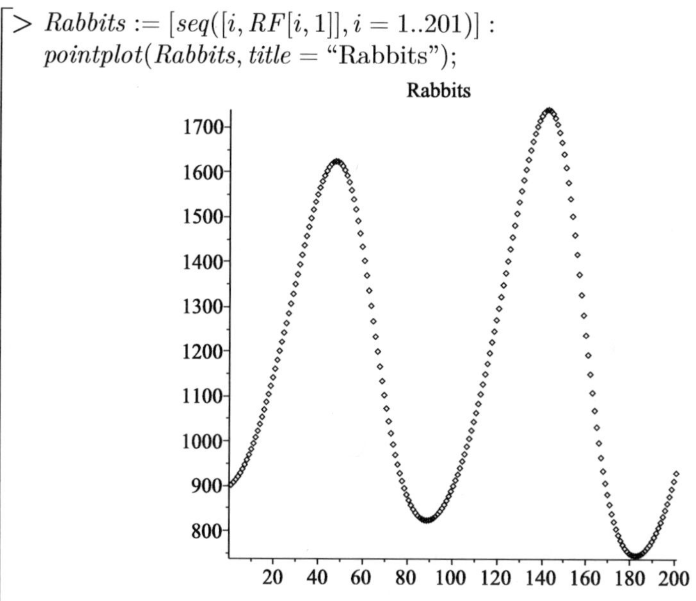

> $Foxes := [seq([i, RF[i,2]], i = 1..201)]:$
> $pointplot(Foxes, title = \text{"Foxes"});$

> $pointplot(RF, labels = [\text{"Rabbits"}, \text{"Foxes"}], title = \text{"Predator-Prey"});$

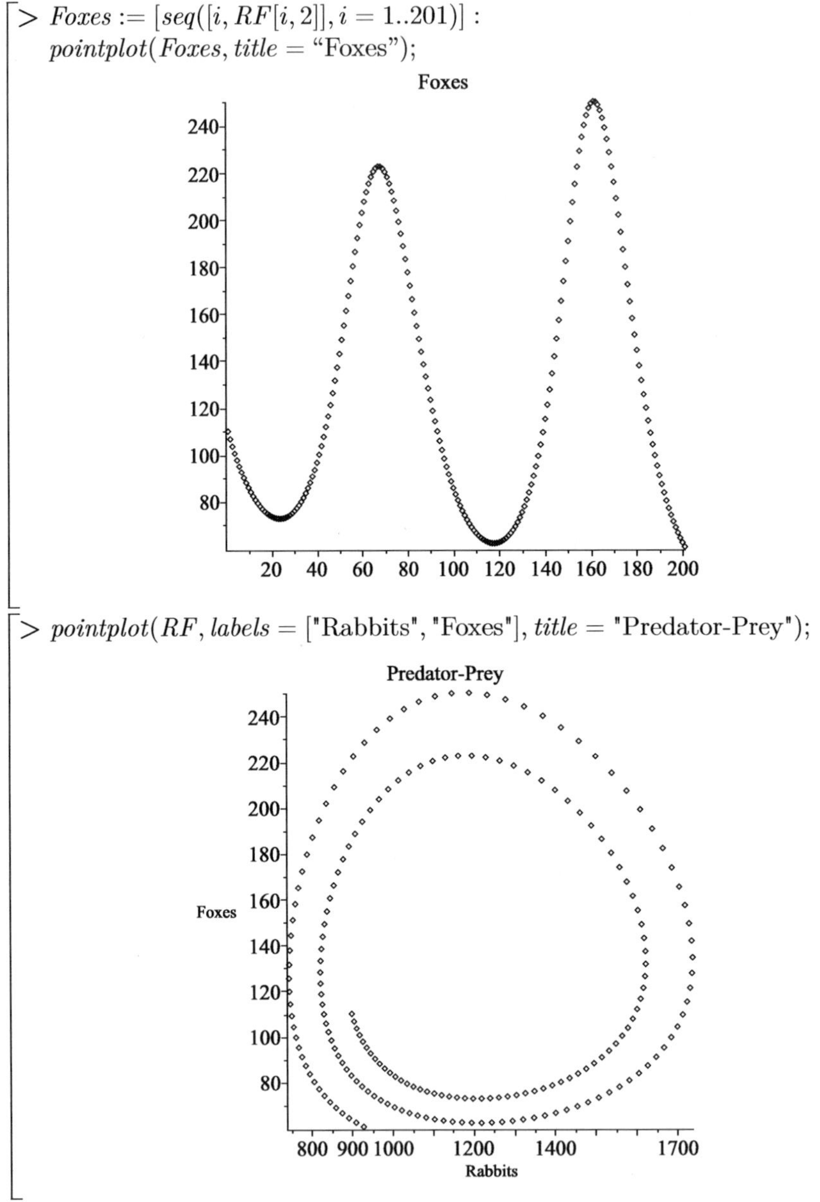

Running this model for even more iterations would show the plot of foxes versus rabbits continuing to spiral. We conclude that the model appears reasonable. We could find the equilibrium values for the system. There are two sets of equilibrium points for rabbits and foxes at $(0,0)$ and $(1200,130)$. The orbits of the spiral indicate that the system is moving away from $(1200,130)$.

We conclude that this system—depending on this set of parameter values—is not stable. Further explorations will appear in the exercise set.

A Discrete SIR Model of an Epidemic

A new flu variant is spreading throughout the Unites States. The Centers for Disease Control and Prevention (CDC) is interested in knowing and experimenting with a model for this new disease prior to it actually becoming a real epidemic.[13] For our model, the population is divided into three categories: susceptible, infected, and removed. We make the following assumptions.

- The community is closed; no one enters or leaves and there is no outside contact.
- Each person is either susceptible to the new flu, infected and can spread the flu; or already has had the flu and now has immunity (includes those who have died from the disease).
- Initially, every person is either susceptible or infected; there is no initial immunity.
- Once someone contracts this flu and recovers, they have immunity and cannot be reinfected.
- The average course of the disease is 2 weeks; during this time, the person is infected and contagious.

The time period for our model will be one week. Define the variables

$$\begin{aligned} n &= \text{ the current week} \\ S(n) &= \text{ the number that are susceptible at week } n \\ I(n) &= \text{ the number that are infected and contagious at week } n \\ R(n) &= \text{ the number that have recovered (or died) and are immune at week } n \end{aligned}$$

Begin by examining $R(n)$. The length of time someone has the flu is 2 weeks. Thus, half the infected people will be removed each week. So

$$R(n+1) = R(n) + 0.5\, I(n).$$

The value 0.5 is called the *removal rate* per week. This value represents the proportion of the infected persons who are 'removed' from infection each week. If real data were available, we would analyze the data to estimate the removal rate parameter.

Now examine $I(n)$. The number infected will both increase by new infections and decrease by removed over time. As the disease lasts for 2 weeks, half

[13]See https://www.cdc.gov/flu/weekly/flusight/.

the number infected are removed each week, $0.5\,I(n)$. The increase is proportional to the number of susceptibles that come into contact with an infected person and subsequently catch the disease, $a\,S(n)\cdot I(n)$. The parameter a, the likelihood that contact leads to infection, the *transmission coefficient.* We realize this is a probabilistic coefficient. We will assume that this rate is a constant that can be estimated from initial conditions.

For illustration, assume the population is 1,000 students in dorms. The Health Center reported that 3 students were infected initially. The next week, 5 students came to the Health Center with flu-like symptoms. We have $I(0) = 3$ and $S(0) = 997$ and 5 new infections. Then

$$5 = a\,S(0)I(0) = a\,997\cdot 3 \implies a = 0.00167$$

The normal course of the disease is 2 weeks, so on average, we expect $0.5\,I(n)$ to recover each week leaving $0.5\,I(n)$ remaining infected. We now have

$$I(n+1) = 0.5\,I(n) + 0.00167\,S(n)I(n)$$

Last, examine $S(n)$. The number of susceptibles is decreased only by the number that become infected. We use the same transmission coefficient as before to obtain the model

$$S(n+1) = S(n) - a\,S(n)I(n)$$

Our coupled SIR model is

$$\begin{aligned} S(n+1) &= S(n) - 0.00167\,S(n)I(n) \\ I(n+1) &= 0.5\,I(n) + 0.00167\,S(n)I(n) \\ R(n+1) &= R(n) + 0.5\,I(n) \end{aligned} \tag{2.2}$$

with $\big(S(0), I(0), R(0)\big) = (997, 3, 0)$.

The Discrete SIR Model (2.2) can be solved iteratively and viewed graphically. Do this with Maple to observe the behavior to gain insights. (Use lowercase letters for the variables since in Maple I is predefined as $I = \sqrt{-1}$.)

```
> DSIR := proc(n :: integer)
    local u, v, s, i, r;
    global s0, i0, r0;
    option remember;
    if n < 1 then
      u := [s0, i0, r0];    # Initial [susceptible, infected, removed]
    else
      v := DSIR(n − 1);
      s := v[1];   # current susceptible
      i := v[2];   # current infected
      r := v[3];   # current infected
      u := [s − 0.00167 · s · i, 0.5 · i + 0.00167 · s · i, r + 0.5 · i];
    end if;
    return(u);
    end proc :
```

```
> s0, i0, r0 := 997, 3, 0 :
```

```
> Flu := Matrix([seq(DSIR(k), k = 0..25)]);
```

$$Flu := \begin{bmatrix} 997 & 3 & 0 \\ 992.00503 & 6.49497 & 1.5 \\ 981.2451483 & 14.00736666 & 4.747485 \\ 958.2915651 & 29.95726650 & 11.75116833 \\ 910.3495461 & 62.92065225 & 26.72980158 \\ 814.6923014 & 127.1175708 & 58.19012770 \\ 641.7442519 & 236.5068349 & 121.7489131 \\ 388.2768258 & 371.7208435 & 240.0023305 \\ 147.2447418 & 426.8925058 & 425.8627523 \\ 42.2724216 & 318.4185731 & 639.3090052 \\ \vdots & \vdots & \vdots \end{bmatrix}$$

26 × 3 Matrix

(Double click the output matrix (blue on your screen) to open the *Matrix Browser* to see all the entries.)

Now for the graphs.

```
> Susceptible := [seq([k, Flu[k, 1]], k = 1..26)] :
  SP := pointplot(Susceptible, title = "Susceptible", color = red);
```

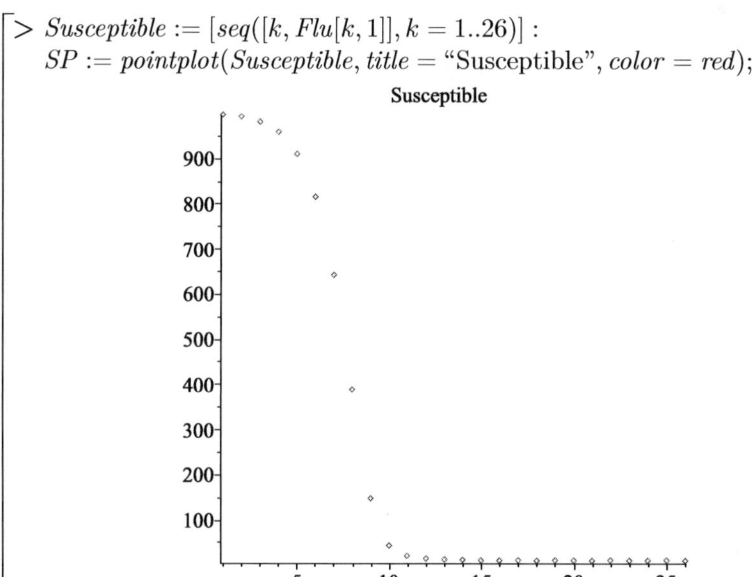

```
> Infected := [seq([k, Flu[k, 2]], k = 1..26)] :
  IP := pointplot(Infected, title = "Infected", color = blue);
```

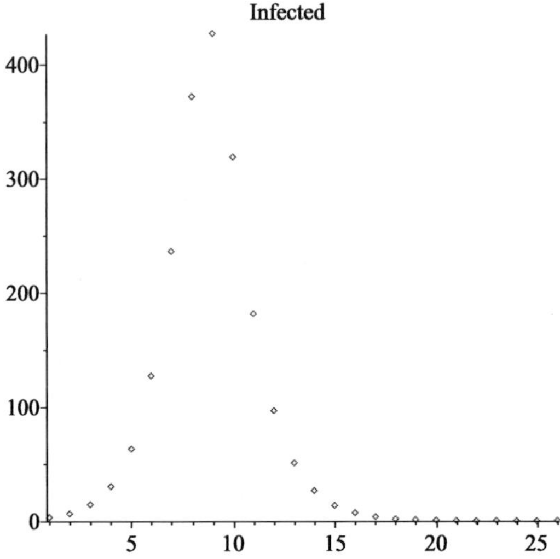


```
> Removed := [seq([k, Flu[k, 3]], k = 1..26)] :
  RP := pointplot(Removed, title = "Removed", color = gold);
```

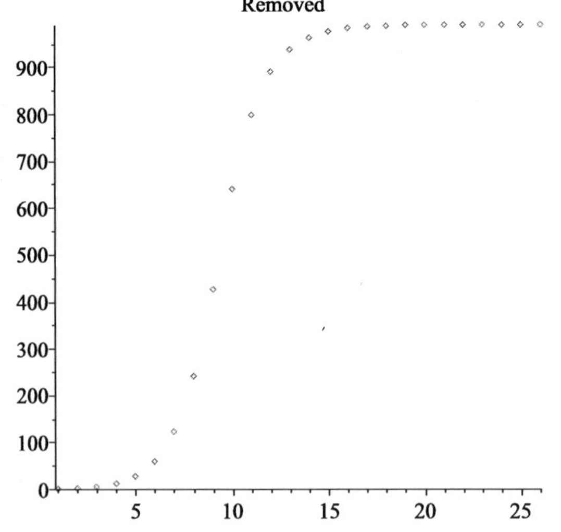

> *display*(*SP*, *IP*, *RP*, *legend* = ["Susceptible", "Infected", "Removed"])

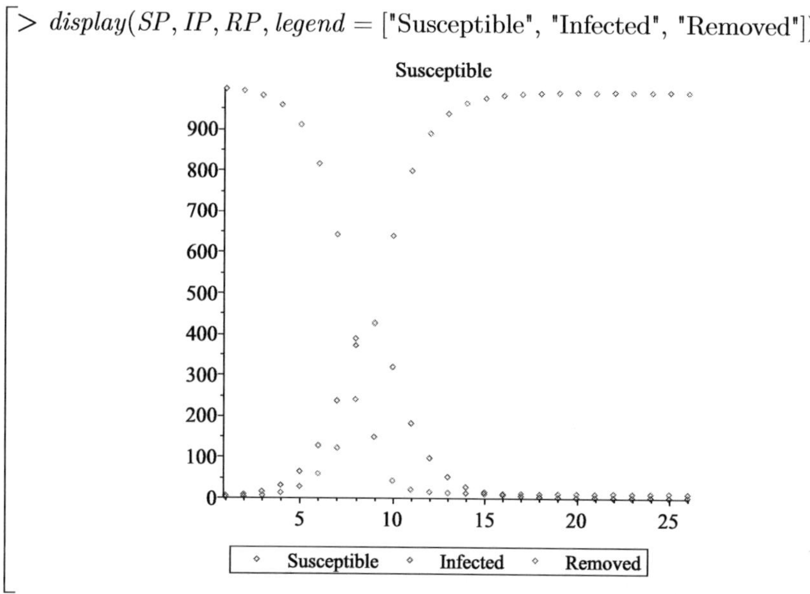

The maximum infections of the flu epidemic occurs around week 8, at the maximum of the *Infected* graph. The maximum number is slightly larger than 400, from the table it is approximately 427. After 25 weeks, slightly more than 9 students never get the flu. You will be asked to check the model for sensitivity to the parameter values in the exercise set.

Use the *spacecurve* function from the *plots* package to see the DDS trajectory for the flu.

```
> pp := pointplot3d(Flu, symbolsize = 16, color = red) :
  sp := spacecure(Flu) :
  display(pp, sp, axes = frame, orientation = [−20, 70],
    tickmarks = [3, 5, 5]);
```

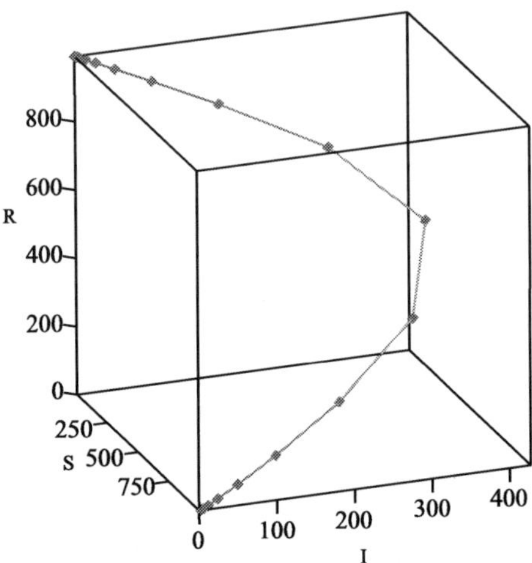

What can you learn about the flu's progression from the graph of the trajectory? Notice the spacing of the data points—remember, the data points are taken at equal time intervals.

Modeling Military Insurgencies

Insurgent forces have a strong foothold in the city of Urbania, a major metropolis in the center of the country of Freedonia. Intelligence estimates they currently have a force of about 1,000 fighters. The local police force has approximately 1,300 officers, many of which have had no formal training in law enforcement methods or in modern tactics for addressing insurgent activity. Based on data collected over the past year, approximately 8% of insurgents switch sides and join the police each week, whereas about 11% of police switch sides joining the insurgents. Intelligence also estimates that around 120 new insurgents arrive from the neighboring country of Sylvania each week. Recruiting efforts in Freedonia yield about 85 new police recruits each week as well. In armed conflict with insurgent forces, the local police are able to capture or kill approximately 10% of the insurgent force each week on average while losing about 3% of their force. See Figure 2.14.

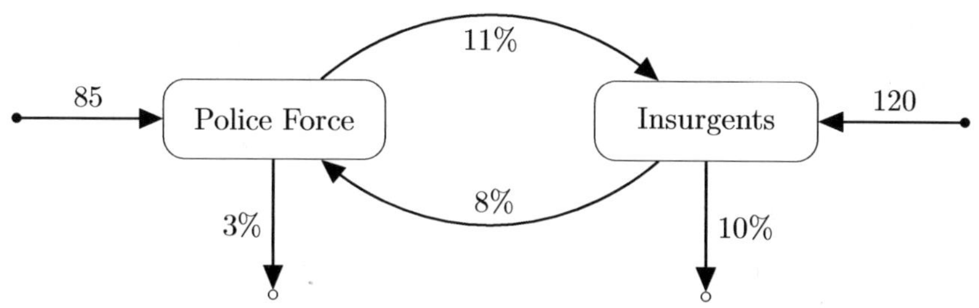

FIGURE 2.14: Diagram of Police versus Insurgents

Problem.

Build a mathematical model of this insurgency. Determine the equilibrium state, if it exists, for the DDS.

Variables.

$$\begin{aligned} n &= \text{ current time period} \\ p(n) &= \text{ number of police in the force at the end of time period } n \\ r(n) &= \text{ number of insurgents/rebels at the end of time period } n \end{aligned}$$

for $n = 0, 1, 2, \ldots$ weeks.

Model.

$$\begin{aligned} p(n+1) &= p(n) - 0.03p(n) - 0.11p(n) + 0.08r(n) + 85 \\ &= 0.86\,p(n) + 0.08\,r(n) + 85 \\ I(n+1) &= r(n) + 0.11p(n) - 0.08r(n) - 0.1r(n) + 120 \\ &= 0.11\,p(n) + 0.82\,r(n) + 120 \end{aligned}$$

with $P(0) = 1300$ and $I(0) = 1000$ and $n = 0, 1, 2, \ldots$.

Use Maple to investigate the *Police-Insurgents* DDS.

```
> Insurgency := proc(n :: integer)
    local u, v, p, r;
    global p0, r0;
    option remember;
    if n < 1 then
      u := [p0, r0];   # Initial [police, insurgents]
    else
      v := Insurgency(n - 1);
      p := v[1];   # current police
      r := v[2];   # current insurgents
      u := [0.86 p + 0.08 r + 85, 0.11 p + 0.82 r + 120];
    end if;
    return(u);
    end proc :
> p0, r0 := 1300, 1000 :
> PR := Matrix([seq(Insurgency(k), k = 0..52)]);
```

$$PR := \begin{bmatrix} 1300 & 1000 \\ 1283.0 & 1083.0 \\ 1275.0200 & 1149.1900 \\ 1273.452400 & 1202.588000 \\ 1276.376104 & 1246.201924 \\ 1282.379603 & 1282.286949 \\ 1290.429415 & 1312.537054 \\ 1299.772261 & 1338.227620 \\ 1309.862354 & 1360.321597 \\ 1320.307352 & 1379.548569 \\ \vdots & \vdots \end{bmatrix}$$

53 × 2 Matrix

(Double click the output matrix (blue on your screen) to open the *Matrix Browser* to see all the entries.)

Now for the graphs.

```
> police := [seq([k, PR[k, 1]], k = 1..53)];
  insurgents := [seq([k, PR[k, 2]], k = 1..53)];
> pol := pointplot(police, symbol = circle);
  ins := pointplot(insurgents, symbol = solidbox);
```

```
> display(pol, ins, labels = ["Week", "Number"],
    title = "Insurgency Model", legend = ["Police", "Insurgents"]);
```

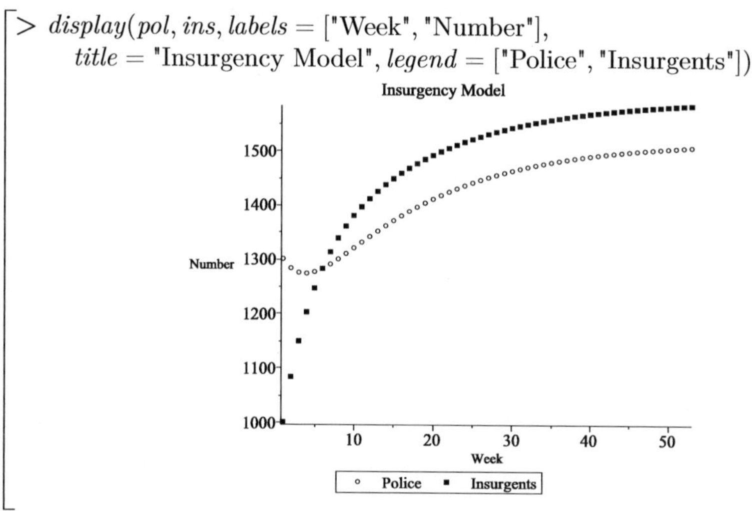

In this insurgency model, we see that under the current conditions, the insurgency overtakes the government after 5 weeks. This is unacceptable to the government, so we must modify conditions that affect the parameters in such a way to obtain a police victory. You will be asked to experiment with DDS's parameters in the exercises.

Exercises

1. Find the equilibrium values for the Predator-Prey model presented.

2. Determine the outcomes in the Predator-Prey model with the following sets of parameters.

   (a) There are 200 foxes and 400 rabbits initially.

   (b) There are 2,000 foxes and 10,00 rabbits initially.

   (c) The birth rate of rabbits increases to 0.1.

3. Find the equilibrium values for the SIR model presented.

4. In the SIR model determine the outcome with the following parameters changed.

   (a) Initially, 5 are sick, and 10 are sick the next week.

   (b) The flu only lasts 1 week.

   (c) The flu lasts 4 weeks.

(d) There are 4,000 students in the dorm. Initially, 5 are infected, and 30 more are sick the next week.

5. Find the equilibrium values analytically, if any, for the *Police-Insurgents* DDS.

6. In the Insurgency model, determine what values of the parameters *police recruiting*, *police operation losses*, and *police conversions to insurgents* enable a reversal of the outcome of insurgents' numbers overtaking police numbers as predicted under the current conditions.

7. Investigate the Maple graph generated by *spacecurve* for the *Police-Insurgents* DDS.

References and Further Reading

[Bauldry2009] William Bauldry, *Introduction to Real Analysis*, Wiley, 2009.

[GFH2014] Frank Giordano, William P. Fox, and Steven Horton, *A First Course in Mathematical Modeling*, 5th ed., Nelson Education, 2014.

[Lawrence1999] Eleanore Lawrence, "Bass are bad news for lake trout," Nature online, Sept. 30, 1999. Available at https://www.nature.com/news/1999/990930/full/news990930-9.html

[Malthus1798] Rev. Thomas Malthus, *An Essay on the Principle of Population as It Affects the Future Improvement of Society, with Remarks on the Speculations of Mr. Godwin, M. Condorcet, and Other Writers.*, Johnson, London, 1798.

[S2002] James Sandefur, *Elementary Mathematical Model: A Dynamic Approach*, Brooks-Cole Pub., 2002.

3

Problem Solving with Single-Variable Optimization

Objectives:

(1) Set up a solution process for single- and multivariable optimization problems.

(2) Recognize when applicable constraints are required.

(3) Perform sensitivity analysis, when applicable and interpret the results.

(4) Know the numerical techniques and when to use them to obtain approximate solutions.

3.1 Single-Variable Unconstrained Optimization

Rigs-R-Us has an oil-drilling rig 9.5 miles offshore. The drilling rig is to be connected to a pumping station via an underwater pipe. A land-based pipe will connect the proposed pumping station to a refinery 15.7 miles down the shoreline from the drilling rig. See Figure 3.1.

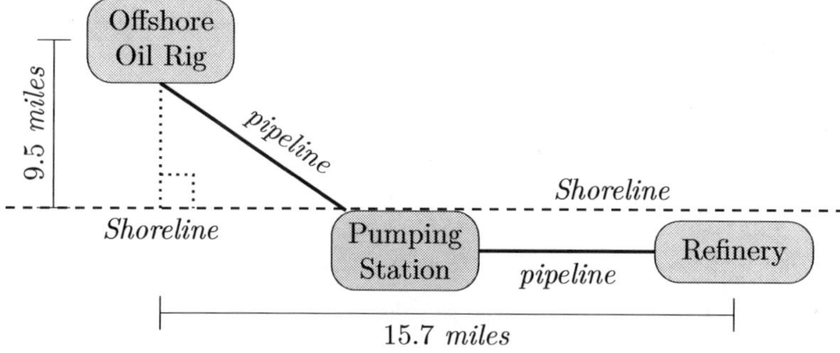

FIGURE 3.1: Oil Pipe Route and Proposed Pumping Station

Underwater pipe costs $32,575 per mile to install, land-based pipe costs $14,442 per mile. Where should *Rigs-R-Us* place the pumping station along the shoreline to minimize the total cost of the pipe?

In this section, we will discuss problems like the pumping station location that we can model and solve using single-variable calculus. First, we'll review the calculus concepts needed for optimization, and then apply them to applications. Maple will handle the computations. We want to solve problems of the form: optimize $f(x)$ (maximize or minimize) for x in an interval.

If $a = -\infty$ and $b = \infty$, then we are looking at $(x, f(x)) \in \mathbb{R}^2$—the entire xy-plane. If either a or b, or both, are restricted, we must consider possible end points in our solution. Define points x_c where $f'(x_c) = 0$ as *critical points* or *stationary points*. The three cases of *extended critical points* x_c are:

Case 1. Points x_c where $a < x_c < b$ and $f'(x_c) = 0$.

Case 2. Points x_c where $a < x_c < b$ and $f'(x_c)$ does not exist.

Case 3. End points $x_c = a$ and $x_c = b$ of the interval $[a, b]$.

A *local extrema* need only be an extreme value over some interval, while a *global extreme value* must be an extreme value over the entire domain of f.

A summary of the relevant definitions and theorems from calculus follow.

Definition. Local and Global Extrema.
A function f has a local maximum (local minimum) at a point c iff $f(x) \leq f(c)$ (respectively, $f(x) \geq f(x)$) for all x in an interval containing c.

A function f has a global maximum (global minimum) at a point c iff $f(x) \leq f(c)$ (respectively, $f(x) \geq f(c)$) for all x in the domain of f.

Theorem. Extreme Value Theorem.
Let f be continuous on a closed interval $[a, b]$. Then f must have both a global maximum and a global minimum over the interval.

Remember, extreme values can occur at an interval's endpoints, and there may be more than one maximum or minimum for any given function. If there are several local maxima, they may have different functional values, and similarly for minima; however, global maxima must all have the same functional value. Consider the extreme values of $f(x) = \sin(x)/x$ and $g(x) = \cos(x)$ for examples.

If f is differentiable, we can do more. Recall your analysis of the shape of f's graph leading to the first and second derivative tests in calculus. As is often customary in calculus classes, we'll state the Second Derivative Test first, and give the First Derivative Test second, with an extension of the Second Derivative Test in between.

Theorem. Second Derivative Test.
Let $f'(x_c) = 0$ for some $x_c \in (a, b)$. Then

- if $f''(x_c) > 0$ (concave up), f has a local minimum at x_c,

- if $f''(x_c) < 0$ (concave down), f has a local maximum at x_c,
- if $f''(x_c) = 0$, f has a possible *inflection point* (change of concavity) at x_c.

Theorem. Extended Derivative Test.
Let $f'(x_c) = f''(x_c) = 0$ for some $x_c \in (a, b)$. Then

- if the first non-zero derivative is even-order (i.e., 4th, 6th, 8th, etc.) and
  - is positive at x_c, then f has a local minimum at x_c,
  - is negative at x_c, then f has a local maximum at x_c,
- if the first non-zero derivative is odd-order (i.e., 3nd, 5th, 7th, etc.), then f does not have an extrema at x_c.

Theorem. First Derivative Test.
Let $f'(x_c) = 0$ for some $x_c \in (a, b)$. Then

- if $f'(x) > 0$ (increasing) below x_c and $f'(x) < 0$ (decreasing) above x_c, then f has a local maximum at x_c,
- if $f'(x) < 0$ (decreasing) below x_c and $f'(x) > 0$ (increasing) above x_c, then f has a local minimum at x_c,
- if $f'(x)$ does not change sign around x_c, then $f(x_c)$ is neither a local maximum nor a local minimum.

Theorem. Nondifferentiable Point Test.
Suppose $f'(x_c)$ does not exist. In this case, test points near x_c: evaluate f at 'nearby' x_1 and x_2 where $x_1 < x_c < x_2$. See Table 3.1.

TABLE 3.1: Testing a Nondifferentiable Point

Relation	Classification
$f(x_1) \le f(x_c)$ and $f(x_c) \le f(x_2)$	then x_c is not a local extrema
$f(x_1) \ge f(x_c)$ and $f(x_c) \le f(x_2)$	then x_c is a local minimum
$f(x_1) \le f(x_c)$ and $f(x_c) \ge f(x_2)$	then x_c is a local maximum
$f(x_1) \ge f(x_c)$ and $f(x_c) \ge f(x_2)$	then x_c is not a local extrema

Theorem. Interval Endpoints Test.
Suppose that f is differentiable on (a, b) and has right and left derivatives at $x = a$ and b, respectively. Then, for $x = a$,

- if $f'(a) > 0$, $f(a)$ is a local minimum.
- if $f'(a) < 0$, $f(a)$ is a local maximum.

For $x = b$, the opposite holds,

- if $f'(b) > 0$, $f(a)$ is a local maximum.
- if $f'(b) < 0$, $f(a)$ is a local minimum.

If $f'(a) = 0 = f'(b)$, then use nearby test points and graphs to determine f's behavior.

Example 3.1. A Set of Simple Examples.

1. Minimize $f(x) = 1.5x^2 + 1$ over $\mathbb{R}$.
   Solution: Since $f'(x) = 3x$ is 0 at $x = 0$ and $f''(0) = +3$, we see that $x = 0$ is a minimum. Since f is always concave up, $x = 0$ must be a global minimum.

2. Maximize $g(x) = x^3 + 1$ over the interval $[-1, 3]$.
   Solution: Since $g'(x) = 3x^2$ is 0 at $x = 0$. Calculating derivatives: $g''(0) = 0$, $g'''(0) = 6$, gives the 3rd derivative as the first nonzero derivative of g; therefore, g has neither maximum, nor minimum at $x = 0$.

   Since the interval is closed, we must check the endpoints $x = -1$ and 3. At the left endpoint, $g'(-1) = +3$; therefore, g is increasing which makes $x = -1$ a minimum. At the right endpoint, $g'(3) = +27$, therefore g is increasing which makes $x = -3$ a maximum.

3. Optimize $h(x) = |x|$ over $\mathbb{R}$.
   Solution: $h'(x) = +1$ if $x > 0$ and -1 if $x < 0$, and has no derivative at $x = 0$. Testing h at $x = -0.1$ and $+0.1$ gives

   $$h(-0.1) = 0.1 > h(0) = 0 \qquad \text{and} \qquad h(0) = 0 < h(+0.1) = 0.1$$

   which indicates $x = 0$ is a minimum. As h always decreases below $x = 0$ and always increases above $x = 0$, we have that $x = 0$ is a global minimum.

For more examples, take any calculus text and work through the optimization exercises in the "Applications of Derivatives" chapter.

We'll use familiar Maple commands for differentiation and solving equations. Recall the commands' usage from their Maple Help Pages.

> **D** – *differential operator*
>
> Calling Sequence
>   **D**(f)
>
> Parameters
>   f – expression which can be applied as a function

diff or Diff – *differentiation or partial differentiation*

Calling Sequence

diff$(f, x1, \ldots, xj)$

$$\frac{\mathrm{d}^j}{\mathrm{d}x_j \ldots \mathrm{d}x_1} f$$

diff$(f, [x1\$n])$

$$\frac{\mathrm{d}^n}{\mathrm{d}x_1^n} f$$

Remark: these calling sequences are also valid with the inert *Diff* command

Parameters

f	– algebraic expression or an equation
x1, x2, ..., xj	– names representing differentiation variables
n	– algebraic expression entering constructions like x$n, representing nth order derivative, assumed to be integer order differentiation

solve – *solve one or more equations*

Calling Sequence

solve(*equations*, *variables*)

Parameters

equations	– equation or inequality, or set or list of equations or inequalities
variables	– (optional) name or set or list of names; unknown(s) for which to solve

fsolve – *solve one or more equations using floating-point arithmetic*

Calling Sequence

fsolve(*equations*, *variables*, complex)

Parameters

equations	– equation, set(equation), expression, set(expression), list(equation)
variables	– (optional) name or set or list of names; unknown(s) for which to solve
complex	– (optional) literal; search for complex solutions

Note: For simple functions $f(x)$, Maple can use $f'(x)$, $f''(x)$, etc., for derivatives.

Basic Applications Calculus Max-Min Theory

Example 3.2. Chemical Manufacturing Company.
A chemical manufacturing company sells sulfuric acid at a price of $200 per ton. The daily total production cost in dollars for x tons is

$$C(x) = 150000 + 145.5x + 0.0025x^2,$$

and the daily production is at most 7,500 tons. How many tons of sulfuric acid should the manufacturer produce to maximize daily profits?

Solution. The profit function is

$$\begin{aligned} \mathit{Profit} &= \mathit{Revenue} - \mathit{Cost} \\ &= (200\,x) - (150000 + 145.5x + 0.0025x^2) \\ &= -150000 + 54.5\,x - 0.0025\,x^2 \quad \text{for } 0 \le x \le 7000 \end{aligned}$$

Profit is twice differentiable, so use Maple for a Second Derivative Test. We'll apply Maple's D operator to the *Profit* function to compute the derivative functions.

$> \mathit{Profit} := x \to -150000 + 54.5 \cdot x - 0.0025 \cdot x^2;$
$\mathit{dProfit} := D(\mathit{Profit});$
$\mathit{d2Profit} := D(\mathit{dProfit});$

$$\mathit{Profit} := x \mapsto -150000 + 54.5 \cdot x + (-1)0.0025 \cdot x^2$$
$$\mathit{dProfit} := x \mapsto 54.5 - 0.0050\,x$$
$$\mathit{d2Profit} := x \mapsto -0.0050$$

$> c := \mathit{solve}(\mathit{dProfit}(x) = 0, x);$

$$c := 10900.$$

$> \mathit{Profit}(c);$

$$147025.0000$$

We must check the endpoints since *Profit* is defined on the interval $[0, 7500]$.

$> [0, \mathit{Profit}(0)], [7500, \mathit{Profit}(7500)];$

$$[0, -150000.], [7500, 118125.0000]$$

Even though $\mathit{Profit}(c)$ is the largest, c = 10,900 tons is greater than the maximum production of 7,500 tons. So the company should produce all it can, 7,500 tons, for a maximum $\mathit{Profit}(7500)$ = \$118,125.00.

Verify the analysis showing the end point solution is selected because the critical point is outside of *Profit's* domain by producing graphs of *Profit* over $[0, 7500]$ and over $[0, 12000]$.

Example 3.3. Computer Production.
Bell Computers spends $250 in *variable costs* to produce an SP6 computer. They have a *fixed cost* of $5500 whenever the plant is in operation and computers are produced. If *Bell* spends x dollars on advertising their new SP6, they

can sell $\sqrt{x}$ SP6's at \$550 per computer. How many SP6 computers should the company produce to maximize profits?

Solution. Since revenue is price times quantity and total cost is fixed cost plus variable costs times quantity, *Bell's* profit function is given by

```
> varcost, fixedcost, price := 250, 5500, 550 :
```

```
> Revenue := x -> price · √x :
  Cost := x -> fixedcost + x + varcost · √x :
```

```
> Profit := Revenue − Cost :   # Note this is an equation of functions!
  Profit(x);
```

$$300\sqrt{x} - 5500 - x$$

Profit is twice differentiable, so again use Maple for a Second Derivative Test.

```
> dProfit := D(Profit) :
  dProfit(x);
  c := solve(dProfit(x) = 0, x);
```

$$\frac{150}{\sqrt{x}} - 1$$

$$c := 22500$$

```
> d2Profit := D(dProfit);
  d2Profit(c);
```

$$\frac{\sqrt{22500}}{6750000}$$

Since *Profit*″(c) is negative, we have the maximum *Profit*(22500) = \$17,000. A graph will verify our analysis.

```
> plot(Profit, 0..30000, thickness = 2, title = "Bell Computer Profit");
```

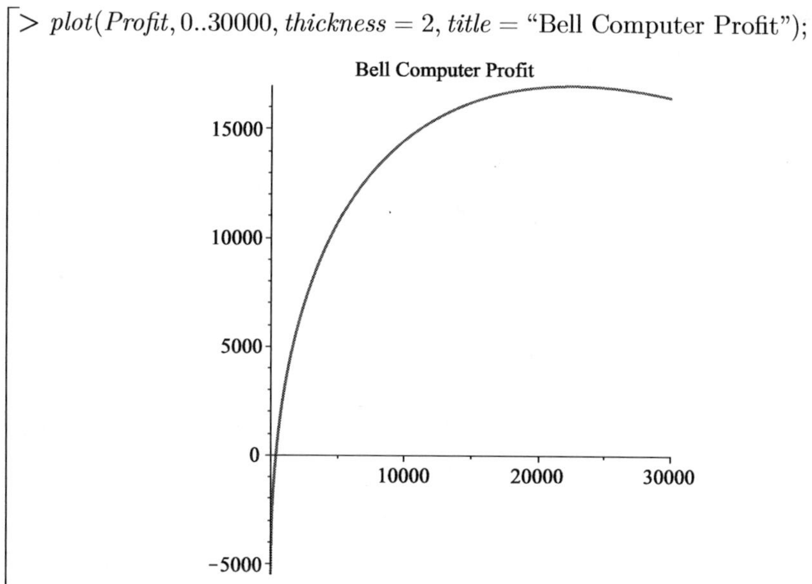

Return to the pumping station example that opened this section.

Example 3.4. Oil Pumping Station Location.
Rigs-R-Us has an oil-drilling rig 9.5 miles offshore. The drilling rig will be connected to a pumping station via an underwater pipe which then connects to a refinery by a land-based pipe. The refinery is 15.7 miles down the shoreline from the drilling rig. See Figure 3.2. Underwater pipeline costs \$32,575 per mile

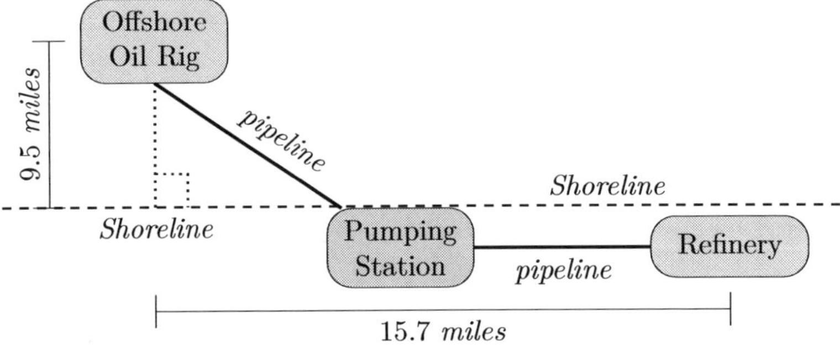

FIGURE 3.2: Oil Pipe Route and Proposed Pumping Station

to install, land-based pipeline costs \$14,442 per mile. Determine a location for the pumping station that minimizes the total cost of the pipeline.

Problem.
Find a relationship between the location of the pumping station and cost of the installation of the pipe, then minimize the cost.

Assumptions.
First, we assume no cost saving for the pipe if we purchase in larger lot sizes. We further assume the costs of preparing the terrain to lay the pipe, both offshore and on land, are captured in given cost per mile figures.

Variables.
$x =$ the location of the Pumping station along the horizontal distance from $x = 0$ to $x = 15.7$ miles
$TC =$ total cost of the pipe for both underwater and on-land piping

Model Construction.
Use the Pythagorean Theorem for the underwater distance of the pipe; that is, the hypotenuse of the right triangle with height 9.5 miles and base x miles along the shoreline. The length of the hypotenuse is $\sqrt{9.5^2 + x^2}$. The length of the pipe on shore is $15.7 - x$. Therefore total cost is

$$TC = 32575 \cdot \sqrt{9.5^2 + x^2} + 14442 \cdot (15.7 - x).$$

We turn to Maple for a Second Derivative Test to determine the minimum cost Pumping Station location.

$> TC := x \to 32575 \cdot \sqrt{9.5^2 + x^2} + 14442 \cdot (15.7 - x) :$

$> dTC := D(TC);$

$$dTC := x \mapsto \frac{32575x}{\sqrt{90.25 + x^2}} - 14442$$

$> c := solve(dTC(x) = 0, x);$

$$c := 4.698818368$$

$> d2TC := D(dTC);$

$$d2TC := x \mapsto \frac{32575}{\sqrt{90.25 + x^2}} - \frac{32575x^2}{\left(90.25 + x^2\right)^{3/2}}$$

$> d2TC(c);$

$$2469.416868$$

Since the second derivative is positive at $c = 4.6988$, we have a minimum total cost. Thus, if the pumping station is located at $15.7 - 4.698 \approx 11.0$ miles from the refinery, we will minimize the total pipeline cost with $TC(c) =$ \$50,4126.27.

Once again, we graph the total cost function to verify our results.

$> plot(TC, 0..8, thickness = 2, title = \text{“Pipeline Cost”});$

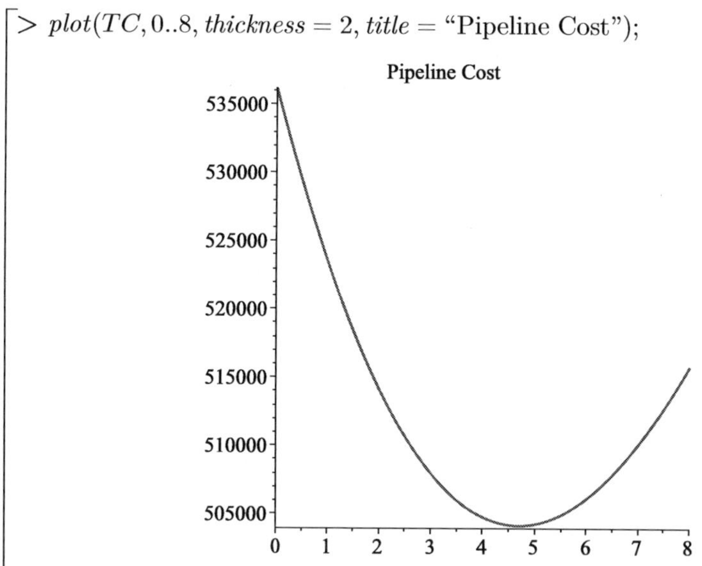

How would the plot command be modified to show a point on the curve at the minimum?

Exercises

1. Each morning during rush hour, 10,000 people travel from New Jersey to New York. The trip lasts 40 minutes taking the subway. If x thousand people drive to New York City, it takes $20 + 5x$ minutes to make the trip.

   (a) Formulate the problem with the objective to minimize the average travel time per person.

   (b) Find the optimal number of people who drive so that the average time per person is minimized.

2. Let $f(x) = 2x^3 - 1$. Use calculus to find the optimal solution to

$$\max_{-1 \le x \le 1} f(x)$$

3. Find all extrema for $f(x) = \sin(2 + \pi x)/x$ on the interval $1 \le x \le 4$.

4. Consider the function $f(x) = 0.5x^3 - 8x^2$.

   a. Use a graph to find and classify all extrema of f.

   b. Use analytical techniques (i.e., calculus) to find and classify all extrema of f.

   c. Define: A differentiable function is

   - *convex* or *concave up* on the interval $[a, b]$ iff f lies above all of its tangents; that is, for all $x_1 < x < x_2$ in $[a, b]$,

$$f'(x) \le \frac{f(x_2) - f(x_1)}{x_2 - x_1}$$

   - *concave* or *concave down* on the interval $[a, b]$ iff f lies below all of its tangents; that is, for all $x_1 < x < x_2$ in $[a, b]$,

$$f'(x) \ge \frac{f(x_2) - f(x_1)}{x_2 - x_1}$$

   A twice differentiable function is

   - *convex* on $[a, b]$ iff fs second derivative is non-negative.
   - *concave* on $[a, b]$ iff fs second derivative is non-positive.

   Apply the definition of *concave* (concave down) to show that f is concave over the interval $[-6, 5]$.

   d. What is the largest interval where f is convex? Concave?

   e. Why is knowing the concavity important in optimization?

5. Dr. E. N. Throat has been taking x-rays of the trachea contracting during coughing. He has found that the trachea appears to contract by 33% (1/3) of its normal size. He has asked the Department of Mathematics to confirm or deny his claim. You perform some initial research and find that under reasonable assumptions about the elasticity of the tracheal wall and about how air near the wall is slowed by friction, the average flow of velocity v can be modeled by the equation

$$v = c(r_0 - r)^2 \text{ cm/sec for } \tfrac{1}{2}r_0 \leq r \leq r_0$$

where c is a positive constant (let $c = 1$) and r_0 is the resting radius of the trachea in centimeters.

   (a) Find the value of r that maximizes v.

   (b) Does your result support or deny Dr. Throat's claim.

3.2 Numerical Search Techniques with Maple

Single-Variable Techniques

When calculus methods aren't feasible, we turn to numerical approximation techniques. The basic approach of most numerical methods in optimization is to produce a sequence of improved approximations to the optimal solution according to a specific scheme. We will examine both elimination methods (Golden section, and Fibonacci) and interpolation methods (Newton's).

In numerical methods of optimization, a procedure is used in obtaining values of the objective function at various combinations of the decision variables, and conclusions are then drawn regarding the optimal solution. The elimination methods can be used to find an optimal solution for even discontinuous functions. An important relationship (assumption) must be made to use these elimination methods. The function must be *unimodal.* A unimodal function—*uni-modal* from "one mode"—is one that has only one maximum (peak) or one minimum (valley), but not both. State this mathematically as

Definition. Unimodal Function.
A function is *unimodal* on the interval $[a, b]$ iff

1. f has a maximum (minimum) x^* in (a, b), and

2. is strictly increasing (decreasing) on $[a, x^*]$, i.e., $x < x^* \Rightarrow f(x) < f(x^*)$, and

3. is strictly decreasing (increasing) on $[x^*, b]$ i.e., if $x > x^* \Rightarrow f(x) < f(x^*)$.

Examples of unimodal and non-unimodal functions are shown in the Figure 3.3. Unimodal functions may or may not be differentiable or even continuous.

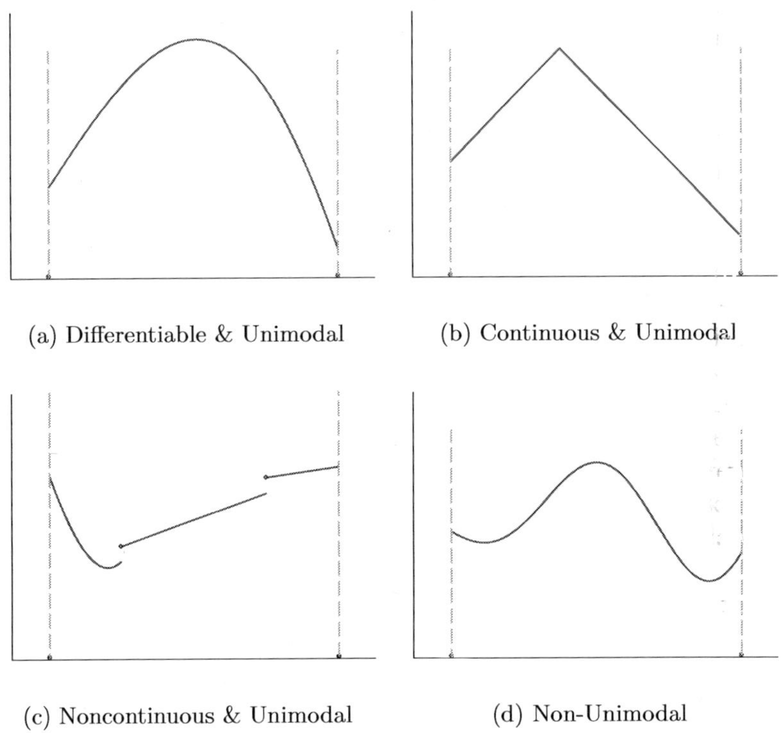

(a) Differentiable & Unimodal

(b) Continuous & Unimodal

(c) Noncontinuous & Unimodal

(d) Non-Unimodal

FIGURE 3.3: Examples of Unimodal and Non-Unimodal Functions

Thus, a unimodal function can be nondifferentiable (have corners) or even discontinuous. If a function is known to be unimodal in a given closed interval, then the optimum (maximum or minimum) can be found.

In this section, we will learn many techniques for numerical searches. For the "elimination methods," we accept an interval answer. If a single value is required, we usually evaluate the function at each end point of the final interval and the midpoint of the final interval, then take the optimum of those three values to approximate our single value.

Golden Section Search

A technique that searches for a value by determining narrower and narrower intervals containing the target value is called a *bracketing method.* A *Golden*

Section search is a type of bracketing method that uses the *golden ratio*[1]. This recent technique was developed in 1953 by Jack Keifer.

To better understand the golden ratio, consider a line segment that is divided into two parts as shown in Figure 3.4. If the ratio of the length of

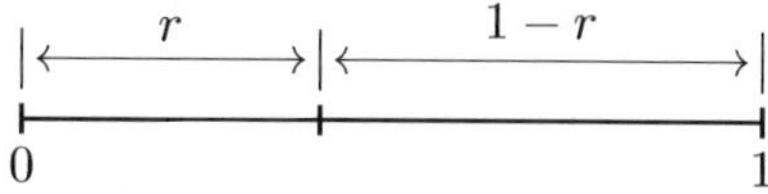

FIGURE 3.4: The Golden Ratio

the whole line to the length of the larger part is the same as the ratio of the length of the larger part to the length of the smaller part, we say the line is divided into the *golden ratio*. Symbolically, taking the length of the original segment as 1, this can be written as

$$\frac{1}{r} = \frac{r}{1-r}$$

Algebraic manipulation of the relationship above yields $r^2 + r - 1 = 0$. The quadratic's positive root $r = (\sqrt{5}-1)/2$ satisfies the ratio requirements for the line segment above. The golden ratio's numerical value is $\phi \approx 0.6180339880$. (The traditional symbol for the golden ratio is ϕ.) This well-known ratio is the limiting value for the ratio of the consecutive terms of the Fibonacci sequences, which we will see in the next method. Another bracketing method, the *Fibonacci search* is often used in lieu of the Golden Section method.

In order to use the Golden Section search procedure, we must ensure that certain assumptions hold. These key assumptions are:

1. The function must be unimodal over a specified interval,

2. the function must have an optimal solution over a known interval of uncertainty, and

3. we will accept an interval solution since the exact optimal value cannot be found by bracketing, only approximated.

Only an *interval solution*, that is, an interval containing the exact optimal value, known as the *final interval of uncertainty*, can be found using this technique. The length of the final interval is controlled by the user and can be made arbitrarily small by selecting a *tolerance value*. Assuming unimodality guarantees that the final interval's length is less than the chosen tolerance.

[1] See http://mathworld.wolfram.com/GoldenRatio.html.

Finding the Maximum of a Function over an Interval with a Golden Section Search

Suppose we know that f is unimodal on an interval I. Break I into two subintervals I_1 and I_2, then f's maximum must be in one or the other. Check a test point in each interval—the higher test point indicates which interval contains the maximum since f is unimodal and has exactly one optimum value. We can reduce the number of times we have to evaluate the function by having the intervals overlap and using the endpoints as test points. We can further reduce the number of function evaluations by cleverly choosing the test points so that the next iteration reuses one of the current test points. These ideas lead to the Golden Section search developed by Prof. Jack Kiefer in his 1953 master's thesis.[2]

A Golden Section search for a maximum is iterative, requiring evaluations of $f(x)$ at test points x_1 and x_2, where

$$x_1 = a + r(b-a) \quad \text{and} \quad x_2 = b - r(b-a).$$

The test points will lie inside the original interval $I_0 = [a,b]$ and are used to determine a new, smaller search interval I_1. Choosing r carefully lets us reuse the function's evaluations in the next iteration: either x_1 or x_2 will be the new interval endpoint, and the other test point will be a test point in the new, reduced interval. For a Golden Section Search, r is chosen to be the golden ratio 0.618. If $f(x_1) > f(x_2)$, the new interval is $[x_2, b]$, and if $f(x_1) < f(x_2)$, the new interval is $[a, x_1]$. Continue to iterate in this manner until the final interval length is less than the chosen tolerance. The final interval I_N contains or *brackets* the optimum solution. The length of the final interval I_N determines the accuracy of the approximate optimum solution. The maximum number of iterations N required to achieve this desired tolerance or final interval length is

$$N = \left\lceil \frac{1}{\ln(0.618)} \cdot \ln\left(\frac{tolerance}{b-a}\right) \right\rceil.$$

When we are required to provide a point solution, instead of a small interval containing the solution, evaluate the function at the end points and the midpoint of the final interval. Select the value of x that yields the largest $f(x)$.

How can the procedure be changed to search for a minimum point?

Figure 3.5 shows the progression of subintervals selected by a Golden Section search.

---

[2]See J. Kiefer, "Sequential minimax search for a maximum," Proc Am Math Soc, 4 (3), 1953, 502–506.

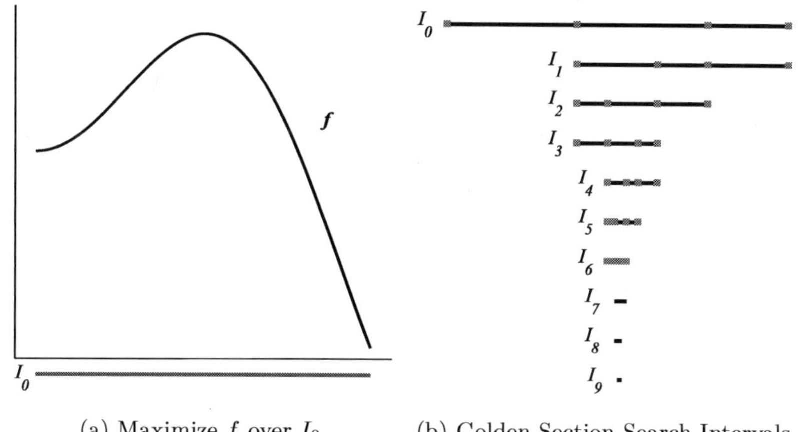

(a) Maximize f over I_0 (b) Golden Section Search Intervals

FIGURE 3.5: Golden Section Search Interval Progression

The Golden Section search is written in algorithmic form below.

Golden Section Search Algorithm

Find the maximum of the unimodal function f on the interval $[a, b]$.

INPUTS:	Function: f
	Endpoints: a, b
	Tolerance: t
OUTPUTS:	Optimal point x^* and maximum value $f(x^*)$

Step 1. Initialize: Set $r = 0.618$, $a_0 = a$, $b_0 = b$, counter $k = 1$, limit $N = \lceil \ln(t/(b-a))/\ln(0.618) \rceil$

Step 2. Calculate the test points $x_1 = a_{k-1} + r(b_{k-1} - a_{k-1})$ and $x_2 = b_{k-1} - r(b_{k-1} - a_{k-1})$.

Step 3. Compute $f(x_1)$ and $f(x_2)$

Step 4. If $f(x_1) \geq f(x_2)$, then $I_k = [x_2, b]$; that is $a = x_2$
If $f(x_1) < f(x_2)$, then $I_k = [a, x_1]$; that is $b = x_1$

Step 5. If $k = N$ or $b_k - a_k < t$, then STOP
Return estimates $x^* =$ midpoint(I_k) and $f(x^*)$.
Otherwise, set $k = k + 1$ and return to Step 2.

Although a Golden Section search can be used with any unimodal function to find the maximum (or minimum) over a specified interval, its main advantage comes when normal calculus procedures fail. Consider the following example.

Example 3.5. Maximizing a Nondifferentiable Function.
Maximize

$$f(x) = -|2-x| - |5-4x| - |8-9x|$$

over the interval $[0, 3]$.

Absolute value functions are not differentiable because they have corner points—graph f to see this. Thus, taking the first derivative and setting it equal to zero is not an option. We'll use a Golden Section search in Maple to solve this problem. Our Maple procedure *GoldenSectionSearch* is in the text's *PSMv2* package. Use *with*(*PSMv2*) to make the procedure available.

```
> with(PSMv2) :
> f := x -> -|2 - x| - |5 - 4x| - |8 - 9x|;
                    f := x ↦ -|2 - x| - |5 - 4x| - |8 - 9x|
> GoldenSectionSearch(f, 0, 3, 0.1) :
  fnormal(%, 5);    #to fit the output to the page width
```

The maximum is f(0.907318)=-2.629270 with tolerance 0.100000

Iteration	x_1	x_2	$f(x_1)$	$f(x_2)$	*Interval*
0	1.8541	1.1459	−11.249	−3.5836	[0.0, 3]
1	1.1459	0.70820	−3.5836	−5.0851	[0.0, 1.8541]
2	1.4164	1.1459	−5.9969	−3.5836	[0.70820, 1.8541]
3	1.1459	0.97871	−3.5836	−2.9149	[0.70820, 1.4164]
4	0.97871	0.87539	−2.9149	−2.7446	[0.70820, 1.1459]
5	0.87539	0.81153	−2.7446	−3.6386	[0.70820, 0.97871]
6	0.91486	0.87539	−2.6594	−2.7446	[0.81153, 0.97871]
7	0.93925	0.91486	−2.7570	−2.6594	[0.87539, 0.97871]
8	0.91486	0.89978	−2.6594	−2.5991	[0.87539, 0.93925]

The midpoint of the final interval is 0.907318 and $f(midpoint) = -2.629270$, so we estimate the maximum of the function is -2.629 at the x value 0.9 (to within 0.1).

Example 3.6. Maximizing a Transcendental Function.
Maximize the function

$$g(x) = 1 - \exp(-x) + \frac{1}{1+x}$$

over the interval $[0, 20]$. (Remember that $\exp(-x) = e^{-x}$.)

```
> g := x → 1 − exp(−x) + 1/(1 + x) :
> GoldenSectionSearch(g, 0, 20, 0.001) :
  fnormal(%, 5);   #to fit the output to the page width
```

The maximum is f(2.512908)=1.203632 with tolerance 0.001000

Iteration	x_1	x_2	$f(x_1)$	$f(x_2)$	*Interval*
0.0	12.361	7.6393	1.0748	1.1153	[0.0, 20]
1	7.6393	4.7214	1.1153	1.1659	[0.0, 12.361]
2	4.7214	2.9180	1.1659	1.2012	[0.0, 7.6393]
3	2.9180	1.8034	1.2012	1.1920	[0.0, 4.7214]
4	3.6068	2.9180	1.1899	1.2012	[1.8034, 4.7214]
5	2.9180	2.4922	1.2012	1.2036	[1.8034, 3.6068]
6	2.4922	2.2291	1.2036	1.2021	[1.8034, 2.9180]
7	2.6548	2.4922	1.2033	1.2036	[2.2291, 2.9180]
8	2.4922	2.3917	1.2036	1.2034	[2.2291, 2.6548]
9	2.5543	2.4922	1.2036	1.2036	[2.3917, 2.6548]
10	2.4922	2.4538	1.2036	1.2036	[2.3917, 2.5543]
11	2.5160	2.4922	1.2036	1.2036	[2.4538, 2.5543]
12	2.5306	2.5160	1.2036	1.2036	[2.4922, 2.5543]
13	2.5160	2.5069	1.2036	1.2036	[2.4922, 2.5306]
14	2.5216	2.5160	1.2036	1.2036	[2.5069, 2.5306]
15	2.5160	2.5125	1.2036	1.2036	[2.5069, 2.5216]
16	2.5125	2.5104	1.2036	1.2036	[2.5069, 2.5160]
17	2.5138	2.5125	1.2036	1.2036	[2.5104, 2.5160]
18	2.5125	2.5117	1.2036	1.2036	[2.5104, 2.5138]
19	2.5130	2.5125	1.2036	1.2036	[2.5117, 2.5138]
20	2.5133	2.5130	1.2036	1.2036	[2.5125, 2.5138]
21	2.5130	2.5128	1.2036	1.2036	[2.5125, 2.5133]

The Golden Section search gives the maximum of the function is 1.204 at the x value $= 2.513$.

Fibonacci Search

A Fibonacci search uses the ratio of Fibonacci numbers to generate a sequence of test points based on the expression

$$F_n = F_{n-1} + F_{n-2} \implies 1 = \frac{F_{n-1}}{F_n} + \frac{F_{n-2}}{F_n}.$$

Since $\lim_{n\to\infty} F_{n-1}/F_n$ equals the golden ratio, A Fibonacci search is a Golden Section search "in the limit." However, a Fibonacci search converges faster than a Golden Section search.

In order to use the Fibonacci search procedure, we must ensure that the key assumptions hold:

1. The function must be unimodal over a specified interval,

2. The function must have an optimal solution over a known interval of uncertainty, and

3. We will accept an interval solution since the exact optimal value cannot be found by bracketing, only approximated.

These are the same assumptions as required for a Golden Section search—the two searches are closely related bracketing methods.

As before, only an interval solution—the final interval of uncertainty—can be found using this or any bracketing technique. The length of the final interval or *tolerance value* is controllable by the user, and can be made arbitrarily small restricted only by the precision of the computations and the computing time available. The final interval is guaranteed to be less than this tolerance value within a specific number of iterations.

Replace the golden ratio r in the test point generating formulas of the Golden Section search with the ratio of Fibonacci numbers to obtain

$$x_1 = a + \frac{F_{n-2}}{F_n} \cdot (b - a)$$

$$x_2 = a + \frac{F_{n-1}}{F_n} \cdot (b - a)$$

These are the Fibonacci search test point generators. The test points must lie inside the original interval $[a, b]$ since $a < x_1 < x_2 < b$, and will determine the new search interval in the same fashion as a Golden Section search.

If $f(x_1) < f(x_2)$, then the new interval is $[x_1, b]$.
If $f(x_1) > f(x_2)$, then the new interval is $[a, x_2]$.

Equality means calculation precision is exceeded or a calculation error has occurred.

The iterations continue until the final interval length is less than our imposed tolerance. Our final interval must contain the optimum solution. The size of the final interval determines the accuracy of the approximate optimum solution and vice versa. The number of iterations required to achieve this accepted interval length is the smallest Fibonacci number that satisfies the inequality

$$\frac{b-a}{tolerance} < F_k.$$

When we require a point solution, instead of an interval solution, the method of selecting a point is to evaluate the function, $f(x)$ at the end points and midpoint of the final interval. For maximization problems, select the endpoint or midpoint that yields the largest $f(x)$ value.

The Fibonacci search algorithm follows.

Fibonacci Search Algorithm

Find the maximum of the unimodal function f on the interval $[a, b]$.

INPUTS: Function: f
Endpoints: a, b
Tolerance: t

OUTPUTS: Optimal point x^* and maximum value $f(x^*)$

Step 1. Initialize: Set $a_0 = a$, $b_0 = b$, counter $k = 1$

Step 2. Calculate the number of iterations n such that F_n is the first Fibonacci number where $(b-a)/t < F_n$

Step 3. Calculate the test points

$$x_1 = a + \frac{F_{n-2}}{F_n} \cdot (b-a)$$
$$x_2 = a + \frac{F_{n-1}}{F_n} \cdot (b-a)$$

Step 4. Compute $f(x_1)$ and $f(x_2)$

Step 5. If $f(x_1) < f(x_2)$, then the new interval is $[x_1, b]$
Set $x_1 = x_2$
Compute the new x_2 (for the new interval)
If $f(x_1) > f(x_2)$, then the new interval is $[a, x_2]$
Set $x_2 = x_1$
Compute the new x_1 (for the new interval)

Step 6. If $n = 2$ or $b_n - a_n < t$, then STOP
Return estimates $x^* =$ midpoint(I_n) and $f(x^*)$.
Otherwise, set $n = n - 1$ and return to Step 4.

Although a Fibonacci search can be used with any unimodal function to find the maximum (or minimum) over a specified interval, its main advantage comes when normal calculus procedures fail, such as when f is not differentiable or continuous. Redo the first example of a Golden Section search.

Example 3.7. Maximizing a Nondifferentiable Function (reprise). Maximize

$$f(x) = -|2-x| - |5-4x| - |8-9x|$$

over the interval $[0, 3]$.

We'll use the *FibonacciSearch* program from the text's *PSMv2* library package.

```
> f := x → −|2 − x| − |5 − 4x| − |8 − 9x|;
```

$$f := x \mapsto -|2-x| - |5-4x| - |8-9x|$$

```
> FibonacciSearch(f, 0, 3, 0.1) :
  fnormal(%, 5);   #to fit the output to the page width
```

The maximum is f(0.905658)=-2.622630 with tolerance 0.100000

Iteration	x_1	x_2	$f(x_1)$	$f(x_2)$	*Interval*
0.0	1.1471	1.8529	−3.5882	−11.235	[0.0, 3]
1	0.70588	1.1471	−5.1176	−3.5882	[0.0, 1.8529]
2	1.1471	1.4160	−3.5882	−5.9916	[0.70588, 1.8529]
3	0.97639	1.1471	−2.9056	−3.5882	[0.70588, 1.4160]
4	0.87395	0.97639	−2.7647	−2.9056	[0.70588, 1.1471]
5	0.80893	0.87395	−3.6749	−2.7647	[0.70588, 0.97639]
6	0.87395	0.91260	−2.7647	−2.6504	[0.80893, 0.97639]
7	0.91260	0.93737	−2.6504	−2.7495	[0.87395, 0.97639]
8	0.89811	0.91260	−2.5924	−2.6504	[0.87395, 0.93737]

How does this answer compare with the optimum found by the Golden Section search?

Now redo the second example from earlier.

Example 3.8. Maximizing a Transcendental Function (reprise). Maximize the function

$$g(x) = 1 - \exp(-x) + \frac{1}{1+x}$$

over the interval $[0, 20]$.

```
> g := x → 1 − exp(−x) + 1/(1 + x) :
```

```
> FibonacciSearch(g, 0, 20, 0.1) :
  fnormal(%, 5);    #to fit the output to the page width
```

The maximum is f(2.523088)=1.203630 with tolerance 0.001000

Iteration	x_1	x_2	$f(x_1)$	$f(x_2)$	*Interval*
0.0	7.6395	12.361	1.1153	1.0748	$[0.0, 20]$
1	4.7210	7.6395	1.1659	1.1153	$[0.0, 12.361]$
2	2.9179	4.7210	1.2012	1.1659	$[0.0, 7.6395]$
3	1.8032	2.9179	1.1920	1.2012	$[0.0, 4.7210]$
4	2.9179	3.6066	1.2012	1.1899	$[1.8032, 4.7210]$
5	2.4920	2.9179	1.2036	1.2012	$[1.8032, 3.6066]$
6	2.2289	2.4920	1.2021	1.2036	$[1.8032, 2.9179]$
7	2.4920	2.6547	1.2036	1.2033	$[2.2289, 2.9179]$
8	2.3916	2.4920	1.2034	1.2036	$[2.2289, 2.6547]$
9	2.4920	2.5542	1.2036	1.2036	$[2.3916, 2.6547]$
10	2.4537	2.4920	1.2036	1.2036	$[2.3916, 2.5542]$
11	2.4920	2.5158	1.2036	1.2036	$[2.4537, 2.5542]$
12	2.5158	2.5304	1.2036	1.2036	$[2.4920, 2.5542]$

The Fibonacci search gives the maximum as $f(2.5) = 1.204$. How does this answer compare with the optimum found by the Golden Section search?

Interpolation with Derivatives: Newton's Method

Finding the Critical Points (Roots of the Derivative) of a Function

Newton's Method can be adapted to solve nonlinear optimization problems. For a twice differentiable function of one variable, the adaptation is straightforward. Newton's Method is applied to the *derivative*, rather than the original function we wish to optimize—the function's critical points occur at roots of the derivative. Replace f with f' in the iterating formula for Newton's Method:

$$N(x) = x - \frac{f(x)}{f'(x)} \qquad \Longrightarrow \qquad M(x) = x - \frac{f'(x)}{f''(x)}$$

Given a specified tolerance $\varepsilon > 0$, the iterations of Newton's Method can be terminated when $|x_{k+1} - x_k| \le \varepsilon$ or when $|f'(x_k)| < \varepsilon$.

In order to use the modified Newton's Method to find critical points, the function's first and second derivatives must exist throughout the neighborhood of interest. Also note that whenever the second derivative at x_k is zero, the next point x_{k+1} cannot be computed. (*Explain why!*) This method finds only approximates the value of critical points; it does not know whether it is finding a maximum or a minimum of the original function. Use the sign of the second derivative to determine whether the critical value is a maximum, minimum, or neither (a likely inflection point).

Example 3.9. A Simple Application.
Maximize the polynomial $p(x) = 5x - x^2$ over the interval $[0, 7]$.

This simple problem is easy to solve using elementary calculus.

1. Differentiate: $p' = 5 - 2x$.
2. The only critical point, the root of p', is $x^* = 5/2$.
3. Differentiate again: $p'' = -2 < 0$.
4. The Second Derivative Test indicates that p has a maximum of $25/4$ at $x^* = 5/2$.

Let's use Maple to implement the modified Newton's method for finding the maximum of p.

$$> p := x \to 5x - x^2 :$$
$$dp := D(p) :$$
$$ddp := D(dp) :$$

$$> Newton := x \to x - \frac{dp(x)}{ddp(x)} :$$

Start with $x_0 = 1.0$.

$$> x_0 := 1.0 :$$
$$x_1 := Newton(x_0);$$
$$x_2 := Newton(x_1);$$

$$x[0] := 1.0$$
$$x[1] := 2.500000000$$
$$x[2] := 2.500000000$$

Starting at any other value also yields $x^* = 2.5$. Since this simple quadratic function has a linear derivative, the linear approximation of the derivative will be exact regardless of the starting point, and the answer will appear at the second iteration. See Figure 3.6. Algebraically simplifying the method's iterating function for p confirms this: $Newton(x) = 2.5$.

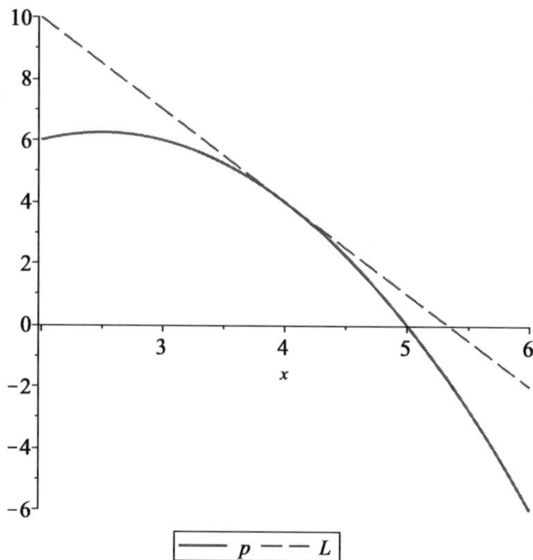

FIGURE 3.6: Plot of $p(x) = 5x - x^2$ with its Linear Approximation at $x = 4$

The slope of the linear approximation of the function at the point is precisely the slope of the function at that point, so the linear approximation is tangent to the function at the point as shown in the figure.

Example 3.10. A Third Degree Polynomial.
Find a maximal point, if possible, for $p(x) = -2x^3 + 10x + 5$ on the interval $[0, 3]$ to a tolerance of $\varepsilon = 0.01$.

We'll use the same process in Maple, this time, with the starting value $x = 1.0$.

$$> p := x \to -2x^3 + 10x + 5 :$$
$$dp := D(p) :$$
$$ddp := D(dp) :$$

$$> Newton := x \to x - \frac{dp(x)}{ddp(x)} :$$

Start again with $x_0 = 1.0$.

$$> x_0 := 1.0 :$$
$$x_1 := Newton(x_0);$$
$$x_2 := Newton(x_1);$$

$$x[0] := 1.0$$
$$x[1] := 1.333333333$$
$$x[2] := 1.291666667$$

$$> abs(x[2] - x[1]) < 0.01$$

$$0.041666666 < 0.01$$

The desired tolerance has not been achieved—we're not finished.

```
> x3 := Newton(x2);
  |x3 - x2| < 0.01;
```

$$x_3 := 1.290994624$$
$$0.000672043 < 0.01$$

We have a result! Checking the second derivative shows that $p(1.29) = 13.6$ is our maximum (to within 0.01).

Maple's *Student[CalculusI]* package contains *NewtonsMethod* for finding the roots of a function. To use *NewtonsMethod* to find critical points, all we have to do is replace the function with its first derivative. We illustrate below.

```
> with(Student[Calculus1]) :
```

```
> p := x -> -2x^3 + 10x + 5 :
  dp := D(p) :
```

```
> NewtonsMethod(dp(x), x = 1.0);
```

$$1.290994449$$

```
> NewtonsMethod(dp(x), x = 1.0, output = sequence);
```

$$1.0, 1.333333333, 1.291666667, 1.290994624, 1.290994449$$

```
> NewtonsMethod(dp(x), x = 0.3, output = plot);
```

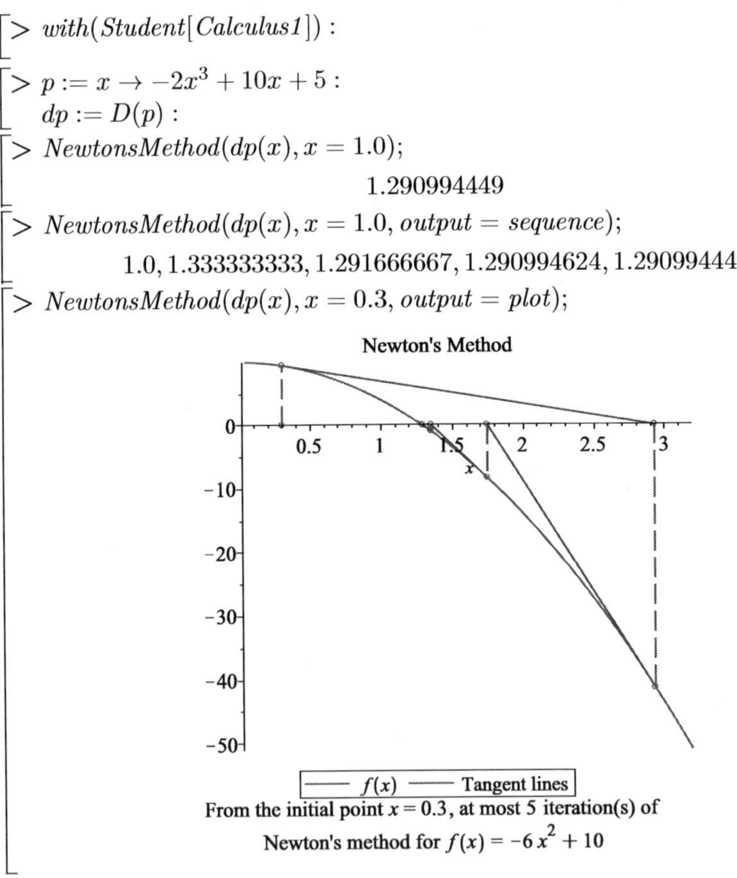

Try *NewtonsMethod* with *output=animation.*

Exercises

Compare the results of using the Golden Section, Fibonacci's, and Newton's methods in the following.

1. Maximize $f(x) = -x^2 - 2x$ on the closed interval $[-2, 1]$. Using a tolerance for the final interval of 0.6. Hint: Start Newton's method at $x = -0.5$.

2. Maximize $f(x) = -x^2 - 3x$ on the closed interval $[-3, 1]$. Using a tolerance for the final interval of 0.6. Hint: Start Newton's method at $x = 1$.

3. Minimize $f(x) = x^2 + 2x$ on the closed interval $[-3, 1]$. Using a tolerance for the final interval of 0.5. Hint: Start Newton's method at $x = -3$.

4. Minimize $f(x) = -x + e^x$ over the interval $[-1, 3]$ using a tolerance of 0.1. Hint: Start Newton's method at $x = -1$.

5. List at least two assumptions required by both Golden Section and Fibonacci's search methods.

6. Consider minimizing $f(x) = -x + e^x$ over the interval $[-1, 3]$. Assume the search method yielded a final interval of $[-0.80, 0.25]$ to within the tolerance of ε. Report a single best value of x to minimize $f(x)$ over the interval. Explain your reasoning.

7. Modify the Golden Section (pg. 109) algorithm to find a minimum value of a unimodal function. Write a Maple procedure to implement your change.

8. Modify the Fibonacci search (pg. 113) algorithm to find a minimum value of a unimodal function. Write a Maple procedure to implement your change.

Projects

Project 3.1. Write a program in Maple that uses a *secant method search.* Apply your program to Exercises 1 through 4 above.

Project 3.2. Write a program in Maple that uses a *Regula-Falsi* search method. Apply your program to Exercises 1 through 4 above.

Project 3.3. Carefully analyze the *rate of convergence* of the different search methods presented.

References and Further Reading

[BSS2013] Mokhtar S. Bazaraa, Hanif D. Sherali, and Chitharanjan M. Shetty, *Nonlinear Programming: Theory and Algorithms*, Wiley, 2013.

[B2016] William C. Bauldry, *Introduction to Computation*, 2016.

[F1992] William P. Fox, "Teaching nonlinear programming with Minitab," COED Journal, Vol. II (1992), no. 1, 80–84.

[F1993] William P. Fox, "Using microcomputers in undergraduate nonlinear optimization," Collegiate Microcomputer, Vol. XI (1993), no. 3, 214–218.

[FA2000] William P. Fox and Jeffrey Appleget, "Some fun with Newton's method," COED Journal, Vol. X (2000), no. 4, 38–43.

[FGMW1987] William P. Fox, Frank Giordano, S. Maddox, and Maurice D. Weir, *Mathematical Modeling with Minitab*, Brooks/Cole, 1987.

[FR2000] William P. Fox and William Richardson, "Mathematical modeling with least squares using Maple," Maple Application Center, Nonlinear Mathematics, 2000.

[FGW1997] William P. Fox, Frank Giordano, and Maurice Weir, *A First Course in Mathematical Modeling*, 2nd Ed., Brooks/Cole, 1997.

[FW2001] William P. Fox and Margie Witherspoon, "Single variable optimization when calculus fails: Golden section search methods in nonlinear optimization using Maple," COED Journal, Vol. XI (2001), no. 2, 50–56.

[GFH2014] Frank Giordano, William P. Fox, and Steven Horton, *A First Course in Mathematical Modeling*, 5th ed., Nelson Education, 2014.

[M2013] Mark M. Meerschaert, *Mathematical Modeling*, Academic Press, 2013.

[PRS1976] D.T. Phillips, A. Ravindran, and J. Solberg, *Operations Research*, Wiley, 1976.

[PFTV2007] William H. Press, Brian P. Flannery, Saul A. Teukolsky, William T. Vetterling, et al., *Numerical Recipes*, Cambridge U. Press, 2007.

[R1979] S.S. Rao, *Optimization: Theory and Applications*, Wiley Eastern Ltd., 1979.

[WVG2003] W.L. Winston, M. Venkataramanan, and J.B. Goldberg, *Introduction to Mathematical Programming*, Duxbury, 2003.

4

Problem Solving with Multivariable Constrained and Unconstrained Optimization

Objectives:

(1) Set up a solution process for multivariable optimization problems.

(2) Recognize when applicable constraints are required.

(3) Perform sensitivity analysis, when applicable, and interpret the results.

(4) Know the numerical techniques to obtain approximate solutions and when to use them.

4.1 Unconstrained Optimization: Theory

A small company is planning to install a central computer with cable links to five new departments. According to their floor plan, the peripheral computers for the five departments will be situated as shown in Figure 4.1 below. The company wishes to locate the central computer to minimize the amount of cable used to link the five peripheral stations because signals degrade with cable length and high quality cable is expensive. Assume that cable is strung above the ceiling panels in a straight line from a point above a peripheral to a point above the central computer; the standard distance formula may be used to determine the length needed to connect a remote to the central computer. Ignore the segments of cable from any device to a point above the ceiling panel immediately over that device. That is, work only with lengths of cable strung over the ceiling panels—the cable segment from the peripheral to the ceiling doesn't change with the different routings. The location coordinates of the five peripheral computers appear in Table 4.1.

TABLE 4.1: Grid of the Five Departments

X	15	25	60	75	80
Y	60	90	75	60	25

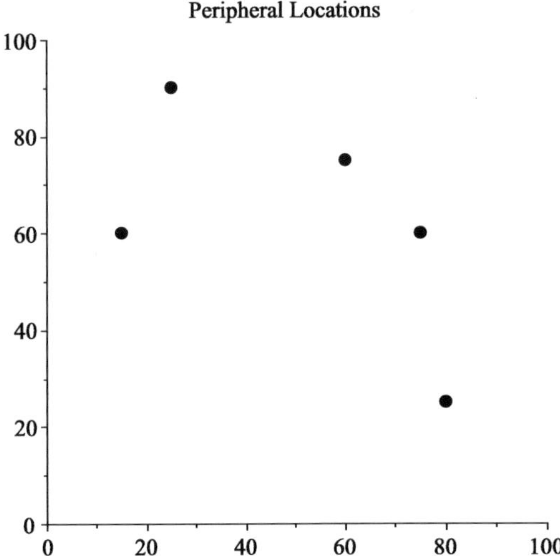

FIGURE 4.1: Grid of the Five Departments

The central computer will be positioned at coordinates (m, n) where m and n are both *integers* in the grid representing office space. Determine the coordinates (m, n) that minimize the total amount of cable needed. Report the total number of feet of cable needed for this placement along with the coordinates (m, n).

To model and solve problems like this, we need *multivariable optimization.*

In this chapter, our goal is to introduce both constrained and unconstrained nonlinear optimization. For a more thorough coverage, we suggest studying complete texts on the subject such as Ruszczyński's *Nonlinear Optimization* (Princeton Univ. Press, 2006).

Basic Theory

We begin with how to find an optimal solution, if one exists, for the unconstrained nonlinear optimization problem

$$\text{Maximize (or minimize) } f(x_1, x_2, \ldots, x_n) \text{ over } \mathbb{R}^n$$

We assume that all the first and second partial derivatives of f exist and are continuous at all points in f's domain. Let the partial derivative of f with respect to x_i be

$$\frac{\partial f(x_1, x_2, \ldots, x_n)}{\partial x_i}.$$

Candidate critical points (stationary points) are found where

$$\frac{\partial f(x_1, x_2, \ldots, x_n)}{\partial x_i} = 0 \text{ for } i = 1, 2, \ldots, n$$

This set of conditions gives a system of equations that when solved yields one or more critical points (if any exist) of f. Each critical point will satisfy each equation $\partial f(\textit{critical point})/\partial x_i = 0$ for $i = 1, \ldots, n$.

Theorem. Local Extremum Characterization.
If $\mathbf{x}^* = (x_1^*, x_2^*, \ldots, x_n^*)$ is a local extremum of the twice continuously differentiable function f, then $\mathbf{x}^*$ satisfies

$$\frac{\partial f(\mathbf{x}^*)}{\partial x_i} = 0 \text{ for } i = 1, 2, \ldots, n.$$

We have previously defined all points $\mathbf{x}$ that satisfy $\partial f(\mathbf{x})/\partial x_i = 0$ for $i = 1, 2, \ldots, n$ as critical points or *stationary points*. Not all critical points (stationary points) are local extrema. If a stationary point is not a local extremum (a maximum or a minimum), then it is called a *saddle point*.

The Hessian Matrix

We used a function's concavity to recognize extrema in the single-variable case. How do we determine the "concavity" of functions of more than one variable?

Definition. Convex Multivariable Function.
A function $f(\mathbf{x})$ is *convex* iff for every pair of points $\mathbf{x}^{(1)}$ and $\mathbf{x}^{(2)}$ in f's domain and any $\lambda \in [0, 1]$, we have

$$f\left((1-\lambda)\mathbf{x}^{(1)} + \lambda\mathbf{x}^{(2)}\right) \le (1-\lambda)f(\mathbf{x}^{(1)}) + \lambda f(\mathbf{x}^{(2)}).$$

Similarly, f is *concave* iff for every pair of points $\mathbf{x}^{(1)}$ and $\mathbf{x}^{(2)}$ in f's domain and any $\lambda \in [0, 1]$, we have

$$f\left((1-\lambda)\mathbf{x}^{(1)} + \lambda\mathbf{x}^{(2)}\right) \ge (1-\lambda)f(\mathbf{x}^{(1)}) + \lambda f(\mathbf{x}^{(2)}).$$

In words, this definition says that f is convex iff whenever we look at an $\mathbf{x}^*$ on the line segment connecting $\mathbf{x}^{(1)}$ and $\mathbf{x}^{(2)}$, the value of the function at $f(\mathbf{x}^*)$ is below the point the same distance along the line segment connecting $f(\mathbf{x}^{(1)})$ and $f(\mathbf{x}^{(2)})$. For concave, change "below" to "above."

Next, we introduce the *Hessian matrix*, named for Ludwig Hesse who studied surfaces in the 1800s, that allows us to determine the convexity of multivariable functions. The Hessian matrix will also provide additional information about the critical points.

Definition. The Hessian Matrix of f.
The *Hessian matrix* of a multivariable function f is the $n \times n$ matrix of second partial derivatives where the (i, j)th entry is $\partial f/(\partial x_i \partial x_j)$

$$H(f) = \begin{bmatrix} \frac{\partial^2 f}{\partial x_1^2} & \frac{\partial^2 f}{\partial x_1 x_2} & \cdots & \frac{\partial^2 f}{\partial x_1 x_n} \\ \frac{\partial^2 f}{\partial x_2 x_1} & \frac{\partial^2 f}{\partial x_2^2} & \cdots & \frac{\partial^2 f}{\partial x_2 x_n} \\ \vdots & \vdots & \ddots & \vdots \\ \frac{\partial^2 f}{\partial x_n x_1} & \frac{\partial^2 f}{\partial x_n x_2} & \cdots & \frac{\partial^2 f}{\partial x_n^2} \end{bmatrix}$$

When f has continuous second partial derivatives, the mixed partials are equal and the Hessian matrix is symmetric.[1]

In the 2-variable case of a function with continuous second partials,

$$H(f) = \begin{bmatrix} \frac{\partial^2 f}{\partial x_1^2} & \frac{\partial^2 f}{\partial x_1 x_2} \\ \frac{\partial^2 f}{\partial x_1 x_2} & \frac{\partial^2 f}{\partial x_2^2} \end{bmatrix}$$

Example 4.1. A Simple 2×2 Hessian.
Let $f(x_1, x_2) = x_1^2 + 3x_2^2$. Find f's Hessian. Since

$$\frac{\partial f}{\partial x_1} = 2x_1, \quad \frac{\partial f}{\partial x_2} = 6x_2$$

then

$$\frac{\partial^2 f}{\partial x_1^2} = 2, \quad \frac{\partial^2 f}{\partial x_1 x_2} = 0, \quad \frac{\partial^2 f}{\partial x_2^2} = 6$$

So f's Hessian is

$$\begin{bmatrix} 2 & 0 \\ 0 & 6 \end{bmatrix}$$

Example 4.2. Another 2×2 Hessian.
Let $g(x_1, x_2) = -x_1^2 + 3x_1x_2 - 3x_2^2$. Find g's Hessian. Since

$$\frac{\partial g}{\partial x_1} = -2x_1 + 3x_2, \quad \frac{\partial g}{\partial x_2} = -6x_2 + 3x_1$$

then

$$\frac{\partial^2 g}{\partial x_1^2} = -2, \quad \frac{\partial^2 g}{\partial x_1 x_2} = 3, \quad \frac{\partial^2 g}{\partial x_2^2} = -6$$

[1]The equality of mixed partials for *smooth* functions is called Schwarz's Theorem or Clairaut's Equality of Mixed Partials Theorem. Euler had discovered this very important result in 1734 when he was 27 years old.

So g's Hessian is

$$\begin{bmatrix} -2 & 3 \\ 3 & -6 \end{bmatrix}$$

The Hessian will play the role of the second derivative in the Second Derivative Test as we extend from single-variable to multivariable functions. We will use *determinants* to analyze the Hessian. Recall, the determinant of a 2×2 matrix A is

$$|A| = \begin{vmatrix} a_{11} & a_{12} \\ a_{21} & a_{22} \end{vmatrix} = a_{11}a_{22} - a_{12}a_{21}.$$

Calculating determinants of larger matrices is defined recursively using *expansion by minors.*[2] We can use *leading principal minors* to analyze the Hessian.

Definition 4.1. Leading Principal Minor.
The ith *leading principal minor* of a square matrix A is the determinant of the $i \times i$ submatrix obtained by deleting $n - i$ rows of A and deleting the same $n - i$ columns of A where $i = 0, 1, 2, \ldots, n - 1$.

Example 4.3. The Leading Principal Minors of a Hessian Matrix.
Given the 3×3 Hessian matrix

$$H = \begin{bmatrix} 2 & 0 & 4 \\ 0 & 1 & 5 \\ 4 & 5 & 3 \end{bmatrix},$$

determine all leading principal minors.

1. There are three 1st order leading principal minors, ($i = 1$: eliminate $3-1 = 2$ rows and the same columns).

    a. Eliminate rows 1 and 2 and columns 1 and 2: $M = \begin{bmatrix} 3 \end{bmatrix}$. Then

    $$|3| = 3$$

    b. Eliminate rows 1 and 3 and columns 1 and 3: $M = \begin{bmatrix} 1 \end{bmatrix}$. Then

    $$|1| = 1$$

    c. Eliminate rows 2 and 3 and columns 2 and 3: $M = \begin{bmatrix} 1 \end{bmatrix}$. Then

    $$|2| = 2$$

    The 1st order leading principal minors are the entries of the main diagonal.

2. There are three 2nd order leading principal minors, ($i = 2$: eliminate $3 - 2 = 1$ row and the same column).

[2]For a full discussion of determinants, see, e.g., Gilbert Strang's *Introduction to Linear Algebra*.

(a) Eliminate row 1 and column 1: $M = \begin{bmatrix} 1 & 5 \\ 5 & 3 \end{bmatrix}$. Then

$$\begin{vmatrix} 1 & 5 \\ 5 & 3 \end{vmatrix} = -22$$

(b) Eliminate row 2 and column 2: $M = \begin{bmatrix} 2 & 4 \\ 4 & 3 \end{bmatrix}$. Then

$$\begin{vmatrix} 2 & 4 \\ 4 & 3 \end{vmatrix} = -10$$

(c) Eliminate row 3 and column 3: $M = \begin{bmatrix} 2 & 0 \\ 0 & 1 \end{bmatrix}$. Then

$$\begin{vmatrix} 2 & 0 \\ 0 & 1 \end{vmatrix} = 2$$

3. There is only one 3rd order leading principal minor, the determinant of H itself.[3] (Eliminate no rows or columns.) Then

$$|H| = -60$$

Notice that the leading principal minors of a matrix are just the determinants of all the submatrices formed along the main diagonal.

How do we use all these determinants, the leading principal minors, of the Hessian matrix to determine the convexity of the multivariate function? The following theorem answers our question.

Theorem. Convexity and the Hessian Matrix.
Suppose $f(x_1, x_2, \ldots, x_n)$ has continuous second partial derivatives at all points of its domain D. Let H be f's Hessian matrix. Then

a. if all the leading principal minors of H are nonnegative, then f is convex;

b. if all the leading principal minors of H match the sign of $(-1)^k$ where k is the minor's order (positive for even order, negative for odd), then f is concave;

c. otherwise, f is neither convex, nor concave.

Example 4.4. Determining Convexity.
Determine the convexity of the function $f(x_1, x_2) = x_1^2 + 3x_2^2$ using its Hessian matrix.

[3]Maple uses the same notation for determinant. Enter the matrix H, then execute $|H|$.

1. The Hessian of f is

$$H(f) = \begin{bmatrix} \frac{\partial^2 f}{\partial x_1^2} & \frac{\partial^2 f}{\partial x_1 x_2} \\ \frac{\partial^2 f}{\partial x_1 x_2} & \frac{\partial^2 f}{\partial x_2^2} \end{bmatrix} = \begin{bmatrix} 2 & 0 \\ 0 & 6 \end{bmatrix}$$

2. a. The 1st order leading principal minors are: $|2| = 2 > 0$ and $|6| = 6 > 0$.
   b. The 2nd order leading principal minor is $|H| = 12 > 0$.

3. Since all leading principal minors are nonnegative, the Convexity and the Hessian Matrix theorem tell us that f is a convex function.

Figure 4.2 shows f's graph.

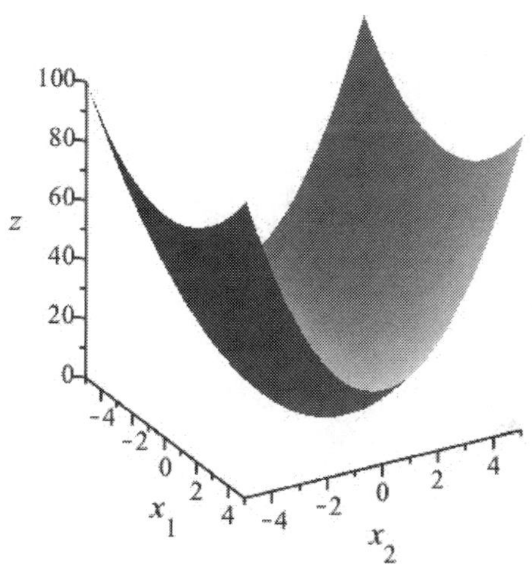

FIGURE 4.2: Graph of $f(x_1, x_2) = x_1^2 + 3x_2^2$

Example 4.5. Determining Concavity Using the Hessian.
Determine the convexity of the function $g(x_1, x_2) = -x_1^2 - 3x_2^2 + x_1 x_2$ using its Hessian matrix.

1. The Hessian of g is

$$H(g) = \begin{bmatrix} \frac{\partial^2 g}{\partial x_1^2} & \frac{\partial^2 g}{\partial x_1 x_2} \\ \frac{\partial^2 g}{\partial x_1 x_2} & \frac{\partial^2 g}{\partial x_2^2} \end{bmatrix} = \begin{bmatrix} -2 & 1 \\ 1 & -6 \end{bmatrix}$$

2. a. The 1st order leading principal minors are: $|-2| = -2 < 0$ & $|-6| = -6 < 0$.
   b. The 2nd order leading principal minor is $|H| = 11 > 0$.

3. Since the 1st order leading principal minors are negative and the 2nd order is positive, the Convexity and the Hessian Matrix theorem tells us that g is a concave function.

Figure 4.3 shows g's graph.

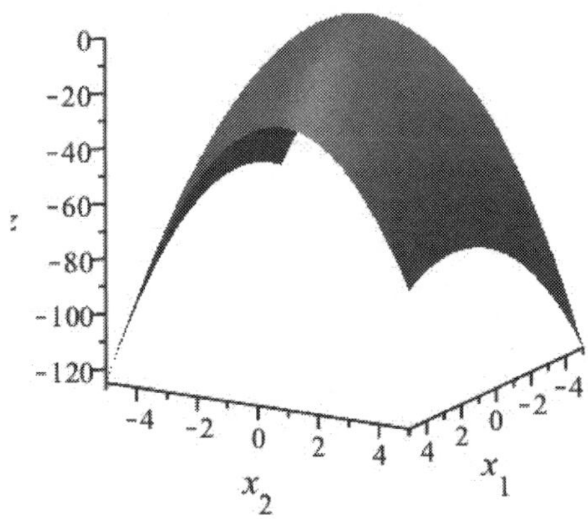

FIGURE 4.3: Graph of $g(x_1, x_2) = -x_1^2 - 3x_2^2 + x_1x_2$

Example 4.6. Determining Nonconvexity Using the Hessian.
Determine the convexity of the function $h(x_1, x_2) = x_1^2 - 3x_2^2 + x_1x_2$ using its Hessian matrix.

1. The Hessian of h is

$$H(h) = \begin{bmatrix} \frac{\partial^2 h}{\partial x_1^2} & \frac{\partial^2 h}{\partial x_1 x_2} \\ \frac{\partial^2 h}{\partial x_1 x_2} & \frac{\partial^2 h}{\partial x_2^2} \end{bmatrix} = \begin{bmatrix} 2 & 1 \\ 1 & -6 \end{bmatrix}$$

2. The 1st order leading principal minors are: $|2| = 2 > 0$ & $|-6| = -6 < 0$. Stop. Since we have opposite signs, h cannot be either convex or concave; h has a *saddle point.*

The function is neither convex nor concave; this Hessian matrix is called *indefinite.*

Figure 4.4 shows h's graph.

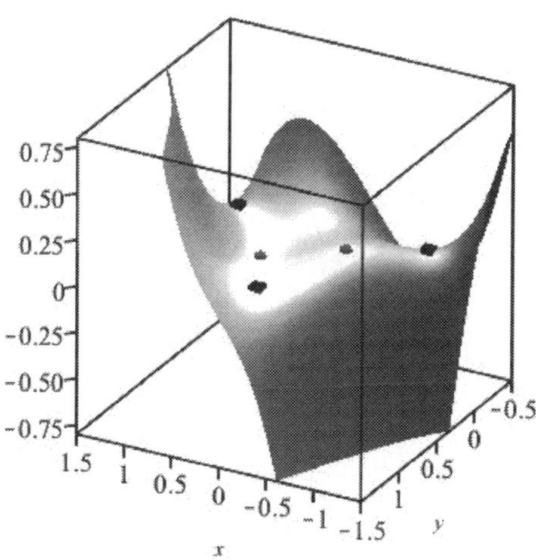

FIGURE 4.4: Graph of $h(x_1, x_2) = x_1^2 - 3x_2^2 + x_1 x_2$

A quick trip to Maple verifies the analysis.

```
> with(Student[VectorCalculus]) :    # for 'Hessian'
  with(LinearAlgebra) :              # for 'IsDefinite'
> h := (x1, x2) -> x1^2 - 3x2^2 + x1 x2 :
> H := Hessian(h(x1, x2)), [x1, x2]);
```

$$\begin{bmatrix} 2 & 1 \\ 1 & -6 \end{bmatrix}$$

```
> IsDefinite(H, query = 'indefinite');
```

true

Part a of the Convexity and the Hessian Matrix Theorem (pg. 126) says a convex function has all nonnegative leading principal minors; the function's Hessian is then called *positive definite.* Part b of the theorem says a concave function has all nonpositive leading principal minors; the function's Hessian is then called *negative definite.* Part c gives us an *indefinite* Hessian. The calculation of definiteness becomes more involved when variables appear in the function's Hessian. Let's use Maple to test and verify definiteness for the functions of our examples. (Remember to load both *Student*[*VectorCalculus*] and *LinearAlgebra* packages.)

```
> f := (x, y) -> x^2 + 3·y^2 :
  g := (x, y) -> -x^2 - 3·y^2 + x·y :
  h := (x, y) -> x^2 - 3·y^2 + x·y :
```

> $Hf := Hessian(f(x_1, x_2), [x_1, x_2]);$
> $IsDefinite(Hf, query = \text{'positive\_definite'});$
> $IsDefinite(Hf, query = \text{'negative\_definite'});$
> $IsDefinite(Hf, query = \text{'indefinite'});$

$$Hf := \begin{bmatrix} 2 & 0 \\ 0 & 6 \end{bmatrix}$$

true
false
false

> $Hg := Hessian(g(x_1, x_2), [x_1, x_2]);$
> $IsDefinite(Hg, query = \text{'positive\_definite'});$
> $IsDefinite(Hg, query = \text{'negative\_definite'});$
> $IsDefinite(Hg, query = \text{'indefinite'});$

$$Hg := \begin{bmatrix} -2 & 1 \\ 1 & -6 \end{bmatrix}$$

false
true
false

> $Hh := Hessian(g(x_1, x_2), [x_1, x_2]);$
> $IsDefinite(Hh, query = \text{'positive\_definite'});$
> $IsDefinite(Hh, query = \text{'negative\_definite'});$
> $IsDefinite(Hh, query = \text{'indefinite'});$

$$Hh := \begin{bmatrix} 2 & 1 \\ 1 & -6 \end{bmatrix}$$

false
false
true

Maple has a number of functions that compute the Hessian of a function. Versions of the *Hessian* command are in the *Student*[*VectorCalculus*], *VectorCalculus*, and *LinearAlgebra* packages. Consult Maple's Help pages to help you decide which version is best to use in a given situation.

In a more general setting, if a function has terms with power higher than 2 or includes transcendental functions, such as $x_1 x_2^2$ or $\sin(x_1 x_2)$, the Hessian will have terms with variables. Then the analysis of definiteness for classifying a stationary point as a maximum, minimum, or saddle point is a bit more complicated.

For two variables, we use the *quadratic form*

$$Q(x_1, x_2) = \begin{bmatrix} x_1 & x_2 \end{bmatrix} \times H \times \begin{bmatrix} x_1 \\ x_2 \end{bmatrix}$$

to test the Hessian matrix. In general, we use $Q(\vec{x}) = \vec{x}^T H \vec{x}$. Then define

Definition 4.2. Definiteness of a Matrix.
Let H be f's Hessian matrix. Then H is

1. *positive definite* iff $H > 0$ for every nonzero vector $z \in \text{domain}(f)$,
2. *positive semi-definite* iff $H \geq 0$ for every nonzero vector $z \in \text{domain}(f)$,
3. *negative definite* iff $H < 0$ for every nonzero vector $z \in \text{domain}(f)$,
4. *negative semi-definite* iff $H \leq 0$ for every nonzero vector $z \in \text{domain}(f)$,
5. *indefinite* otherwise.

We can relate definiteness to the signs of the leading principal minors.

Positive Definite: All leading principal minors are positive.

Positive Semi-definite: All leading principal minors are non-negative (some may be zero).

Negative Definite: The leading principal minors follow the signs of $(-1)^k$, where k is the order of the leading principal minor. Even order leading principal minors are positive, and odd order leading principal minors are negative.

Negative Semi-definite: All nonzero leading principal minors follow the signs of $(-1)^k$, where k is the order of the leading principal minor. Even order nonzero leading principal minors are positive, and odd order nonzero leading principal minors are negative. Some may be zero.

Indefinite: Some leading principal minors do not follow any of the rules above.

If the Hessian is also a function of the independent variables, its definiteness might vary from one point to another. To test the definiteness of the Hessian at a point x^*, evaluate the Hessian at the point. For example, in the matrix

$$H(x_1, x_2) = \begin{bmatrix} 2x_1 & x_2 \\ x_2 & 4 \end{bmatrix},$$

the specific values of x_1 and x_2 determine whether H is positive definite, positive semi-definite, negative definite, negative semi-definite, or indefinite *at* $[x_1, x_2]$. *Give an example of each case!*

Table 4.2 summarizes the relationship between the Hessian matrix definiteness and the classification of stationary points (extrema) as maximum, minimum, saddle points, or 'inconclusive.'

TABLE 4.2: Summary of Hessian Definiteness

Determinants: H_k $= k$ order leading principal minors	Results	Conclusions About the Stationary Point
All $H_k > 0$	Positive Definite, f is convex	minimum
All $H_k \geq 0$	Positive Semi-Definite, f is convex	local minimum
$\text{sgn}(H_k) = \text{sgn}((-1)^k)$	Negative Definite, f is concave	maximum
$\text{sgn}(H_k) = \text{sgn}((-1)^k)$ or some, not all, $H_k = 0$	Negative Semi-Definite, f is concave	local maximum
H_k not all 0, but none of the above	Indefinite, f is neither	saddle point
$H_k = 0$ for all k	Indefinite	*inconclusive*

Example 4.7. Using the Summary Table.

1. Suppose the Hessian of the function $f(x_1, x_2)$ is given by

$$H(x_1, x_2) := \begin{bmatrix} -2 & -1 \\ -1 & -4 \end{bmatrix}$$

   The 1st order leading principal minors are -2 and -4. The 2nd order leading principal minor is $|H(x_1, x_2)| = 7$. Since the first orders are negative and the second order is positive, the leading principal minors follow the sign of $(-1)^k$. The summary table indicates that f is concave and the stationary point found is a maximum. This Hessian matrix is negative definite.

$$> H := Matrix([[-2, -1], [-1, -4]]);$$

$$H := \begin{bmatrix} -2 & -1 \\ -1 & -4 \end{bmatrix}$$

$$> IsDefinite(H, query = \text{'negative\_definite'});$$

$$true$$

2. Suppose the Hessian of the function f is given by

$$H := \begin{bmatrix} 3 & 2 & 1 \\ 6 & 5 & 4 \\ 9 & 8 & 7 \end{bmatrix}$$

The 1st order leading principal minors are 3, 5, and 7. The 2nd order leading principal minors are 3, 12, and 3. The 3nd order leading principal minor, the determinant of H, is 0. All leading principal minors are positive or zero. The summary table indicates that f is convex and any corresponding stationary points found would be local minima. This Hessian matrix is positive semi-definite.

> $H := Matrix([[3, 2, 1], [6, 5, 4], [9, 8, 7]]);$

$$H := \begin{bmatrix} 3 & 2 & 1 \\ 6 & 5 & 4 \\ 9 & 8 & 7 \end{bmatrix}$$

> $IsDefinite(H, query = \text{'positive\_definite'});$

$$false$$

> $IsDefinite(H, query = \text{'positive\_semidefinite'});$

$$true$$

3. Suppose the Hessian of the function f is given by

$$H := \begin{bmatrix} -2 & -4 \\ -4 & -3 \end{bmatrix}$$

The 1st order leading principal minors are -2 and -3. The 2nd order leading principal minor is $|H| = -10$. Since both the first orders and the second order are negative, the summary table indicates the Hessian is indefinite and the stationary point found is a saddle point.

> $H := Matrix([[-2, -4], [-4, -3]]);$

$$H := \begin{bmatrix} -2 & -4 \\ -4 & -3 \end{bmatrix}$$

> $IsDefinite(H, query = \text{'indefinite'});$

$$true$$

4.2 Unconstrained Optimization: Examples

The definitions and theorems from the previous section are put to work to solve a set of unconstrained optimization problems in the following examples. In the Maple sessions below, remember to start with a fresh document and to load the *Student*[*VectorCalculus*] and *Student*[*LinearAlgebra*] packages.

Example 4.8. Finding Extrema, I.
Find and classify all the stationary points of

$$f(x_1, x_2) = 55x_1 - 4x_1^2 + 135x_2 - 15x_2^2 - 100.$$

The computations for f are relatively straightforward, so we'll do them by hand.

The partial derivates are

$$\frac{\partial f}{\partial x} = 55 - 8x \quad \text{and} \quad \frac{\partial f}{\partial y} = 135 - 30,$$

and so the second partials are

$$\frac{\partial^2 f}{\partial x^2} = -8, \quad \frac{\partial^2 f}{\partial x\, \partial y} = 0, \quad \text{and} \quad \frac{\partial^2 f}{\partial y^2} = -30$$

Solving the system $\{f_x(x_1, x_2) = 0, f_y(x_1, x_2) = 0\}$ gives the single critical point $\mathbf{x}^* = (55/8, 135/30)$.

The Hessian of f is

$$H = \begin{bmatrix} -8 & 0 \\ 0 & -30 \end{bmatrix}.$$

The 1st order leading principal minors (LPM) for f are -8 and -30. The 2nd order LMP is $|H| = 240 > 0$. The 1st order LPMs are negative, the 2nd order is positive: by Table 4.2, the Summary of Hessian Definiteness, we see that f is concave at $\mathbf{x}^*$; therefore $f(55/8, 135/30) = 392.8125$ is the maximum.

Example 4.9. Finding Extrema, II.
Find and classify all stationary points of the function

$$g(x_1, x_2) = 2x_1x_2 + 4x_1 + 6x_2 - 2x_1^2 - 2x_2^2.$$

We'll go straight to Maple to analyze g. To avoid indexing issues, define g as a function of x and y. Then we can switch to x_1 and x_2 immediately. Maple's D operator uses "numeric" notation: $\partial/\partial x_1 = D_1$ and $\partial/\partial x_2 = D_2$, etc.

```
> g := (x, y) → 2·x·y + 4x + 6y − 2x^2 − 2y^2;
  dx1 := D1(g);
  dx2 := D2(g);
  ddx1 := D1(dx1);
  ddx2 := D2(dx2);
  ddx1x2 := D2(dx1);
```

$$g := (x, y) \mapsto 2xy + 4x + 6y + (-2x^2) + (-2y^2)$$
$$dx1 := (x, y) \mapsto -4x + 2y + 4$$
$$dx2 := (x, y) \mapsto 2x - 4y + 6$$
$$ddx1 := (x, y) \mapsto -4$$
$$ddx2 := (x, y) \mapsto -4$$
$$ddx1x2 := (x, y) \mapsto 2$$

```
> solve([dx1(x1, x2) = 0, dx2(x1, x2) = 0], [x1, x2]);
  CP := rhs~ (%1);
```

$$\left[\left[x_1 = \frac{7}{3}, x_2 = \frac{8}{3}\right]\right]$$

$$CP := \left[\frac{7}{3}, \frac{8}{3}\right]$$

```
> H := Hessian(g(x1, x2), [x1, x2]);
```

$$H := \begin{bmatrix} -4 & 2 \\ 2 & -4 \end{bmatrix}$$

```
> IsDefinite(H, query = 'negative_definite');
```

true

We see that the 1st order LPMs are -4 and -4, which follow $(-1)^1$, and the 2nd order LPM is 12, which also follows $(-1)^2$. Thus, H is negative definite (at the point $(7/3, 8/3)$), and so g is concave. It follows that $g(7/3, 8/3) \approx 12.667$ is a global maximum. See Figure 4.5.

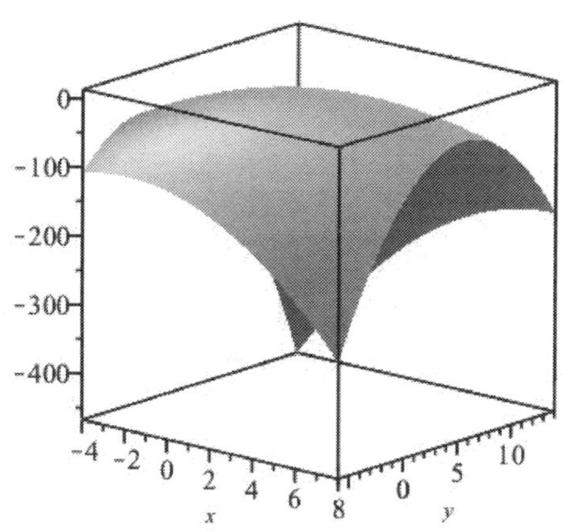

FIGURE 4.5: Graph of $g(x_1, x_2)$

Example 4.10. Finding Extrema, III.
Find and classify all the stationary points of

$$h(x_1, x_2) = x^3 y + x\, y^2 - x\, y.$$

We will use Maple for the computations this time also.

$> h := (x, y) \to x \cdot y^2 + x^3 \cdot y - x \cdot y$

$> dx_1 := D_1(h);$
$dx_2 := D_2(h);$

$$dx_1 := (x, y) \mapsto 3x^2y + y^2 - y$$
$$dx_2 := (x, y) \mapsto x^3 + 2yx - x$$

Our next step is to solve the system $\{\partial g/\partial x_1 = 0, \partial g/\partial x_2 = 0\}$ to find critical points. In order to avoid complex roots and Maple's *RootOf*s that occur in higher powers equations, we'll solve in the *RealDomain*; i.e., using only real numbers.

> **use** *RealDomain* **in**
  $CriticalPoints := solve([dx_1(x_1, x_2) = 0, dx_2(x_1, x_2) = 0], [x_1, x_2]);$
  end use :
$CP := map(rhs\sim, CriticalPoints)$

$$CP := \left[\left[\frac{\sqrt{5}}{5}, \frac{2}{5}\right], \left[-\frac{\sqrt{5}}{5}, \frac{2}{5}\right], [0, 1], [-1, 0], [1, 0], [0, 0]\right]$$

We have 6 stationary points that need to be tested with the Hessian to determine if they are local maxima, local minima, or saddle points.

$> Hessian(h(x_1, x_2), [x_1, x_2]);$
$H := unapply(\%, x):$ # *H will be a function of the vector* $x = \langle x_1, x_2 \rangle$

$$\begin{bmatrix} 6\,x_2 x_1 & 3\,{x_1}^2 + 2\,x_2 - 1 \\ 3\,{x_1}^2 + 2\,x_2 - 1 & 2\,x_1 \end{bmatrix}$$

Now evaluate the Hessian and its determinant at each of the critical points.

We'll use a **for** loop to compute the Hessian at the different critical points successively. Recall that the syntax of this type of loop is

for *var* **from** *start_value* **to** *stop_value* **do**
  statements
  end do

If "**from** *start_value*" is omitted, the *start_value* is assumed to be 1.

On to the analysis. In each line, list the critical point, the Hessian at that point, and the Hessian's determinant, which is the 2nd order LPM, at that critical point.

```
> for i to 6 do
    CP_i, H(CP_i), Determinant(H(CP_i));
  end do;
```

$$\left[\frac{\sqrt{5}}{5}, \frac{2}{5}\right], \begin{bmatrix} \frac{12\sqrt{5}}{25} & \frac{2}{5} \\ \frac{2}{5} & \frac{2\sqrt{5}}{5} \end{bmatrix}, \frac{4}{5}$$

$$\left[-\frac{\sqrt{5}}{5}, \frac{2}{5}\right], \begin{bmatrix} -\frac{12\sqrt{5}}{25} & \frac{2}{5} \\ \frac{2}{5} & -\frac{2\sqrt{5}}{5} \end{bmatrix}, \frac{4}{5},$$

$$[0,1], \begin{bmatrix} 0 & 1 \\ 1 & 0 \end{bmatrix}, -1$$

$$[-1,0], \begin{bmatrix} 0 & 2 \\ 2 & -2 \end{bmatrix}, -4$$

$$[1,0], \begin{bmatrix} 0 & 2 \\ 2 & 2 \end{bmatrix}, -4$$

$$[0,0], \begin{bmatrix} 0 & -1 \\ -1 & 0 \end{bmatrix}, -1$$

We used Maple's *Determinant* command from the *Student*[*LinearAlgebra*] package to compute $|H|$ giving us the 2nd order LPMs for each critical point. The *Minor* command will give H's 1st order LPMs as seen in

```
> Minor(H(CP_1), 1, 1);
  Minor(H(CP_1), 2, 2);
```

$$\frac{2\sqrt{5}}{5}$$

$$\frac{12\sqrt{5}}{5}$$

Determinant, *IsDefinite*, and *Minor* are all in the *Student*[*LinearAlgebra*] package. How can the **for** loop above be modified to determine the definiteness of the Hessian at each critical point?

Table 4.3 shows the data collected for h in a chart for easier analysis.

TABLE 4.3: Analysis of h's Critical Points

Critical Point	Leading Principal Minors		Classification and Results
	1st Order LPMs	2nd Order LPM	
$(0,0)$	$0,\ 0;\ [\geq 0]$	-1	neither, saddle point
$(1,0)$	$0,\ 0;\ [\geq 0]$	-4	neither, saddle point
$(0,1)$	$0,\ 0;\ [\geq 0]$	-1	neither, saddle point
$(-1,0)$	$0,\ -2;\ [\sim (-1)^1]$	-4	neither, saddle point
$\left(\frac{\sqrt{5}}{5}, \frac{2}{5}\right)$	$\frac{12\sqrt{5}}{5},\ \frac{2\sqrt{5}}{5};\ [\geq 0]$	$\frac{4}{5}$	convex, local minimum
$\left(-\frac{\sqrt{5}}{5}, \frac{2}{5}\right)$	$-\frac{12\sqrt{5}}{5},\ -\frac{2\sqrt{5}}{5};\ [\sim (-1)^1]$	$\frac{4}{5};\ [\sim (-1)^2]$	concave, local maximum

Using Maple to verify the definiteness of each Hessian is easy.

```
> IsDefinite(H(CP1), query = 'positive_definite');
  IsDefinite(H(CP2), query = 'negative_definite');
                          true
                          true
> seq(IsDefinite(H(CPk), query = indefinite'), k = 3..6);
                 true, true, true, true
```

We find the maximum and minimum points are $h(-\sqrt{5}/5, 2/5) \approx 0.072$ and $h(\sqrt{5}/5, 2/5) \approx -0.072$. Finally, a graph of h is shown in Figure 4.6 with the maximum and minimum points marked as spheres and the saddle points as boxes. Carefully zoom in on each critical point in a graph of h to verify the behaviors claimed.

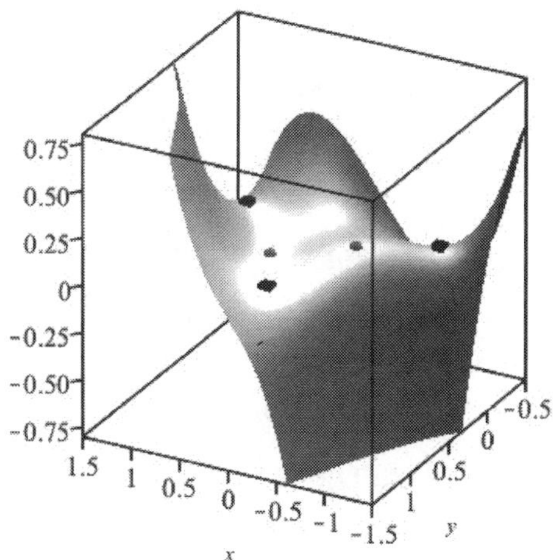

FIGURE 4.6: Graph of $h(x_1, x_2)$ and its Critical Points

We could also use Maple's *Minor* command from the *Student*[*LinearAlgebra*] package in the form

$[> Minor(Minor(M, 1, 1, output = matrix), 1, 1);$

to compute LPMs of larger matrices M; however, the syntax rapidly gets complicated as the number of leading principal minors gets large quickly.

Tides change the sea level around an island. Local tidal variation is influenced by a number of factors many of which are based on local topography. An island in a harbor is a navigational hazard. In the next example, we look for an island using a function based on bathymetric data of the sea bed combined with land height data.

Example 4.11. Finding an Island.
The bathymetric and topographic data sets combined lead to the sea and land surface model

$$f(x_1, x_2) = -300\,x_2^3 - 695\,x_2^2 + 7\,x_2 - 300\,x_1^3 - 679\,x_1^2 - 235\,x_1 + 570$$

for $(x, y) \in [-2..2] \times [-2..2]$.

Going right to Maple using our templates from the previous examples, we find:

$$> f := (x,y) \to -300\,y^3 - 695\,y^2 + 7\,y - 300\,x^3 - 679\,x^2 - 235\,x + 570 :$$
$$dx := D_1(f);$$
$$dy := D_2(f);$$

$$dx := (x,y) \mapsto -900x^2 - 1358x - 235$$
$$dy := (x,y) \mapsto -900y^2 - 1390y + 7$$

$$> Hessian(f(x,y),[x,y]);$$
$$H := unapply(\%,[x,y]) :$$

$$\begin{bmatrix} -1800\,x - 1358 & 0 \\ 0 & -1800\,y - 1390 \end{bmatrix}$$

$>$ **use** *RealDomain* **in**
$CriticalPoints := evalf_5(solve([dx(x,y)=0, dy(x,y)=0],[x,y])) :$
end use :
$CP := map(rhs\sim, CriticalPoints);$

$$CP := [[-0.19940,-1.5495],[-1.3095,-1.5495],[-0.19940,0.00504], [-1.3095,0.00504]]$$

$>$ **for** i **to** 4 **do**
$CP_i, f(op(CP_i)), H(op(CP_i));$
end do;

$$[-0.19940,-1.5495], 28.8149803, \begin{bmatrix} -999.08000 & 0 \\ 0 & 1399.1000 \end{bmatrix}$$

$$[-1.3095,-1.5495], -176.379930, \begin{bmatrix} 999.1000 & 0 \\ 0 & 1399.1000 \end{bmatrix}$$

$$[-0.19940,0.00504], 592.2577678, \begin{bmatrix} -999.08000 & 0 \\ 0 & -1399.07200 \end{bmatrix}$$

$$[-1.3095,0.00504], 387.0628571, \begin{bmatrix} 999.1000 & 0 \\ 0 & -1399.07200 \end{bmatrix}$$

$> IsDefinite(H(op(CP_3)), query = \text{'negative\_definite'});$

$$true$$

We find that at CP_3, $x = -0.1994$ and $y = 0.0050$, the height is $f(-0.1994, 0.0050) \approx 592.2578$. The Hessian $H(-0.1994, 0.0050)$ is negative definite at that point indicating that we have found the maximum. Assuming that $f(x,y) = 0$ is sea level, we have found that an island which is 592.2578 feet above sea level exists in the harbor. That's a significant island for ships to avoid! See Figure 4.7.

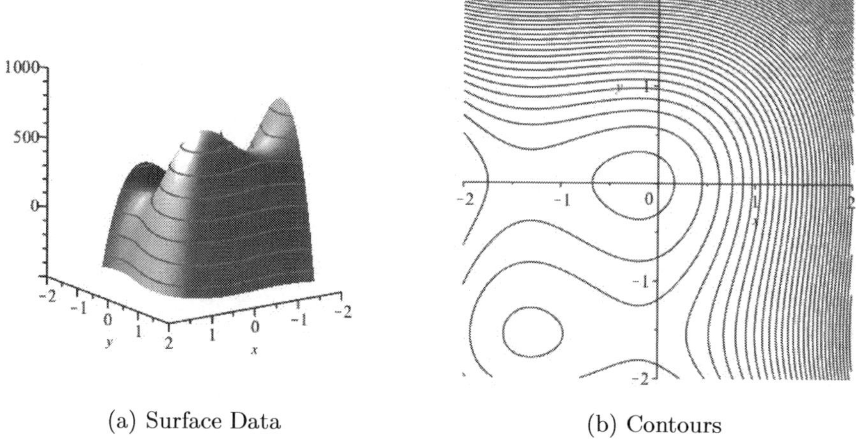

(a) Surface Data (b) Contours

FIGURE 4.7: Topographical Plot

For our last example in this section, we turn to an interesting and quite useful application of multivariate optimization: finding the least squares regression line fitting a data set. We'll begin by theoretically finding the regression line using what we've developed for optimization. Then we'll step through the computations in Maple to find the least squares regression line for a small data set.

Example 4.12. Finding the Least Squares Regression Line.
In using least squares regression to fit a line $y = a + bx$ to a set of n data points $\{(x_i, y_i)\}$, we minimize the sum of squared errors

$$SSE = \sum_{i=1}^{n} \big(y_i - (a + b\,x_i)\big)^2.$$

the variables are a and b.

Begin by finding the system of equations that will determine the critical points. First,

$$\frac{\partial f}{\partial a} = \sum_{i=1}^{n} 2\big(y_i - (a + b\,x_i)\big)(-1)$$

$$\frac{\partial f}{\partial b} = \sum_{i=1}^{n} 2\big(y_i - (a + b\,x_i)\big)(-x_i)$$

Then $\{\partial f/\partial a = 0, \partial f/\partial b = 0\}$ holds when the *Normal Equations for Least Squares* hold

$$n\,a + b\sum_{i=1}^{n} x_i = \sum_{i=1}^{n} y_i$$

$$a\sum_{i=1}^{n} x_i + b\sum_{i=1}^{n} x_i^2 = \sum_{i=1}^{n} x_i y_i$$

Remember, the summations are all constants once given a data set.

The Hessian is

$$H = \begin{bmatrix} 2n & 2\sum_{i=1}^{n} x_i \\ 2\sum_{i=1}^{n} x_i & 2\sum_{i=1}^{n} x_i^2 \end{bmatrix}$$

So the 1st order LPMs are $2n$ and $2\sum_{i=1}^{n} x_i^2$ which are both greater than 0. The 2nd order LPM is

$$|H| = 4n\sum_{i=1}^{n} x_i^2 - \left[\sum_{i=1}^{n} x_i\right]^2$$

which can also be shown to be greater than zero. Therefore, H is positive definite, and the critical point (a, b) must be a minimum.

Let's look at this optimization computation in Maple with a very small data set:

X	1	2	3
Y	2	4.8	7

Once more, we go right to Maple.

```
> X, Y := [1, 2, 3], [2, 4.8, 7] :
  n := 3 :
  sse := sum((Y_i − (a + b · X_i))^2, i = 1..n);
  SSE := unapply(sse, [a, b]) :
        sse := (−a − b + 2)^2 + (−a − 2b + 4.8)^2 + (−a − 3b + 7)^2
> da := D_1(SSE);
  db := D_2(SSE);
        da := (a, b) ↦ −27.6 + 6a + 12b
        db := (a, b) ↦ −65.2 + 12a + 28b
> solve([da(a, b) = 0, db(a, b) = 0], [a, b]);
  CP := %_1;
        [[a = −0.4000000000, b = 2.500000000]]
        CP := [a = −0.4000000000, b = 2.500000000]
```

```
> H := Hessian(SSE(a, b), [a, b]);
```

$$\begin{bmatrix} 6 & 12 \\ 12 & 28 \end{bmatrix}$$

```
> IsDefinite(H, query = 'positive_definite');
```

$$true$$

```
> L := subs(CP, a + b x);
```

$$L := 2.500000000\,x - 0.4000000000$$

The Hessian matrix is positive definite, so we have found the minimum for *SSE*, the sum of the square errors. The linear regression line for the small data set is $y = 2.5x - 0.4$. The graph in Figure 4.8 shows how well the regression line fits the data.

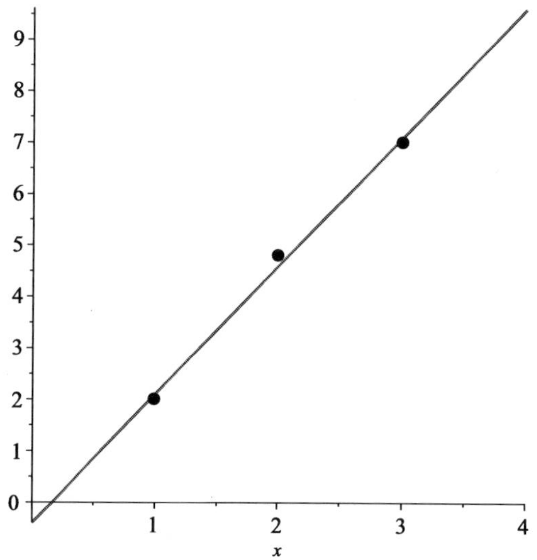

FIGURE 4.8: Regression Line with Data

Exercises

1. Indicate both the definiteness: positive definite, positive semi-definite, negative definite, negative semi-definite, or indefinite of the given Hessian matrices and the concavity of the function f from which these Hessians were derived.

a. $H = \begin{bmatrix} 2 & 3 \\ 3 & 5 \end{bmatrix}$ b. $H = \begin{bmatrix} 4 & 3 \\ 3 & 2 \end{bmatrix}$ c. $H = \begin{bmatrix} -2 & 1 \\ 1 & -2 \end{bmatrix}$

d. $H = \begin{bmatrix} -3 & 4 \\ 4 & -5 \end{bmatrix}$ e. $H = \begin{bmatrix} 6x & 0 \\ 0 & 2x \end{bmatrix}$ f. $H = \begin{bmatrix} x^2 & 0 \\ 0 & 2y \end{bmatrix}$

g. $H = \begin{bmatrix} 2 & 2 & 3 \\ 2 & 6 & 4 \\ 3 & 4 & 4 \end{bmatrix}$ h. $H = \begin{bmatrix} 1 & 2 & 2 \\ 2 & 1 & 2 \\ 2 & 2 & 1 \end{bmatrix}$ i. $H = \begin{bmatrix} 12x & 2 & 2 \\ 2 & -12y & -1 \\ 2 & -1 & 18z \end{bmatrix}$

2. Determine the convexity of the given function using the Hessian matrix H. Find and classify the function's critical points.

   a. $f(x, y) = x^2 + 3xy - y^2$
   b. $f(x, y) = x^2 + y^2$
   c. $f(x, y) = -x^2 - xy - 2y^2$
   d. $f(x, y) = 3x + 5y - 4x^2 + y^2 - 5xy$
   e. $f(x, y, z) = 2x + 3y + 3z - xy + xz - yz - x^2 - 3y^2 - z^2$

3. Determine values of a, b, and c such that $g(x, y) = ax^2 + bxy + cy^2$ is convex. Determine the values where g is concave.

4. Find and classify all the extreme points for the following functions.

   (a) $f(x, y) = x^2 + 3xy - y^2$
   (b) $f(x, y) = x^2 + y^2$
   (c) $f(x, y) = -x^2 - xy - 2y^2$
   (d) $f(x, y) = 3x + 5y - 4x^2 + y^2 - 5xy$
   (e) $f(x, y) = 3x^3 + 7x^2 + 2x + 3y^3 + 7y^2 - y - 5$
   (f) $f(x, y, z) = 2x + 3y + 3z - xy + xz - yz - x^2 - 3y^2 - z^2$

5. Find and classify all critical points of $f(x, y) = e^{(x-y)} + x^2 + y^2$.

6. Find and classify all critical points of $f(x, y) = (x^2 + y^2)^{1.5} - 4(x^2 + y^2)$.

7. Three oil wells are located at coordinates $(0, 0)$, $(12, 6)$, and $(10, 20)$. Each well produces an equal amount of oil. A pipeline is to be laid from each oil well to a central refinery located at (a, b). Where should the refinery be located to minimize the total squared Euclidean distance?

$$d^2 = \sum_{i=1}^{3} (x - a_i)^2 + (y_i - b)^2$$

   That is, determine the optimal location (a, b).

8. A small company is planning to install a central computer with cable links to five departments. According to the floor plan, the peripheral computers for the five departments will be situated as shown by Figure 4.1 (pg. 122) based on the coordinates listed in Table 4.1. The company wishes to locate the central computer to minimize the total amount of cable used to link to the five peripheral computers. Assuming that cable may be strung over the ceiling panels in a straight line from a point above any peripheral to a point above the central computer, the standard distance formula may be used to determine the length of cable needed. Ignore all lengths of cable from the computer itself to a point directly above the ceiling panel over that computer. That is, work only with straight lengths of cable strung above the ceiling panels. Find the optimal location for the central computer.

9. The first partials of a function $f(x, y)$ evaluated at $(0, 0)$ are $\partial f/\partial x = -5$ and $\partial f/\partial y = 1$. The Hessian of f at $(0, 0)$ is

$$H_f(0,0) = \begin{bmatrix} 6 & -1 \\ -1 & 2 \end{bmatrix}$$

Use your knowledge of partial derivatives and Hessians to find the point (x^*, y^*) that minimizes f. What is the minimum value? Show all work.

10. The water depth in a region where we want to build a port is given by the function

$$k(x, y) = \frac{x - 2y}{1 + 2x^2 + y^2}$$

Identify the general location of any entry points that will cause problems for ships entering into the region.

11. The water depth in a region where we want to build a port is given by the function

$$\begin{aligned} f(x, y) &= 10(3 + x)^3 + 3(6 + x) - 20(2 + y)^2 - \frac{10(2 + y)^5}{11 - 30x - 25x^2} \\ &\quad - 20(3 + y) - 30\cos(10 + (6 + y)^2) + 4\sin(28 + 3x + 6y + xy) \end{aligned}$$

Identify the general location of any entry points that will cause problems for ships entering into the region.

4.3 Unconstrained Optimization: Numerical Methods

We've been investigating analytical techniques to solve unconstrained multivariable optimization problems

$$\max_{\mathbf{x} \in \mathbb{R}^n} f(x_1, x_2, \ldots, x_n). \tag{4.1}$$

In many problems (*most real problems!*), it is quite difficult to find the stationary or critical points, and then use them to determine the nature of the stationary point. In this section, we'll discuss numerical techniques to maximize or minimize a multivariable function.

Gradient Search Methods

We want to solve the *unconstrained nonlinear programming problem* (NLP) given in Equation 4.1 above. Calculus tells us that if a function f is concave, then the optimal solution—if there is one—will occur at a stationary point $\mathbf{x}^*$; i.e., at a point $\mathbf{x}^*$ where

$$\frac{\partial f}{\partial x_1}(\mathbf{x}^*) = \frac{\partial f}{\partial x_2}(\mathbf{x}^*) = \cdots = \frac{\partial f}{\partial x_n}(\mathbf{x}^*) = 0$$

Finding the stationary points in many problems is quite difficult. The methods of *Steepest Ascent* (maximization) and *Steepest Descent* (minimization) offer an alternative by finding an approximation to a stationary point. We'll focus on the gradient method, the Steepest Ascent.

Examine the function shown in Figure 4.9.

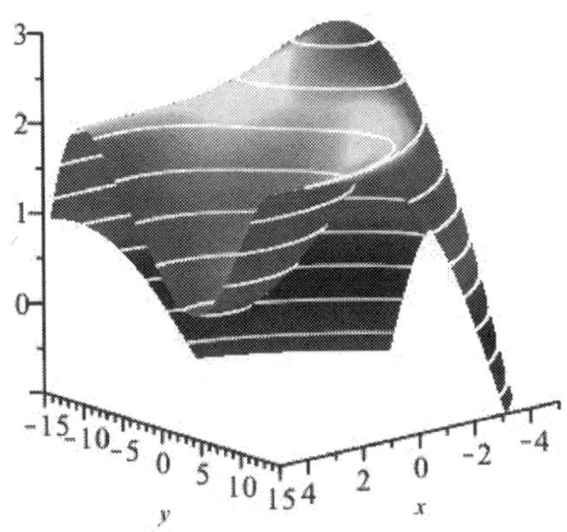

FIGURE 4.9: A Surface Defined by a Function

We want to find the maximum point on the surface. If we started at the bottom of the hill, we might proceed by finding the gradient vector, since the gradient is the vector that points "up the hill" in the direction of maximum increase. The gradient vector is defined as

$$\nabla f(\mathbf{x}) = \left[\frac{\partial f(\mathbf{x})}{\partial x_1}, \frac{\partial f(\mathbf{x})}{\partial x_2}, \ldots, \frac{\partial f(\mathbf{x})}{\partial x_n}\right].$$

(The symbol ∇ is called "del" or "nabla".[4])

If we were lucky, the gradient would point all the way to the maximum of the function, but the contours of functions do not always cooperate—actually, they rarely do. The gradient "points uphill," but for how far? We need to find the distance along the line given by the gradient to travel that maximizes the height of the function in that direction. From that new point, we compute a new gradient vector to find a new direction that "points uphill" in the direction of maximum increase. Continue this method until reaching to the "top of the hill," the maximum of f.

To summarize: from a given starting point, we move in the direction of the gradient as long as f's value continues to increase. At that point, recalculate the gradient and move in the new direction as far as f continues to improve. This process continues until we achieve a maximum value within some specific tolerance (margin of acceptable error). The algorithm for the Method of Steepest Ascent using the gradient is:

Method of Steepest Ascent Algorithm

Find the unconstrained maximum of the function $f : \mathbb{R}^n \to \mathbb{R}$.

INPUTS: Function: f
Starting point: $\mathbf{x}_0$
Tolerance: ε

OUTPUTS: Maximal point $\mathbf{x}^*$ and maximum value $f(\mathbf{x}^*)$

Step 1. Initialize: Set $\mathbf{x} = \mathbf{x}_0$

Step 2. Calculate the gradient $\mathbf{g} = \nabla f(\mathbf{x})$

Step 3. Compute the maximum t^* of the 1-variable function $\phi(t) = f(\mathbf{x} + t \cdot \mathbf{g})$

Step 4. Find the new point $\mathbf{x}_{new} = \mathbf{x} + t^* \cdot \mathbf{g} = \mathbf{x} + t^* \cdot \nabla f(\mathbf{x})$

Step 5. If $\|\mathbf{x} - \mathbf{x}_{new}\| < \varepsilon$ OR $\|\nabla f(\mathbf{x})\| < \varepsilon$, then STOP and return estimates $\mathbf{x}^* = \mathbf{x}_{new}$ and $f(\mathbf{x}_{new})$.
Otherwise, set $\mathbf{x} = \mathbf{x}_{new}$ and return to Step 2.

Remember that $\|\mathbf{y}\| = \|\langle y_1, y_2, \ldots, y_n\rangle\| = \sqrt{y_1^2 + y_2^2 + \cdots + y_n^2}$.
It's time for several examples using the method of steepest ascent.

[4]Wikipedia https://en.wikipedia.org/wiki/Nabla_symbol has a humorous history of ∇.

Example 4.13. Steepest Ascent Example, I.
Maximize $f(\mathbf{x}) = 2x_1x_2 + 2x_2 - x_1^2 - 2x_2^2$ to within $\varepsilon = 0.01$.

We start with $\mathbf{x}_0 = \langle 0, 0\rangle$.
ITERATION 1.
The gradient of $f(x_1, x_2)$, ∇f, is found using the partial derivatives; f's gradient is the vector $\nabla f(x_1, x_2) = \langle 2x_2 - 2x_1, 2x_1 + 2 - 4x_2\rangle$. Then $\nabla f(0, 0) = \langle 0, 2\rangle$. From $\langle 0, 0\rangle$, we move along (up) the x_2-axis in the direction of $\langle 0, 2\rangle$; that is, along the line $L(t) = \mathbf{x}_0 + \nabla f(\mathbf{x}_0) \cdot t = \langle 0, 0\rangle + t\langle 0, 2\rangle = \langle 0, 2t\rangle$. How far do we go?

We need to maximize the function $\phi(t) = f(0, 2t) = 4t - 8t^2$ starting at $t = 0$ to find how far along the line $L(t)$ to step. This function can be maximized by using any of the one-dimensional search techniques for single-variable optimization from Chapter 3 or by simple calculus.

$$\frac{d\phi}{dt} = 4 - 16t = 0 \quad \text{when} \quad t^* = 0.25.$$

Then $L(t^*)$ gives the new point $\mathbf{x}_1 = L(0.25) = \langle 0, 0.5\rangle$.

The magnitude of the difference $(\mathbf{x}_0 - \mathbf{x}_1)$ is 0.5 which is not less than our desired tolerance of 0.01 *(chosen arbitrarily at the beginning of the solution)*. Since we are not optimal, we continue. Repeat the calculations from the new point $\langle 0, 0.5\rangle$.
ITERATION 2.
The gradient vector $\mathbf{g} = \nabla f(\langle 0, 0.5\rangle) = \langle 1, 0\rangle$. Now $L(t) = \langle 0, 0.5\rangle + \langle 1, 0\rangle t$. Again, how far do we travel along L to maximize $\phi(t) = f(\langle t, 0.5\rangle) = -t^2 + t + 0.5$? Simple calculus gives

$$\frac{d\phi}{dt} = -2t + 1 = 0 \quad \text{when} \quad t^* = 0.5.$$

Then, the new $\mathbf{x}$ is $\mathbf{x}_2 = L(0.5) = \langle 0.5, 0.5\rangle$.

The magnitude of the difference $(\mathbf{x}_1 - \mathbf{x}_2)$ is 0.5 which is still not less than our desired tolerance of 0.01. The magnitude of $\nabla f(\mathbf{x}_1) = 1$ which is also not within our tolerance 0.01. We continue to iterate.

Maple will continue the iterations for us using the function *SteepestAscent* which is in the book's *PSMv2* package. Load the package via *with*(*PSMv2*).

Define the function using vectors.

```
> f := x → 2x₁x₂ + 2x₂ − x₁² − 2x₂² :
```

The syntax of our *SteepestAscent* function is

$$SteepestAscent(f, \langle x_0, y_0\rangle, \varepsilon)$$

Adding a third argument tells *SteepestAscent* to produce a graph of the path $\mathbf{x}_0$, $\mathbf{x}_1$, ..., $\mathbf{x}_n$. The output of *SteepestAscent* is a *DataFrame* containing the generated points, gradients, function value, and step-size.

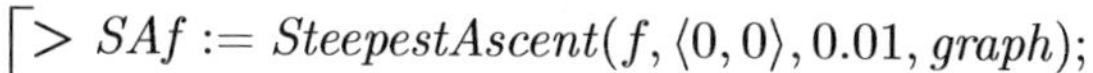

$> SAf := SteepestAscent(f, \langle 0,0 \rangle, 0.01, graph);$

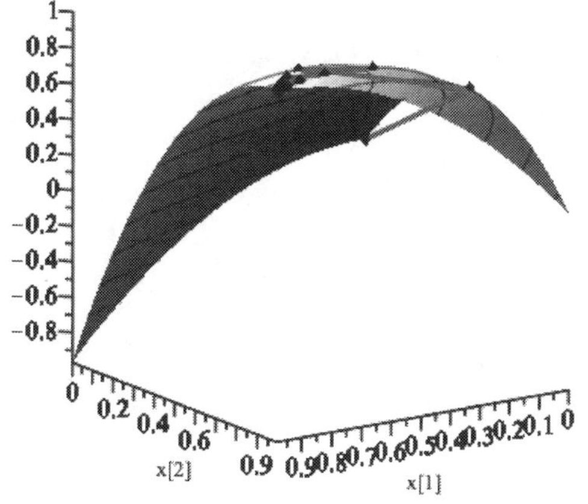

$SAf :=$

	Pts	Gradient	Fcn	Grad_Step
1	$\begin{bmatrix}0.\\0.\end{bmatrix}$	$\begin{bmatrix}0.\\2.\end{bmatrix}$	0.	0.2500
2	$\begin{bmatrix}0.\\0.5000\end{bmatrix}$	$\begin{bmatrix}1.000\\0.\end{bmatrix}$	0.5000	0.5000
3	$\begin{bmatrix}0.5000\\0.5000\end{bmatrix}$	$\begin{bmatrix}0.\\1.000\end{bmatrix}$	0.7500	0.2500
4	$\begin{bmatrix}0.5000\\0.7500\end{bmatrix}$	$\begin{bmatrix}0.500\\0.\end{bmatrix}$	0.8750	0.5000
5	$\begin{bmatrix}0.7500\\0.7500\end{bmatrix}$	$\begin{bmatrix}0.\\0.500\end{bmatrix}$	0.9375	0.2500
6	$\begin{bmatrix}0.7500\\0.8750\end{bmatrix}$	$\begin{bmatrix}0.250\\0.\end{bmatrix}$	0.9699	0.5000
7	$\begin{bmatrix}0.8750\\0.8750\end{bmatrix}$	$\begin{bmatrix}0.\\0.250\end{bmatrix}$	0.9844	0.2500
8	$\begin{bmatrix}0.8750\\0.9375\end{bmatrix}$	$\begin{bmatrix}0.125\\0.\end{bmatrix}$	0.9922	0.5000
9	$\begin{bmatrix}0.9375\\0.9375\end{bmatrix}$	$\begin{bmatrix}0.\\0.125\end{bmatrix}$	0.9961	0.2500
10	$\begin{bmatrix}0.9375\\0.9688\end{bmatrix}$	$\begin{bmatrix}0.063\\0.\end{bmatrix}$	0.9980	0.5000
⋮	⋮	⋮	⋮	⋮

14 × 4 DataFrame

To see the list of points generated, use *SAf['Pts']*; to see just the last point and its function value, use *SAf['Pts'][-1]*, *SAf['Fcn'][-1]*.

```
> SAf['Pts'][−1], SAf['Fcn'][−1];
```

$$\begin{bmatrix} 0.9844 \\ 0.9922 \end{bmatrix}, 0.9999$$

The multivariable calculus solution is straightforward to compute by solving the system $\{\partial f/\partial x_1 = 0, \partial f/\partial x_2 = 0\}$, and checking f's Hessian at the critical point.

$$\left.\begin{aligned} \frac{\partial f}{\partial x_1} &= 2x_2 - 2x_1 = 0 \\ \frac{\partial f}{\partial x_2} &= 2x_1 + 2 - 4x_2 = 0 \end{aligned}\right. \quad \Longrightarrow \mathbf{x}^* = \langle x_1, x_2 \rangle = \langle 1, 1 \rangle$$

The Hessian

$$\begin{bmatrix} -2 & 2 \\ 2 & -4 \end{bmatrix}$$

is negative definite, so the point $\mathbf{x}^* = \langle 1, 1 \rangle$ is a maximum with $f(\mathbf{x}^*) = 1$.

To get a closer approximation with the steepest ascent method, we would make our tolerance smaller. A look at f's contour plot confirms a hill at approximately $(1, 1)$ in Figure 4.10.

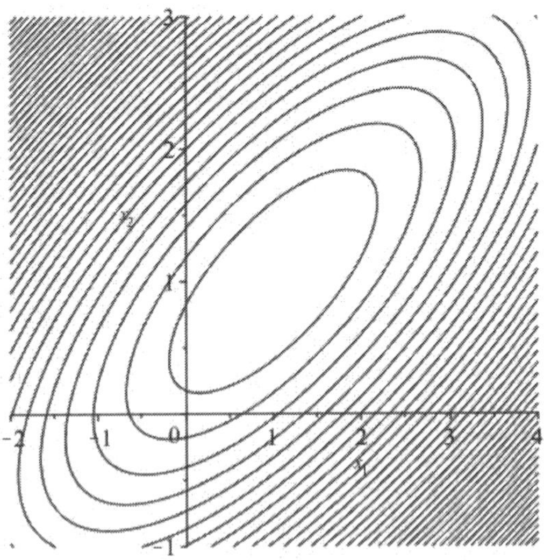

FIGURE 4.10: Contour Plot of f

Example 4.14. Steepest Ascent Example, II.
Maximize $g(\mathbf{x}) = 55x_1 - 4x_1^2 + 135x_2 - 15x_2^2 - 100$ using the Steepest Ascent method starting at the point $\langle 1, 1\rangle$ to within a tolerance of $\varepsilon = 0.01$.

Figure 4.11 provides a visual reference for the maximum.

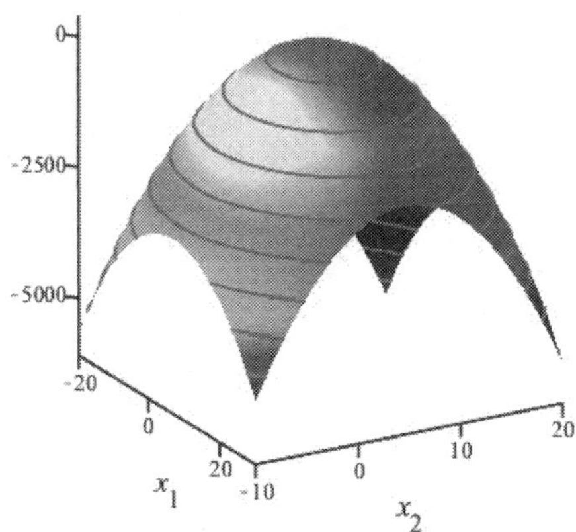

FIGURE 4.11: Surface Plot of g Showing Contours

The gradient is $\nabla g(\mathbf{x}) = \langle 55 - 8x_1, 135 - 30x_2\rangle$.

ITERATION 1.
We begin with $\mathbf{x}_0 = \langle 1, 1\rangle$. Then $\nabla g(\mathbf{x}_0) = \langle 47, 105\rangle$. From $\langle 1, 1\rangle$, we move in the direction of $\langle 47, 105\rangle$. How far do we go along the line $L(t) = \langle 1, 1\rangle + \langle 37, 105\rangle t$? We need to maximize the function

$$\begin{aligned} g(L(t)) &= g(\mathbf{x}_0 + t \cdot \nabla g(\mathbf{x}_0)) \\ &= -4(1 + 47t)^2 - 15(1 + 105t)^2 + 90 + 16760t \\ &= -174211t^2 + 13234t + 71. \end{aligned}$$

This function can also be maximized by simple single-variable calculus:

$$\frac{d}{dt}\, g(L(t)) = 0 \quad \Longrightarrow \quad t^* = 0.03798$$

The new point $\mathbf{x}_1$ is found by evaluating $L(t^*) = \mathbf{x}_0 + t^* \cdot \nabla g(\mathbf{x}_0)$.

$$\mathbf{x}_1 = \langle 1, 1\rangle + t^* \cdot \langle 37, 105\rangle = \langle 2.785, 4.988\rangle$$

Since $\|\mathbf{x}_1 - \mathbf{x}_0\| > 0.01$, iterate again.

ITERATION 2.

Now compute $\nabla g(\mathbf{x}_1) = \langle 32.719, -14.645\rangle$. We move from $\langle 2.785, 4.988\rangle$ in the direction of $\langle 32.719, -14.645\rangle$. How far do we go along the line $L(t) = \langle 2.785, 4.988\rangle + \langle 32.719, -14.645\rangle t$? We need to maximize the function

$$\begin{aligned} g(L(t)) &= g(\mathbf{x}_1 + t \cdot \nabla g(\mathbf{x}_1)) \\ &= 322.33 + 1285.0t - 7499.3t^2. \end{aligned}$$

This function can also be maximized by simple single-variable calculus:

$$\frac{d}{dt}\, g(L(t)) = 0 \qquad \Longrightarrow \qquad t^* = 0.08567$$

The new point $\mathbf{x}_2$ is found by evaluating $L(t^*) = \mathbf{x}_1 + t^* \cdot \nabla g(\mathbf{x}_1)$.

$$\mathbf{x}_2 = \langle 5.588, 3.733\rangle$$

Since $\|\mathbf{x}_2 - \mathbf{x}_1\| > 0.01$, we must iterate again.

Use our *SteepestAscent* function to have Maple finish the iterations. Define g as a function of the vector $\mathbf{x}$.

```
> g := x → −4x_1^2 − 15x_2^2 + 55x_1 + 135x_2 − 100 :
```

> $SAg := SteepestAscent(g, \langle 1, 1\rangle, 0.01, graph);$

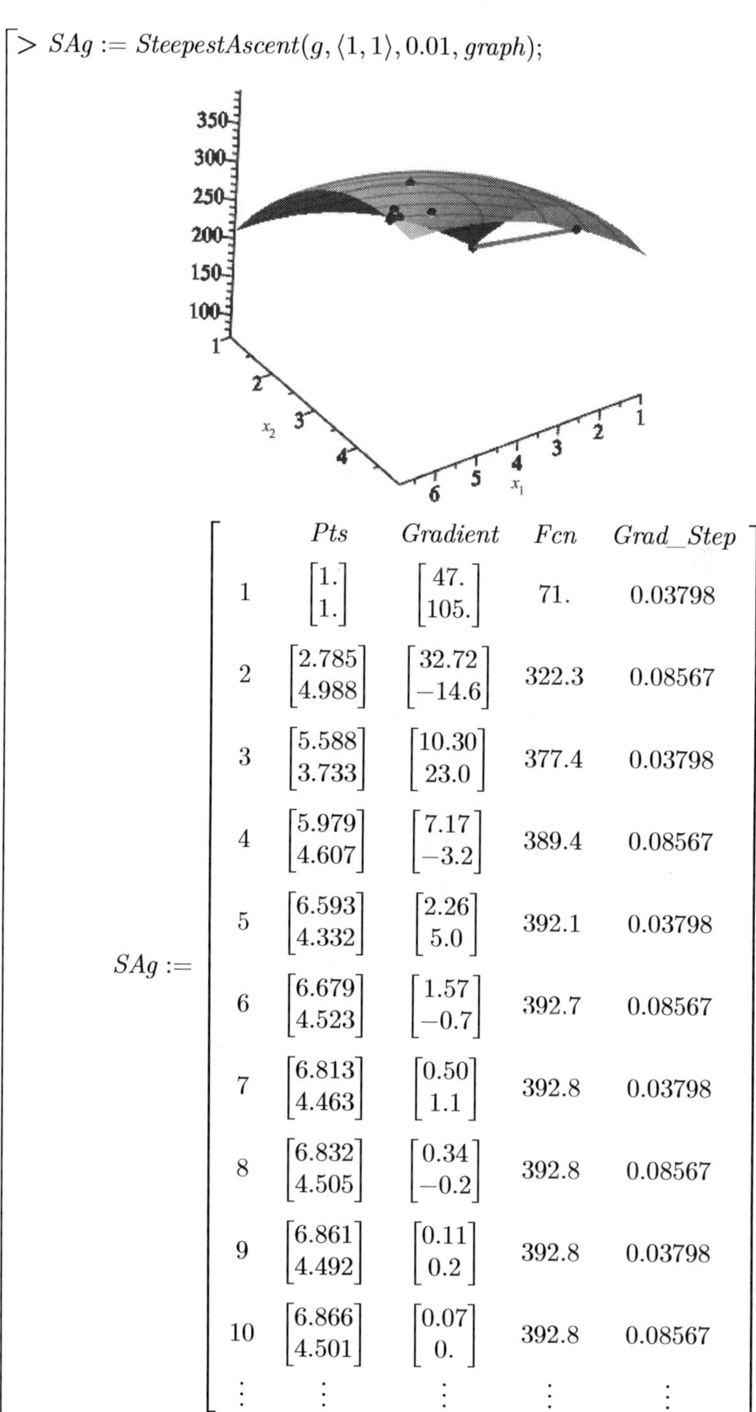

SAg :=

	Pts	Gradient	Fcn	Grad_Step
1	$\begin{bmatrix}1.\\1.\end{bmatrix}$	$\begin{bmatrix}47.\\105.\end{bmatrix}$	71.	0.03798
2	$\begin{bmatrix}2.785\\4.988\end{bmatrix}$	$\begin{bmatrix}32.72\\-14.6\end{bmatrix}$	322.3	0.08567
3	$\begin{bmatrix}5.588\\3.733\end{bmatrix}$	$\begin{bmatrix}10.30\\23.0\end{bmatrix}$	377.4	0.03798
4	$\begin{bmatrix}5.979\\4.607\end{bmatrix}$	$\begin{bmatrix}7.17\\-3.2\end{bmatrix}$	389.4	0.08567
5	$\begin{bmatrix}6.593\\4.332\end{bmatrix}$	$\begin{bmatrix}2.26\\5.0\end{bmatrix}$	392.1	0.03798
6	$\begin{bmatrix}6.679\\4.523\end{bmatrix}$	$\begin{bmatrix}1.57\\-0.7\end{bmatrix}$	392.7	0.08567
7	$\begin{bmatrix}6.813\\4.463\end{bmatrix}$	$\begin{bmatrix}0.50\\1.1\end{bmatrix}$	392.8	0.03798
8	$\begin{bmatrix}6.832\\4.505\end{bmatrix}$	$\begin{bmatrix}0.34\\-0.2\end{bmatrix}$	392.8	0.08567
9	$\begin{bmatrix}6.861\\4.492\end{bmatrix}$	$\begin{bmatrix}0.11\\0.2\end{bmatrix}$	392.8	0.03798
10	$\begin{bmatrix}6.866\\4.501\end{bmatrix}$	$\begin{bmatrix}0.07\\0.\end{bmatrix}$	392.8	0.08567
⋮	⋮	⋮	⋮	⋮

11 × 4 DataFrame

The final point $\mathbf{x}_{11}$ and the g's maximum value are:

> $SAg['Pts'][-1], SAg['Fcn'][-1];$

$$\begin{bmatrix} 6.872 \\ 4.498 \end{bmatrix}, 392.8$$

This time, the solution process zig-zagged and converged more slowly to an optimal solution as illustrated in Figure 4.12.

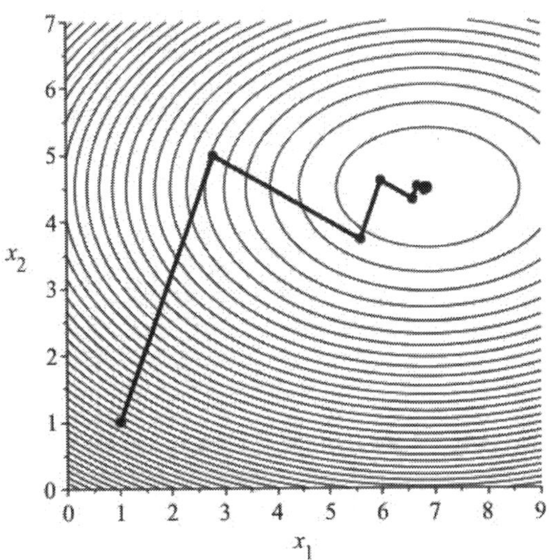

FIGURE 4.12: Zig-Zagging to a Solution

We'll now look at maximizing a transcendental multivariable function where the critical points cannot be found analytically in a closed form, but must be approximated numerically.

Example 4.15. Steepest Ascent Example, III.
Maximize $h(\mathbf{x}) = 2x_1x_2 + 2x_2 - e^{x_1} - e^{x_2} + 10$ using the Steepest Ascent method starting at the point $\langle 0.5, 0.5\rangle$ to within a tolerance of $\varepsilon = 0.01$.

Verify that h has a local maximum near $\langle 1, 1.25\rangle$ by looking at a graph. The gradient of h is

$$\nabla h(\mathbf{x}) = \langle 2x_2 - e^{x_1}, 2x_1 + 2 - e^{x_2}\rangle$$

ITERATION 1.
We begin with $\mathbf{x}_0 = \langle 0.5, 0.5\rangle$. Then $\nabla h(\mathbf{x}_0) = \langle -0.649, 1.351\rangle$. From $\langle 0.5, 0.5\rangle$, we move in the direction of $\langle -0.6487, 1.3513\rangle$. How far do we go along the line $L(t) = \langle 0.5, 0.5\rangle + \langle -0.6487, 1.3513\rangle t$? We need to maximize the

function

$$\begin{aligned} h(L(t)) &= g(\mathbf{x}_0 + t \cdot \nabla g(\mathbf{x}_0)) \\ &= 11.5 + 3.405t - 1.753t^2 - e^{(0.5-0.6487t)} - e^{(0.5+1.351t)}. \end{aligned}$$

Maple's *fsolve* gives $t^* = 0.2875$. Therefore

$$\mathbf{x}_1 = \mathbf{x}_0 + t^* \cdot \nabla h(\mathbf{x}_0) = \langle 0.3136, 0.8883 \rangle.$$

Since $\|\mathbf{x}_1 - \mathbf{x}_0\| > 0.01$, we continue. But let Maple do the work.

$> h := x \to 2x_1 x_2 + 2x_2 - exp(x_1) - exp(x_2) + 10 :$

$> SAh := SteepestAscent(h, \langle 0.5, 0.5 \rangle, 0.01);$

$$SAh := \begin{bmatrix} & Pts & Gradient & Fcn & Grad\_Step \\ 1 & \begin{bmatrix} 0.5 \\ 0.5 \end{bmatrix} & \begin{bmatrix} -0.649 \\ 1.351 \end{bmatrix} & 8.203 & 0.2874 \\ 2 & \begin{bmatrix} 0.3136 \\ 0.8883 \end{bmatrix} & \begin{bmatrix} 0.409 \\ 0.196 \end{bmatrix} & 8.534 & 1.704 \\ 3 & \begin{bmatrix} 1.009 \\ 1.222 \end{bmatrix} & \begin{bmatrix} -0.299 \\ 0.624 \end{bmatrix} & 8.773 & 0.2001 \\ 4 & \begin{bmatrix} 0.9496 \\ 1.347 \end{bmatrix} & \begin{bmatrix} 0.109 \\ 0.053 \end{bmatrix} & 8.822 & 0.7345 \\ 5 & \begin{bmatrix} 1.030 \\ 1.386 \end{bmatrix} & \begin{bmatrix} -0.029 \\ 0.061 \end{bmatrix} & 8.827 & 0.1868 \\ 6 & \begin{bmatrix} 1.025 \\ 1.397 \end{bmatrix} & \begin{bmatrix} 0.007 \\ 0.007 \end{bmatrix} & 8.828 & 0. \end{bmatrix}$$

The final point $\mathbf{x}_6$ and the h's local maximum value are:

$> SAh['Pts'][-1], SAh['Fcn'][-1];$

$$\begin{bmatrix} 1.025 \\ 1.397 \end{bmatrix}, 8.828$$

Modified Newton's Method

The zig-zag pattern we saw in Figure 4.12 shows that steepest ascent doesn't always go directly to an optimum value. The Newton-Raphson[5] iterative root finding technique using the partial derivatives of the function provides an alternative numerical search method for an optimum value when modified

[5]Newton invented the method in 1671, but it wasn't published until 1736; Raphson independently discovered the method and published it in 1690.

appropriately. Given the right conditions, this numerical method is more efficient and converges faster to an approximate optimal solution: *quadratic convergence* versus the *linear convergence* of steepest ascent.

Newton's Method for multivariable optimization searches is based on the single-variable root-finding algorithm. Modify the procedure to look for roots of the first derivative rather than roots of the original function:

$$x_{n+1} = x_n + \frac{f(x_n)}{f'(x_n)} \implies x_{n+1} = x_n + \frac{f'(x_n)}{f''(x_n)},$$

and iterate until $|x_{n+1} - x_n| < \varepsilon$ for our chosen tolerance $\varepsilon > 0$.

Extend the Modified Newton's Method to several variables by using gradients and the Hessian, the matrix of second partial derivatives.

$$x_{n+1} = x_n + \big(f''(x_n)\big)^{-1} \cdot f'(x_n) \implies \mathbf{x}_{n+1} = \mathbf{x}_n + \big(H(\mathbf{x}_n)\big)^{-1} \cdot \nabla f(\mathbf{x}_n)$$

now iterating until $\|\mathbf{x}_{n+1} - \mathbf{x}_n\| < \varepsilon$ for our chosen tolerance $\varepsilon > 0$. Applying a little bit of linear algebra gives us the method as an algorithm.

Modified Newton's Method Algorithm

Find the unconstrained maximum of the function $f : \mathbb{R}^n \to \mathbb{R}$.

INPUTS: Function: f
Starting point: $\mathbf{x}_0$
Tolerance and Maximum Iterations: ε, N

OUTPUTS: Maximal point $\mathbf{x}^*$ and maximum value $f(\mathbf{x}^*)$

Step 1. Initialize: Set $\mathbf{x} = \mathbf{x}_0$

Step 2. Calculate the gradient $\mathbf{g} = \nabla f(\mathbf{x})$ and $H = \text{Hessian}(\mathbf{x})$

Step 3. Compute $d = |H|$ and create new matrices

$$H_1 = \text{substitute } \mathbf{g} \text{ for column 1 of } H$$
$$H_2 = \text{substitute } \mathbf{g} \text{ for column 2 of } H$$

Step 4. Compute: $\Delta x_1 = |H_1|/d$ and $\Delta x_2 = |H_2|/d$.

Step 5. Find the new point $\mathbf{x}_{new} = \langle x_1 + \Delta x_1, x_2 + \Delta x_2\rangle$

Step 6. If $\|\langle \Delta x_1, \Delta x_2\rangle\| < \varepsilon$, then STOP
and return estimates $\mathbf{x}^* = \mathbf{x}_{new}$ and $f(\mathbf{x}_{new})$.
Otherwise, set $\mathbf{x} = \mathbf{x}_{new}$ and return to Step 2.

Again, remember that $\|\mathbf{y}\| = \|\langle y_1, y_2\rangle\| = \sqrt{y_1^2 + y_2^2}$.

Let's repeat our examples for the steepest ascent method using Newton's method. We'll use the function *ModifiedNewtonMethod* which is in the book's

PSMv2 package. Load the package via *with*(*PSMv2*). The syntax of the *ModifiedNewtonMethod* function is

$$ModifiedNewtonMethod(f, \langle x_0, y_0\rangle, \varepsilon, N)$$

where f is the function, $\langle x_0, y_0\rangle$ is the starting vector, ε is the tolerance, and N is the maximum number of iterations allowed. The output of *ModifiedNewtonMethod* is a *DataFrame* containing the generated points, function values, Hessians, and the definiteness of the Hessian.

Example 4.16. Modified Newton's Method Example, I.
Maximize $f(\mathbf{x}) = 2x_1x_2 + 2x_2 - x_1^2 - 2x_2^2$ to within $\varepsilon = 0.01$ starting at $\mathbf{x}_0 = \langle 0, 0\rangle$.

Limit the number of iterations to $N = 20$.

> $f := x \to 2x_1x_2 + 2x_2 - x_1^2 - 2x_2^2$:

> $MNf := ModifiedNewtonMethod(f, \langle 0, 0\rangle, 0.01, 20)$;

$$SAf := \begin{bmatrix} & x_k & F(x_k) & Hessian & definiteness \\ 1 & \begin{bmatrix} 0. \\ 0. \end{bmatrix} & 0. & \begin{bmatrix} -2. & 2. \\ 2. & 4 \end{bmatrix} & negative\_definite \\ 2 & \begin{bmatrix} 1. \\ 1. \end{bmatrix} & 0. & \begin{bmatrix} -2. & 2. \\ 2. & 4 \end{bmatrix} & negative\_definite \\ 3 & \begin{bmatrix} 1. \\ 1. \end{bmatrix} & 0. & \begin{bmatrix} -2. & 2. \\ 2. & 4 \end{bmatrix} & negative\_definite \end{bmatrix}$$

We have a maximum of f of 1 at $\langle 1, 1\rangle$ since the Hessian is negative definite there.

On to the second example.

Example 4.17. Modified Newton's Method Example, II.
Maximize $g(\mathbf{x}) = 55x_1 - 4x_1^2 + 135x_2 - 15x_2^2 - 100$ starting at the point $\langle 1, 1\rangle$ to within a tolerance of $\varepsilon = 0.01$ limiting iterations to $N = 20$.

> $g := x \to 55x_1 - 4x_1^2 + 135x_2 - 15x_2^2 - 100$:

> $MNg := ModifiedNewtonMethod(g, \langle 1, 1\rangle, 0.01, 20)$;

$$SAg := \begin{bmatrix} & x_k & F(x_k) & Hessian & definiteness \\ 1 & \begin{bmatrix} 1. \\ 1. \end{bmatrix} & 71. & \begin{bmatrix} -8. & 0. \\ 0. & -30 \end{bmatrix} & negative\_definite \\ 2 & \begin{bmatrix} 6.875 \\ 4.500 \end{bmatrix} & 392.9 & \begin{bmatrix} -8. & 0. \\ 0. & -30 \end{bmatrix} & negative\_definite \\ 3 & \begin{bmatrix} 6.875 \\ 4.500 \end{bmatrix} & 392.9 & \begin{bmatrix} -8. & 0. \\ 0. & -30 \end{bmatrix} & negative\_definite \end{bmatrix}$$

We have a maximum of g of 392.9 at $\langle 6.875, 4.500\rangle$ since the Hessian is negative definite there.

Now, for the third example.

Example 4.18. Modified Newton's Method Example, III.
Maximize $h(\mathbf{x}) = 2x_1x_2 + 2x_2 - e^{x_1} - e^{x_2} + 10$ starting at the point $\langle 0.8, 0.8\rangle$ to within a tolerance of $\varepsilon = 0.01$ with no more than $N = 20$ iterations.

$$> h := x \to 2x_1x_2 + 2x_2 - e^{x_1} - e^{x_2} + 10 :$$

$$> \mathit{MNg} := \mathit{ModifiedNewtonMethod}(h, \langle 0.8, 0.8\rangle, 0.01, 20);$$

$$\mathit{SAg} := \begin{bmatrix} & x_k & F(x_k) & \mathit{Hessian} & \mathit{definiteness} \\ 1 & \begin{bmatrix} 0.8 \\ 0.8 \end{bmatrix} & 8.428 & \begin{bmatrix} -2.226 & 2. \\ 2. & -2.226 \end{bmatrix} & \mathit{negative\_definite} \\ 2 & \begin{bmatrix} 2.224 \\ 2.697 \end{bmatrix} & 3.306 & \begin{bmatrix} -9.241 & 2. \\ 2. & -14.83 \end{bmatrix} & \mathit{negative\_definite} \\ 3 & \begin{bmatrix} 1.669 \\ 2.057 \end{bmatrix} & 7.851 & \begin{bmatrix} -5.306 & 2. \\ 2. & -7.820 \end{bmatrix} & \mathit{negative\_definite} \\ 4 & \begin{bmatrix} 1.288 \\ 1.642 \end{bmatrix} & 8.723 & \begin{bmatrix} -3.264 & 2. \\ 2. & -5.164 \end{bmatrix} & \mathit{negative\_definite} \\ 5 & \begin{bmatrix} 1.088 \\ 1.450 \end{bmatrix} & 8.825 & \begin{bmatrix} -2.968 & 2. \\ 2. & -4.265 \end{bmatrix} & \mathit{negative\_definite} \\ 6 & \begin{bmatrix} 1.034 \\ 1.404 \end{bmatrix} & 8.829 & \begin{bmatrix} -2.813 & 2. \\ 2. & -4.073 \end{bmatrix} & \mathit{negative\_definite} \\ 7 & \begin{bmatrix} 1.031 \\ 1.402 \end{bmatrix} & 8.827 & \begin{bmatrix} -2.803 & 2. \\ 2. & -4.061 \end{bmatrix} & \mathit{negative\_definite} \end{bmatrix}$$

We have a maximum of h of 8.827 at $\langle 1.031, 1.402\rangle$ since the Hessian is negative definite there.

Even though Newton's method is faster and more direct, we must be cautious. The method requires a relatively good starting point. Try searching for a maximum for h starting at $\langle 0.5, 0.5\rangle$ and looking at just the last entry in the *DataFrame* report.

$$> \mathit{MNh2} := \mathit{ModifiedNewtonMethod}(h, \langle 0.5, 0.5\rangle, 0.01, 20) :$$
$$\mathit{MNh2}[-1, ..];$$

$$\begin{bmatrix} x_k & \begin{bmatrix} -0.2667 \\ 0.3829 \end{bmatrix} \\ F(x_k) & 8.329 \\ \mathit{Hessian} & \begin{bmatrix} -0.7659 & 2. \\ 2. & -1.467 \end{bmatrix} \\ \mathit{definite\_ness} & \mathit{indefinite} \end{bmatrix}$$

We started at $\langle 0.5, 0.5\rangle$, not far from our original $\langle 0.8, 0.8\rangle$. But the Hessian being indefinite tells us that we've found an approximate saddle point, not a maximum. Modified Newton's method is very sensitive to the initial value chosen.

Comparisons of Methods

We compared these two routines finding that Modified Newton's Method converges faster than the gradient method. Table 4.4 shows the comparison.

TABLE 4.4: Comparison of the Steepest Ascent and Modified Newton's Methods

f	Initial Value	Iterations	Solution	$\max f(\mathbf{x})$
Steepest Ascent	$\langle 0, 0\rangle$	16	$\langle 0.9922, 0.9961\rangle$	1.0
Modified Newton	$\langle 0, 0\rangle$	2	$\langle 1.0, 1.0\rangle$	1.0

g	Initial Value	Iterations	Solution	$\max g(\mathbf{x})$
Steepest Ascent	$\langle 0, 0\rangle$	4	$\langle -0.266, 0.385\rangle$	8.3291
Modified Newton	$\langle 0, 0\rangle$	2	$\langle -0.269, 0.381\rangle$	8.3291

h	Initial Value	Iterations	Solution	$\max h(\mathbf{x})$
Steepest Ascent	$\langle 0.5, 0.5\rangle$	6	$\langle 1.025, 1.397\rangle$	8.828
Modified Newton	$\langle 0.8, 0.8\rangle$	7	$\langle 1.031, 1.402\rangle$	8.827

As a final comparison, add the modified Newton's method points to the contour map showing the steepest ascent points for g of Figure 4.12 to obtain Figure 4.13 below.

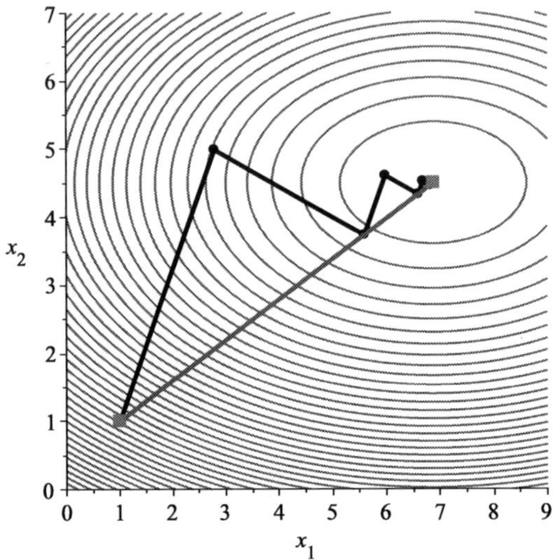

FIGURE 4.13: Steepest Ascent and Modified Newton's Method Solutions

When given a good starting value, a Modified Newton's method search is much faster and more direct than a steepest ascent gradient method.

Exercises

1. Maximize $f(\mathbf{x}) = 2x_1x_2 - 2x_1^2 - 2x_2^2$ to within a tolerance of 0.1.

   a. Start at the point $\mathbf{x} = \langle 1, 1 \rangle$. Perform 2 complete iterations of the steepest ascent gradient search. For each iteration clearly show $\mathbf{x}_n$, $\mathbf{x}_{n+1}$, $\nabla f(\mathbf{x}_n)$, and t^*. Justify that the process will eventually find the approximate maximum.

   b. Use Newton's method to find the maximum. Starting at $\mathbf{x} = \langle 1, 1 \rangle$. Clearly show $\mathbf{x}_n$, $\mathbf{x}_{n+1}$, $\nabla f(\mathbf{x}_n)$, and the Hessian for each iteration. Indicate when the stopping criterion is achieved.

2. Maximize $f(\mathbf{x}) = 3x_1x_2 - 4x_1^2 - 4x_2^2$ to within a tolerance of 0.1.

   a. Start at the point $\mathbf{x} = \langle 1, 1 \rangle$. Perform 2 complete iterations of the steepest ascent gradient search. For each iteration clearly show $\mathbf{x}_n$, $\mathbf{x}_{n+1}$, $\nabla f(\mathbf{x}_n)$, and t^*. Justify that the process will eventually find the approximate maximum.

   b. Use Newton's method to find the maximum. Starting at $\mathbf{x} = \langle 1, 1 \rangle$. Clearly show $\mathbf{x}_n$, $\mathbf{x}_{n+1}$, $\nabla f(\mathbf{x}_n)$, and the Hessian for each iteration. Indicate when the stopping criterion is achieved.

3. Apply the modified Newton's method to find the following:

   a. Maximize $f(x, y) = -x^3 + 3x + 84y - 6y^2$ starting with the initial value $\langle 1, 1\rangle$. Why can't we start at $\langle 0, 0\rangle$?

   b. Minimize $f(x, y) = -4x + 4x^2 - 3y + y^2$ starting at $\langle 0, 0\rangle$.

   c. Perform 3 modified Newton's method iterations to minimize $f(x, y) = (x - 2)^4 + (x - 2y)^2$ starting at $\langle 0, 0\rangle$. Why is this problem not converging as quickly as b?

4. Use a gradient search to approximate the minimum to $f(\mathbf{x}) = (x_1 - 2)^2 + x_1 + x_2^2$. Start at $\langle 2.5, 1.5\rangle$.

Projects

Project 4.1. Modify the Steepest Ascent Algorithm (pg 147) to approximate the minimum of a function. This technique is called the *Steepest Descent Algorithm.*

Project 4.2. Write a program in Maple that uses the one-dimensional Golden Section search algorithm instead of calculus to perform iterations of a gradient search. Use your code to find the maximum of

$$f(x, y) = xy - x^2 - y^2 - 2x - 2y + 4$$

Project 4.3. Write a program in Maple that uses the one-dimensional Fibonacci search algorithm instead of calculus to perform iterations of a gradient search. Use your code to find the maximum of

$$f(x, y) = xy - x^2 - y^2 - 2x - 2y + 4$$

References and Further Reading

[B2016] William C. Bauldry, *Introduction to Computation*, 2016.

[BSS2013] Mokhtar S. Bazaraa, Hanif D. Sherali, and Chitharanjan M. Shetty, *Nonlinear Programming: Theory and Algorithms*, Wiley, 2013.

[F1992] William P. Fox, "Teaching nonlinear programming with Minitab," COED Journal, Vol. II (1992), no. 1, 80–84.

[F1993] William P. Fox, "Using microcomputers in undergraduate nonlinear optimization," Collegiate Microcomputer, Vol. XI (1993), no. 3, 214–218.

[FGMW1987] William P. Fox, Frank Giordano, S. Maddox, and Maurice D. Weir, *Mathematical Modeling with Minitab*, Brooks/Cole, 1987.

[FR2000] William P. Fox and William Richardson, "Mathematical modeling with least squares using Maple," Maple Application Center, Nonlinear Mathematics, 2000.

[FGW1997] William P. Fox, Frank Giordano, and Maurice Weir, *A First Course in Mathematical Modeling*, 2nd Ed., Brooks/Cole, 1997.

[FW2001] William P. Fox and Margie Witherspoon, "Single variable optimization when calculus fails: Golden section search methods in nonlinear optimization using Maple," COED Journal, Vol. XI (2001), no. 2, 50–56.

[M2013] Mark M. Meerschaert, *Mathematical Modeling*, Academic Press, 2013.

[PRS1976] D.T. Phillips, A. Ravindran, and J. Solberg, *Operations Research*, Wiley, 1976.

[PFTV2007] William H. Press, Brian P. Flannery, Saul A. Teukolsky, William T. Vetterling, et al., *Numerical Recipes*, Cambridge U. Press, 2007.

[R1979] S.S. Rao, *Optimization: Theory and Applications*, Wiley Eastern Ltd., 1979.

4.4 Constrained Optimization: The Method of Lagrange Multipliers

Equality Constraints: Method of Lagrange Multipliers

A company manufactures new phones that are projected to take the market by storm. The two main input components of the new phone are the circuit board and the LCD Touchscreen that make the phone faster, smarter, and easier to use. The number of phones to be produced E is estimated to equal $E = 250\,a^{1/4}b^{1/2}$ where a and b are the number of circuit board production hours and the number of LCD Touchscreen production hours available, respectively. Such a formula is known to economists as a *Cobb-Douglas function.* Our laborers are paid by the type of work they do: the circuit board labor cost is \$5 an hour and the LCD Touchscreen labor cost is \$8 an hour. What is the maximum number of phones that can be made if the company has \$175,000 allocated for these components in the short run?

Problems such as this can be modeled using constrained optimization. We begin our discussion with optimization with equality constraints, then we move to optimization with inequality constraints in the next section.

Lagrange multipliers[6] can be used to solve nonlinear optimization problems (NLPs) in which all the constraints are equations. Consider the NLP given by

$$\begin{aligned}
&\text{Maximize (Minimize) } z = f(\mathbf{x}) \\
&\text{subject to} \\
&\qquad g_1(\mathbf{x}) = b_1 \\
&\qquad g_2(\mathbf{x}) = b_2 \\
&\qquad \vdots \\
&\qquad g_m(\mathbf{x}) = b_m
\end{aligned} \tag{4.2}$$

where $m \leq n$.

We can build an equality constrained model for our phone problem: maximize the number of phones made using all available production hours.

$$\begin{aligned}
&\text{Maximize } E = 250\,a^{1/4}b^{1/2} \\
&\text{subject to} \\
&\qquad 5a + 8b = 175000
\end{aligned}$$

Lagrange Multipliers: Introduction and Basic Theory

In order to solve NLPs in the form of (4.2), we associate a Lagrange multiplier λ_i with the ith constraint, and form the Lagrangian function

$$L(\mathbf{x}, \boldsymbol{\lambda}) = f(\mathbf{x}) + \sum_{i=1}^{m} \lambda_i \big(g_i(\mathbf{x}) - b_i\big) \tag{4.3}$$

The computational procedure for Lagrange multipliers requires that all the partials of the Lagrange function (4.3) must equal zero. The partials all equaling zero at $\mathbf{x}^*$ form the *necessary conditions* that $\mathbf{x}^*$ is a solution to the NLP problem; i.e., these conditions are required for $\mathbf{x} = \langle x_1, x_2, \ldots, x_n \rangle$ to be a solution to (4.2). We have

Proposition. Lagrange Multiplier Necessary Conditions.
For $\mathbf{x}^*$ to be a solution of the Lagrange function, all partials must satisfy

$$\frac{\partial L}{\partial x_i} = 0 \quad \text{for } i = 1, 2, \ldots, n \ (\textit{variables})$$

$$\frac{\partial L}{\partial \lambda_j} = 0 \quad \text{for } j = 1, 2, \ldots, m \ (\textit{constraints})$$

at $\mathbf{x}^*$.

[6]For a nice history of Lagrange's technique, see Bussotti, "On the Genesis of the Lagrange Multipliers," J Optimization Theory & Applications, Vol 117, No 3, pp 453-459, June 2003.

The points $\mathbf{x}$ we will consider as candidates for optimal are

Definition. Regular Point.
$\mathbf{x}$ is a *regular point* iff the set $\{\nabla g_i(\mathbf{x}) : i = 1..m\}$ is linearly independent.

The main theorem for Lagrange's technique is

Theorem. Lagrange Multipliers.
Let $\mathbf{x}$ be a point satisfying the *Lagrange multiplier necessary conditions.* Then

a. If f is concave and each g_i is linear, then $\mathbf{x}$ is an optimal solution of the maximization problem (4.2).

b. If f is convex and each g_i is linear, then $\mathbf{x}$ is an optimal solution of the minimization problem (4.2).

Previously, we used the Hessian matrix to determine if a function was convex, concave, or neither. Also note that the theorem limits constraints to linear functions.

What if the constraints are nonlinear? We can use the *bordered Hessian* in the sufficient conditions. For a bivariate Lagrangean function with one constraint

$$L(\langle x_1, x_2\rangle, \lambda) = f(\langle x_1, x_2\rangle) + \lambda\big(g(\langle x_1, x_2\rangle) - b\big),$$

define the bordered Hessian matrix as

$$BdH = \begin{bmatrix} 0 & g_1 & g_2 \\ g_1 & f_{11} - \lambda g_{11} & f_{12} - \lambda g_{12} \\ g_2 & f_{21} - \lambda g_{21} & f_{22} - \lambda g_{22} \end{bmatrix}$$

The determinant of this bordered Hessian matrix is

$$|BdH| = g_1 g_2 \cdot (f_{21} - \lambda g_{21}) + g_2 g_1 \cdot (f_{12} - \lambda g_{12}) - g_2^2 \cdot (f_{11} - \lambda g_{11}) - g_1^2 \cdot (f_{22} - \lambda g_{22})$$

The necessary condition for a *maximum* in the bivariate case with one constraint is that the determinant of its bordered Hessian is positive when evaluated at the critical point.

The necessary condition for a *minimum* in the bivariate case with one constraint is the determinant of its bordered Hessian is negative when evaluated at the critical point.

If $\mathbf{x}$ is a regular point and $g_i(\mathbf{x}) = 0$ (all constraints are satisfied at $\mathbf{x}$), then

$$M_{\mathbf{x}} = \{\mathbf{y} \mid \nabla g_i(\mathbf{x}) \bullet \mathbf{y} = 0\}$$

defines a plane tangent to the feasible region at $\mathbf{x}$. (The '$\bullet$' is the usual dot product.)

Lemma.
If $\mathbf{x}$ is regular, $g_i(\mathbf{x}) = 0$, and $\nabla g_i(\mathbf{x}) \bullet y = 0$, then $\nabla f(\mathbf{x}) \bullet y = 0$.

Note that the Lagrange Multiplier conditions are exactly the same for a minimization problem as for a maximization problem. This is the reason that these conditions alone are not sufficient conditions. Thus, a given solution can either be a maximum or a minimum. In order to determine whether the point found is a maximum, minimum, or saddle point we will use the Hessian.

The value of the Lagrange multiplier itself has an important modeling interpretation. The multiplier λ is the "shadow price" for scarce resources. Thus, λ_i is the shadow price of the ith constraint. If the right-hand side of the constraint is increased by a small amount, say Δ, then the optimal solution will change by $\lambda_i \cdot \Delta$. We will illustrate shadow prices both graphically and computationally.

Graphical Interpretation of Lagrange Multipliers

We can best understand the method of Lagrange multipliers by studying its geometric interpretation. This geometric interpretation involves the gradients of both the function and the constraints. Initially, consider only one constraint, $g(\mathbf{x}) = b$, then the Lagrangian equation simplifies to

$$\nabla f = \lambda \nabla g.$$

The solution is the point $\mathbf{x}$ where the gradient vector $\nabla g(\mathbf{x})$ is perpendicular to contours on the surface. The gradient vector ∇f always points in the direction f increases fastest. At either a maximum or a minimum, this direction must be perpendicular to contour lines on f's surface S. Thus, since both ∇f and ∇g point along the same perpendicular line, then $\nabla f = \lambda \nabla g$. Further, g's curve must be tangent to f's contours at optimal points. See Figure 4.14.

The geometrical arguments are similar for the case of multiple constraints.

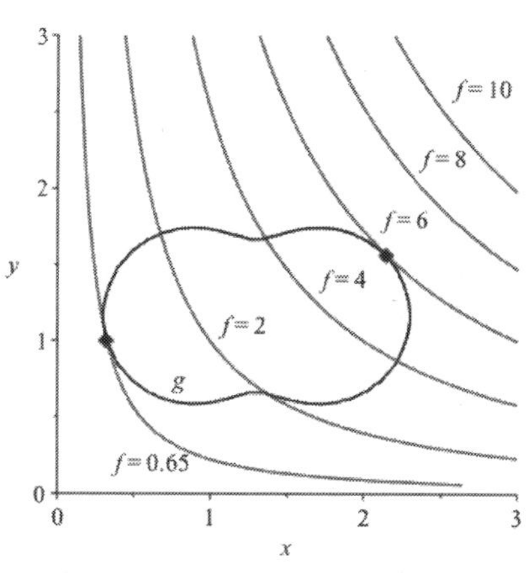

(a) Contours and Constraint

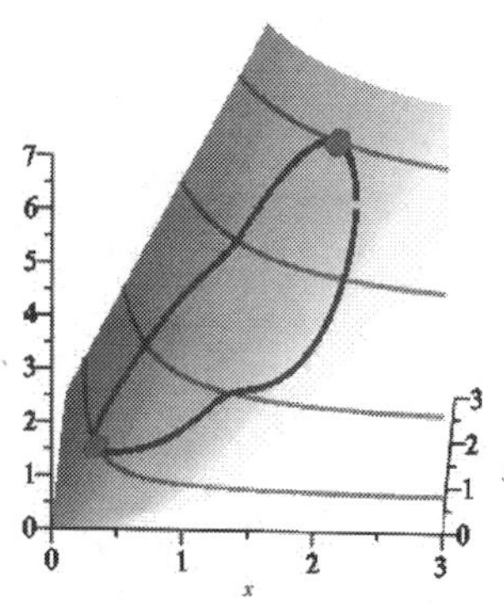

(b) Surface and Constraint

FIGURE 4.14: Lagrange Multipliers Geometrically: One Equality Constraint

Let's preview a graphical solution to an example.

$$\text{Maximize } z = -2x^2 - 2y^2 + xy + 8x + 3y$$
$$\text{subject to}$$
$$3x + y = 6$$

Generate a contour plot of $z = f(\mathbf{x})$ with Maple, and overlay the single constraint onto the contour plot. See Figure 4.15. What information can we obtain from this graphical representation? First, we note that the unconstrained optimum does not lie on the constraint. We estimate the unconstrained maximum as $(x^*, y^*) = (2.3, 1.3)$. The optimal constrained solution lies at the point where the constraint is tangent to a contour of $z = f(\mathbf{x})$. This point is approximately $(1.8, 1.0)$ on the graph. We see clearly that the constraint does not pass through the unconstrained maximum, and thus, it can be modified/adjusted (if feasible) until the line passes through the unconstrained solution. At that point, we would no longer add (or subtract) any more constrained resources (see Figure 4.16). Valuable insights about the problem come from plotting the information, when possible.

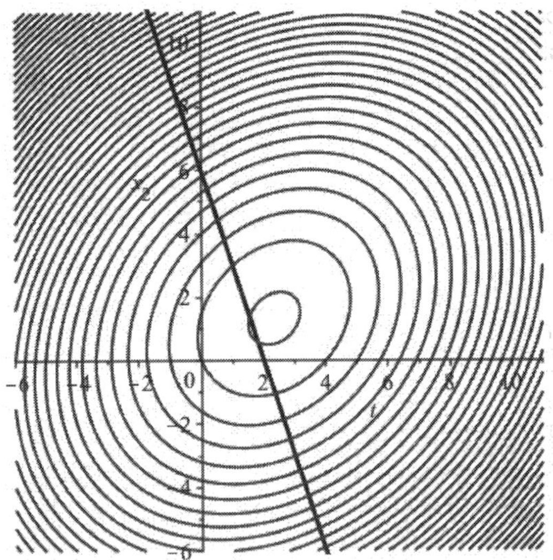

FIGURE 4.15: Contour Plot of f with Linear Constraint g

Interpreting the value of λ as a "shadow price" leads us to consider what happens when the amount of a resource governed by a constraint is changed. Figure 4.16 shows this concept graphically.

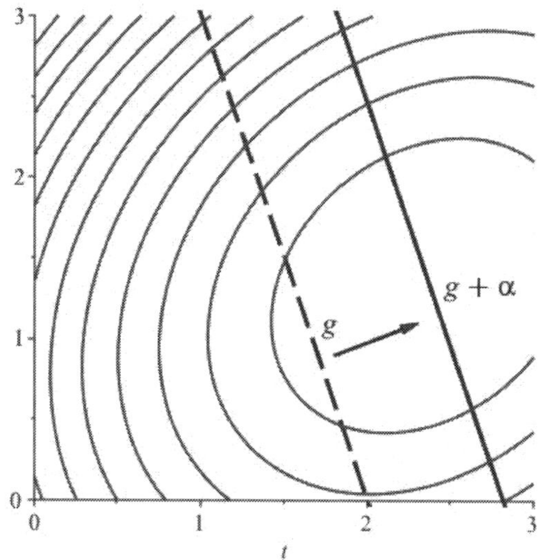

FIGURE 4.16: Resource Constraint g is Increased by Constant α

Let's now turn to the calculations.

Lagrange Multipliers Computations with Maple

The set of equations in (4.2), pg. 163, gives a system of $n+m$ equations in the $n+m$ unknowns $\{x_i, \lambda_j\}$. Generally speaking, this system presents a very difficult problem to solve without a computer except for simple problems. Also, since the Lagrange Multipliers are necessary conditions only, not sufficient, we may find solutions (x_i, λ_j) that are not optimal for our specific NLP. We need to be able to classify the points found in the solution. Commonly used methods for determining optimality include

a. the definiteness of the Hessian matrix

b. the definiteness of the bordered Hessian via $\det(BdH)$

We'll illustrate these, when feasible, in the following examples with Maple.

Example 4.19. Lagrange Multipliers with Maple.

$$\text{Maximize } z = -2x^2 - 2y^2 + xy + 8x + 3y$$
$$\text{subject to}$$
$$3x + y = 6$$

First, define the f and g, and then display a plot.

```
> f := (x, y) → −2x^2 − 2y^2 + xy + 8x + 3y :
  g := t → 6 − 3t :
```

For the plot, we'll embed the contour plot into a 3D plot to make a better visual representation of f. To embed the plot, use the *transform* command from the *plots* package.

```
> T := transform((x, y) → [x, y, −500]) :
```

Then T will take a point (x, y) on a 2D contour plot and embed it in a 3D plot on the plane $z = -500$.

```
> rng := (x = −10..10, y = −10..10) :
  Surface := plot3d(f(x, y), rng, style = patchcontour, contours = 15) :
  Contours := contourplot(f(x, y), rng, contours = 15) :
```

Now, put the graphs together.

```
> display(Surface, T(Contours));
```

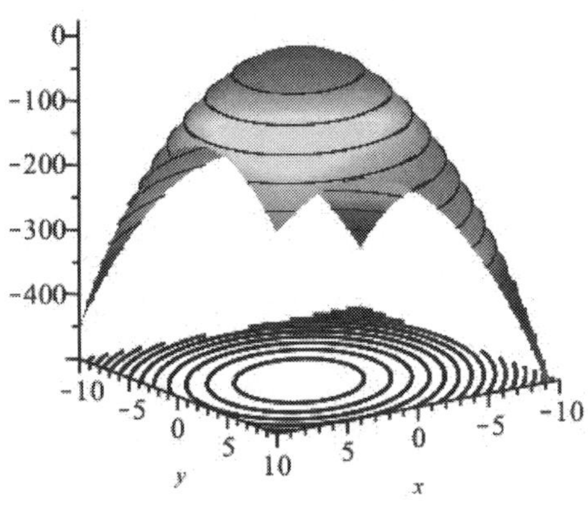

Rotate the 3D figure and inspect it from many different perspectives. Being able to rotate a 3D image is an incredibly useful feature.

Back to the computations.

The Lagrangian function is $L(x, y, \lambda) = f(x, y) + \lambda(g(x, y) - b)$.

```
> L := (x, y, λ) → f(x, y) + λ(3x + y − 6);
```

Calculate the system of necessary conditions.

```
> NecessCond := {diff(L(x, y, λ), x) = 0, diff(L(x, y, λ), y) = 0,
      diff(L(x, y, λ), λ) = 0};
NecessCond := {−4x + y + 8 + 3λ = 0, x − 4y + 3 + λ = 0, 3x + y − 6 = 0}
```

Solve the system *NecessCond* to find potential optimal points.

> $opt := fsolve(NecessCond, \{\lambda, x, y\});$

$$opt := \{\lambda = -0.7608695652, x = 1.673913043, y = 0.9782608696\}$$

> $subs(opt, L(x, y, \lambda)), subs(opt, f(x, y));$

$$10.44565217, 10.44565217$$

Explain why these two values are equal!

It will be no surprise that Maple has several commands for finding solutions to a constrained optimization problem; choose which command to use depending on the form of the output needed. Table 4.5 shows the most commonly used commands.

TABLE 4.5: Maple Commands for Constrained Optimization

Package	Command
Student[MultivariateCalculus]	LagrangeMultipliers
Basic Form: LagrangeMultipliers(ObjectiveFcn, [constraints], [vars], options)	
Options: return more detailed report and/or plots	
Optimization	NLPSolve
Basic Form: NLPSolve(ObjectiveFcn, [constraints], options)	
Options: set solving method, starting point, detailed output, etc.	
Optimization	Maximize/Minimize
Basic Form: Maximize(ObjectiveFcn, [constraints], options)	
Minimize(ObjectiveFcn, [constraints], options)	
Options: recommended: set a starting point	

Remember, in "real world problems," a critical element of the solution using the Lagrange multiplier method is the value and interpretation of the multiplier λ.

Using *LagrangeMultipliers*

> $with(Student[MultivariateCalculus]):$

> $Obj := f(x, y)$

$$Obj := -2x^2 + xy - 2y^2 + 8x + 3y$$

> $Cnstr := 3x + y - 6:$

> $LagrangeMultipliers(Obj, [Cnstr], [x, y], output = detailed);$

$$\left[x = \frac{77}{46}, y = \frac{45}{46}, \lambda_1 = \frac{35}{46}, -2x^2 + xy - 2y^2 + 8x + 3y = \frac{961}{92}\right]$$

The plot output can be very useful.

> $LagrangeMultipliers(Obj, [Cnstr], [x, y], output = plot);$

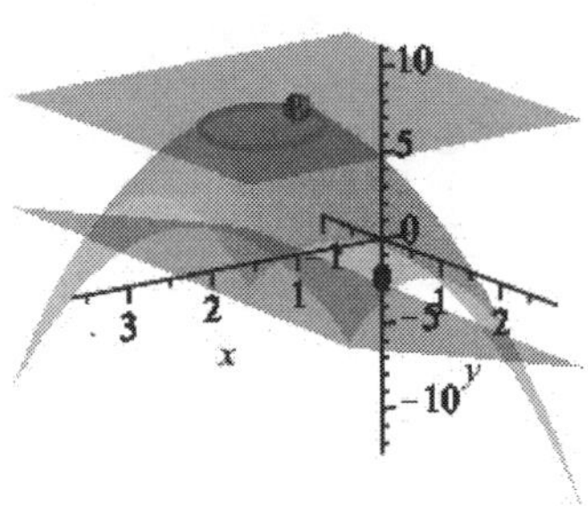

The intersection of the surface $f(x, y) = -2x^2 + xy - 2y^2 + 8x + 3y$ and one or more planes of the form $[10.44565217] = constant$.

Using *NLPSolve*

Constraints must now be written as equations (or inequalities).

> $with(Optimization):$

> $NLPSolve(Obj, \{3x + y = 6\}, maximize);$

[10.4456521739130430, [$x = 1.67391304347826, y = 0.978260869565217$]]

We've lost the value of λ since *NLPSolve* used a different method. Trying to use the Lagrangian equation for the objective function doesn't capture the correct value.

> $NLPSolve(Obj + \lambda \cdot (3x + y - 6), \{3x + y = 6\}, maximize);$

[10.4456521739130430, [$\lambda = 1.00000000000000, x = 1.67391304347826, y =$ 0.978260869565217]]

Using *Maximize*

> $with(Optimization):$

> $Maximize(Obj, \{3x + y = 6\});$

$$[10.4456521739130430, [x = 1.67391304347826, y = 0.978260869565217]]$$

Still no λ.

> $Maximize(Obj + \lambda \cdot (3x + y - 6), \{3x + y = 6\}, maximize);$

Warning, problem appears to be unbounded

$$[10.4456521739130, [\lambda = 0., x = 1.67391304347826, y = 0.978260869565217]]$$

And we did not capture the correct value of λ again.

If we need complete information, it's best to use either direct computation of the method or the *LagrangeMultipliers* function to solve these problems.

Combine the information we've found. The solution is

$$f(\mathbf{x}^*) = 10.4457 \quad \text{with} \quad \mathbf{x}^* = \langle 1.6739, 0.9783 \rangle \quad \text{and} \quad \lambda = 0.76087$$

We have a solution, but we need to know whether this solution represents a maximum or a minimum.

We can use either the Hessian or the bordered Hessian to justify that we have found the correct solution to our constrained optimization problem.

> $with(Student[VectorCalculus]):$
> $with(LinearAlgebra):$

> $H := Hessian(f(x, y), [x, y]);$

$$H := \begin{bmatrix} -4 & 1 \\ 1 & -4 \end{bmatrix}$$

> $IsDefinite(H, query = \text{'negative\_definite'});$

$$true$$

The Hessian is negative definite for all values of $\mathbf{x}$, so the regular point, also called the stationary point, $\mathbf{x}$ is a maximum.

The bordered Hessian gives the same result.

> $BdH := Hessian(f(x, y) + \lambda \cdot (3x + y - 6), [\lambda, x, y]);$

$$BdH := \begin{bmatrix} 0 & 3 & 1 \\ 3 & -4 & 1 \\ 1 & 1 & -4 \end{bmatrix}$$

> $|BdH|$

$$46$$

Since the determinant is positive we have found the maximum at the critical point.

Either method works in this example to determine that we have found the maximum for our constrained optimization problem.

Now, let's interpret the shadow price $\lambda = 0.76$. If the right-hand side of the constraint is increased by a small amount Δ, then the function will increase by approximately $\lambda \cdot \Delta = 0.76 \cdot \Delta$. Since this is a maximization problem, we would add to the restricting resource if possible because it would improve the value of the objective function.

From a graph, it can be seen that the incremental change must be small or the objective function could begin to decrease. Look back at Figure 4.16. If we increase the right side of $g(x) = b$ by one unit so the constraint becomes $3x + y = 7$, the solution at the new point (x^{**}, y^{**}) should yield a new maximum functional value $f(x^{**}, y^{**})$ approximately equal to the old maximum plus λ times the change Δ; that is $f(x^{**}, y^{**}) \approx f(x^*, y^*) + \lambda\,\Delta$, here $10.4457 + 0.7609 = 11.2065$. In actuality, changing the constraints yields a new solution of 11.04347826. (*Verify this!*) The actual increase is about 0.5978.

Example 4.20. Lagrange Multipliers with Multiple Constraints. Minimize $w = x^2 + y^2 + 3z$ subject to $x + y = 3$ and $x + 3y + 2z = 7$.

First, we directly solve the problem using Maple.

$> with(Student[MultivariateCalculus]):$

$> f := (x, y, z) \to x^2 + y^2 + 3z:$

$> Constraints := (3 - (x + y), 7 - (x + 3y + 2z)):$

Now, define the Lagrangian function L and find its gradient.

$> L := (x, y, z, \lambda, \mu) \to f(x, y, z) + \lambda \cdot Constraints_1 + \mu \cdot Constraints_2:$

$> grad := Gradient(L(x, y, z, \lambda, \mu), [x, y, z, \lambda, \mu]);$

$$grad := \begin{bmatrix} 2x + \lambda + \mu \\ 2y + \lambda + 3\mu \\ 3 + 2\mu \\ x + y - 3 \\ x + 3y + 2z - 7 \end{bmatrix}$$

Build the system of equations and solve it to find the potential solution.

$> LMsystem := [seq(G = 0, G \textbf{ in } grad)];$

$$LMsystem := [2x + \lambda + \mu = 0, 2y + \lambda + 3\mu = 0, 3 + 2\mu = 0, x + y - 3 = 0, x + 3y + 2z - 7 = 0]$$

$> fsolve(LMsystem):$
$fnormal(\%, 4);$
$subs(\%, f(x, y, z));$

$$\{\lambda = -0., \mu = -1.500, x = 0.7500, y = 2.250, z = -0.2500\}$$
$$4.87500000$$

Check the solution! (We'll use the "long-form," that is "full name," of the commands so as to not load the packages.)

> *Student*:-*VectorCalculus*:-*Hessian*$(f(x, y, z), [x, y, z])$;
> *LinearAlgebra*:-*IsDefinite*(%, *query* = '*positive_semidefinite*');

$$\begin{bmatrix} 2 & 0 & 0 \\ 0 & 2 & 0 \\ 0 & 0 & 0 \end{bmatrix}$$

true

This Hessian is always positive semi-definite. The function is convex, and so our critical point is a minimum.

Now, we'll solve the problem using Maple' *LagrangeMultipliers* function.

> *with*(*Student*[*MultivariateCalculus*]) :

> *LagrangeMultipliers*$(f(x, y, z)$, [*Constraints*], $[x, y, z]$, *output* = *detailed*);

$$\left[x = \frac{3}{4}, y = \frac{9}{4}, z = -\frac{1}{4}, \lambda_1 = 0, \lambda_2 = -\frac{3}{2}, x^2 + y^2 + 3z = \frac{39}{8}\right]$$

The same Hessian shows the answer is our desired minimum.

Interpret the shadow prices we found, $\lambda = 0$ and $\mu = -3/2$. If we only had the funds to increase one of the two constraint resources, which one should we choose? Since the shadow price of the first constraint is 0, we expect no improvement in the objective function's value; we would not spend extra funds to increase the first resource. Since the shadow price of the second constraint is -1.5, we expect to change the objective function's value by -1.5Δ (*Improving the objective since we are minimizing!*) if we increase that resource by Δ; we would spend extra funds to increase the second resource as long as the cost of increasing resource 2 was less than 1.5Δ. *Why does the cost need to be less than* 1.5Δ*?*

Applications using Lagrange Multipliers

Many applied constrained optimization problems can be solved with Lagrange multipliers. We'll start by revisiting the opening problem of this section (from page 162).

Example 4.21. The Cobb-Douglas Function.

A company manufactures new phones that are projected to take the market by storm. The two main input components of the new phone are the circuit board and the LCD Touchscreen that make the phone faster, smarter, and easier to use. The number of phones to be produced P is estimated to equal $P = 250\, a^{1/4} b^{1/2}$ where a and b are the number of circuit board production hours and the number of LCD Touchscreen production hours available, respectively. This type of formula is known to economists as a *Cobb-Douglas production function*[7]. Laborers are paid by the type of work they do: the circuit board labor cost is \$5 an hour and the LCD Touchscreen labor cost is \$8 an hour. What is the maximum number of phones that can be made if the company has \$175,000 allocated for these components in the short run?

Let's use Maple.

$> CD := (a, b) \to 250 \cdot a^{0.15} \cdot b^{0.5} :$
$Constraint := 5a + 8b - 175000 :$

$> L := (a, b, \lambda) \to CD(a, b) + \lambda \cdot Constraint :$

$> grad := Gradient(L(a, b, \lambda), [a, b, \lambda]);$

$$grad := \begin{bmatrix} 62.50\, a^{-0.75} b^{0.50} + 5\,\lambda \\ 125.0\, a^{0.25} b^{-0.50} + 8\,\lambda \\ 5\,a + 8\,b - 175000 \end{bmatrix}$$

$> LMsystem := [seq(G = 0, Gingrad)];$

$> soln := fsolve(LMsystem, a, b, \lambda);$

$$soln := \{a = 11666.66667, b = 14583.33333, \lambda = -1.344709033\}$$

$> subs(soln, L(a, b, \lambda));$

$$313765.4410$$

Of course, *LagrangeMultipliers* gives the same results.

$> LagrangeMultipliers(CD(a, b), [Constraint], [a, b], output = detailed) :$
$fnormal(\%, 6);$

$$[a = 11666.7, b = 14583.3, \lambda_1 = 1.34471, 250a^{0.25}\sqrt{b} = 313765.]$$

We find we can make $\approx$ 313,765 phones using 11,667.67 production hours for circuit boards and 14,583.33 production hours for LCD touchscreens.

A plot of the Cobb-Douglas function centered at the (11,668, 14,583) showing the constraint in Figure 4.17 finishes our analysis.

[7]Originated in Cobb & Douglas, "A Theory of Production," Am Econ Review, **18**, 1928, pg. 139-165.

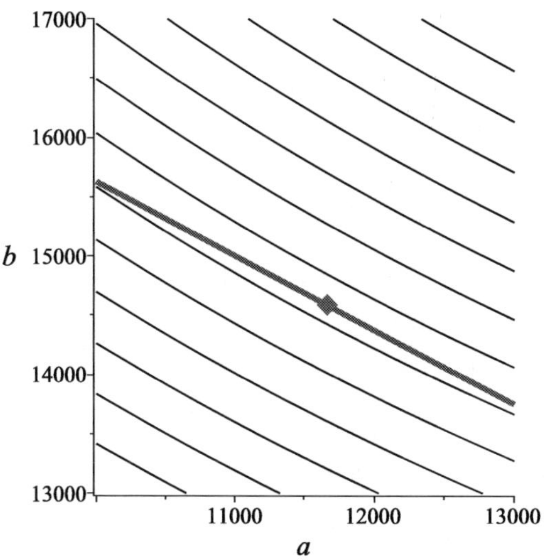

FIGURE 4.17: Cobb-Douglas Phone Production Function

Example 4.22. Oil Transfer Company Storage Facilities.
The management of a small oil transfer company desires a minimum cost policy taking into account the restrictions on available tank storage space. A formula has been derived from historical data records that describes system costs.

$$f(\mathbf{x}) = \sum_{n=1}^{N} \frac{a_n b_n}{x_n} + \frac{h_n x_n}{2}$$

where

a_n is the fixed costs for the nth item,

b_n is the withdrawal rate per unit time for the nth item,

h_n is the holding costs per unit time for the nth item.

The tank space constraint is given by:

$$g(\mathbf{x}) = \sum_{n=1}^{N} t_n x_n = T$$

where t_n is the space required for the nth item (in 1,000s bbl) and T is the available tank space (in 1,000s bbl). The parameter values shown in Table 4.6 are based on the company's data collected over several years.

TABLE 4.6: Oil Storage data Parameter Estimates

Item n	a_n (\$1,000s)	b_n (k bbl/hr)	h_n (\$1,000s)	t_n (1,000s bbl)
1	9.6	3	0.47	1.4
2	4.27	5	0.26	2.62
3	6.42	4	0.61	1.71

The storage tanks have only 22 (1,000s bbl) space available. Find the optimal solution as a minimum cost policy.

We'll first solve the unconstrained problem. This solution will provide an upper bound for the constrained solution and help us gain insight into the dynamics of the problem.

Define the parameters, then define the functions.

$$> N := 3 :$$
$$a := [9.6, 4.27, 6.42] :$$
$$b := [3, 5, 4] :$$
$$h := [0.47, 0.26, 0.61] :$$
$$t := [1.4, 2.62, 1.71] :$$

$$> f := x \rightarrow sum\left(\frac{a_n \cdot b_n}{x_n} + \frac{h_n \cdot x_n}{2}, n = 1..N\right) :$$
$$g := x \rightarrow sum(t_n \cdot x_n, n = 1..N) :$$

$$> \text{`}f\text{`}(x) = f(x);$$
$$\text{`}g\text{`}(x) = g(x);$$

$$f(x) = \frac{28.8}{x_1} + 0.23500\, x_1 + \frac{21.35}{x_2} + 0.13000\, x_2 + \frac{25.68}{x_3} + 0.30500\, x_3$$

$$g(x) = 1.4\, x_1 + 2.62\, x_2 + 1.71\, x_3$$

Build the system of equations $\{\partial f/\partial x_i = 0\}$ and solve it. (Remember to load *Student*[*MultivariateCalculus*] for *Gradient*.)

$$> grad := Gradient(f(x), [x_1, x_2, x_3]);$$

$$grad := \begin{bmatrix} -\dfrac{28.8}{x_1^2} + 0.23500 \\ -\dfrac{21.35}{x_2^2} + 0.13000 \\ -\dfrac{25.68}{x_3^2} + 0.30500 \end{bmatrix}$$

$$> sys := [seq(G = 0, G \textbf{ in } grad)] :$$

```
> solns := solve(sys, [x1, x2, x3]) :
  Matrix(solns);
```

$$\begin{bmatrix} x_1 = 11.07037450 & x_2 = 12.81525533 & x_3 = 9.175877141 \\ x_1 = 11.07037450 & x_2 = -12.81525533 & x_3 = 9.175877141 \\ x_1 = 11.07037450 & x_2 = 12.81525533 & x_3 = -9.175877141 \\ x_1 = 11.07037450 & x_2 = -12.81525533 & x_3 = -9.175877141 \\ x_1 = -11.07037450 & x_2 = 12.81525533 & x_3 = 9.175877141 \\ x_1 = -11.07037450 & x_2 = -12.81525533 & x_3 = 9.175877141 \\ x_1 = -11.07037450 & x_2 = 12.81525533 & x_3 = -9.175877141 \\ x_1 = -11.07037450 & x_2 = -12.81525533 & x_3 = -9.175877141 \end{bmatrix}$$

The only useful solution is where $x_1, x_2, x_3 > 0$, so we choose the first row.

```
> Soln := solns[1];
```

$$Soln := [x_1 = 11.07037450, x_2 = 12.81525533, x_3 = 9.175877141]$$

This solution $\mathbf{x}^* = \langle 11.07, 12.82, 9.18 \rangle$ provides an unconstrained upper bound since it does not satisfy the constraint $g(\mathbf{x}) = 22$.

```
> g(rhs~(Soln)) = 22;
```

$$64.76524317 = 22$$

Now we solve the constrained optimization problem knowing that the solution will be less than the unconstrained value we just found.

```
> L := (x, λ) → f(x) + λ · (g(x) − 22) :
```

```
> Lgrad := Gradient(L(x, λ), [x1, x2, x3, λ]) :
  Lsys := [seq(G = 0, G in Lgrad)] :
```

Using *solve* as we did above gives a plethora of complex and negative values that we don't want. We'll use *fsolve* with a *starting value* of the unconstrained solution and $\lambda = 1$.

```
> StartingPt := {op(Soln), λ = 1} :
  Lsolns := fnormal(fsolve(Lsys, StartingPt), 4);
```

$$Lsolns := \{\lambda = 0.7397, x_1 = 4.761, x_2 = 3.213, x_3 = 4.045\}$$

Do we have the minimum?

The Hessian matrix H (*Remember to load Student[VectorCalculus]*) is

> $H := Hessian(f(x), [x_1, x_2, x_3]);$

$$H := \begin{bmatrix} \frac{57.6}{x_1^3} & 0 & 0 \\ 0 & \frac{42.7}{x_2^3} & 0 \\ 0 & 0 & \frac{51.36}{x_3^3} \end{bmatrix}$$

> $|H|;$

$$\frac{126320.9472}{x_1^3 x_2^3 x_3^3}$$

Since the Hessian is positive for all positive $\mathbf{x}$, the matrix is positive definite. We have found a minimum for the constrained problem at $\mathbf{x}^* = \langle 4.761, 3.213, 4.045 \rangle$.

Should we recommend the company add storage space? We know from the unconstrained solution that, if possible, we would add storage space to decrease the costs. We have found the shadow price $\lambda = 0.74$ which suggests that any small increase Δ in the RHS of the constraint causes the objective function to decrease by approximately $0.74 \cdot \Delta$. The cost of the extra storage tank would have to be less than the savings incurred by adding the tank.

Exercises

1. Solve the following constrained problems using the Lagrangian approach.
   (a) Minimize $z = x^2 + y^2$ subject to $x + 2y = 4$.
   (b) Maximize $z = (x-3)^2 + (y-2)^2$ subject to $x + 2y = 4$.
   (c) Maximize $z = x^2 + 4xy + y^2$ subject to $x^2 + y^2 = 1$.
   (d) Maximize $z = x^2 + 4xy + y^2$ subject to $x^2 + y^2 = 4$ and $x + 2y = 4$.

2. Find and classify the extrema for $f(x, y, z) = x^2 + y^2 + z^2$ subject to $x^2 + 2y^2 - z^2 = 1$.

3. Two manufacturing processes, f_1 and f_2, both use a resource with b units available. Maximize $f_1(x_1) + f_2(x_2)$ subject to $x_1 + x_2 = b$.

   If $f(1(x_1) = 50 - (x_1 - 2)^2$ and $f_2(x_2) = 50 - (x_2 - 2)^2$, analyze the manufacturing processes to

   (a) determine the amount of x_1 and x_2 to use to maximize $f_1 + f_2$,
   (b) determine the amount b of the resource to use.

4. Maximize $Z = -2x^2 - y^2 + xy + 8x + 3y$ subject to $3x + y = 10$ and $x^2 + y^2 = 16$.

5. Use the Method of Lagrange Multipliers to maximize $f(x, y, z) = xyz$ subject to $2x + 3y + 4w = 36$.
   Determine how much $f(x, y, z)$ would change if one more unit was added to the constraint.

4.5 Constrained Optimization: Kuhn-Tucker Conditions

Inequality Constraints and the Kuhn-Tucker Conditions

Previously, we investigated procedures for solving problems with equality constraints. However, in most realistic problems, many constraints are expressed as inequalities. The inequality constraints form the boundaries of a set containing the solution.

One method for solving NLPs with inequality constraints is by using the *Kuhn-Tucker Conditions* (KTC) for optimality, sometimes called the Karush-Kuhn-Tucker conditions.[8] In this section, we'll describe this set of conditions first graphically, then analytically. We'll discuss both necessary and sufficient conditions for $\mathbf{x} = \langle x_1, x_2, \ldots, x_n \rangle$ to be an optimal solution to an NLP with inequality constraints. We'll also illustrate how to use Maple to compute solutions. Last, we'll close with example applications.

Basic Theory of Constrained Optimization

The generic form of the NLPs we will study in this section is

$$\begin{aligned} &\text{Maximize (Minimize) } z = f(\mathbf{x}) \\ &\text{subject to} \\ &\qquad g_i(\mathbf{x}) \begin{Bmatrix} \le \\ = \\ \ge \end{Bmatrix} b_i, \quad \text{for } i = 1, 2, \ldots, m \end{aligned} \tag{4.4}$$

(Note: Since $a = b$ is equivalent to $(a \le b \wedge a \ge b)$ and $a \ge b$ is equivalent to $-a \le -b$, we could focus only on less-than inequalities; however, the technique is more easily understood by allowing all three forms.)

Recall that the optimal solution to an NLP with only equality constraints had to fall on one constraint or at an intersection of several constraints. With inequality constraints, the solution no longer must lie on a constraint or at an intersection point of constraints. We need a method that describes the position of the optimal solution relative to each constraint.

[8] First formulated in Kuhn & Tucker, "Nonlinear programming," Proc 2nd Berkeley Sym, U Cal Press, 1951, pp. 481-492.

The technique based on the Kuhn-Tucker conditions involves defining a Lagrangian function of the decision variables $\mathbf{x}$, the Lagrange multipliers λ_i, and the nonnegative *slack* or *surplus variables* μ_i^2. The nonnegative slack variable μ_i^2 is a variable added to the ith 'less-than or equal' constraint to transform it to an equality: $g_i(\mathbf{x}) \leq b_i \mapsto g_i(\mathbf{x}) + \mu_i^2 = b_i$; the nonnegative variable μ_i^2 "picks up the slack" in the inequality. The surplus variable μ_j^2 is a variable subtracted from the jth 'greater-than or equal' constraint to make it an equality: $g_j(\mathbf{x}) \geq b_j \mapsto g_j(\mathbf{x}) - \mu_j^2 = b_j$; the variable μ_j^2 "holds the surplus" in the inequality. In this formulation, the shadow price for the ith constraint is $-\lambda_i$.

The Lagrangian function for our generic NLP (4.4) is

$$L(\mathbf{x}, \boldsymbol{\lambda}, \boldsymbol{\mu}) = f(\mathbf{x}) + \sum_{i=1}^{m} \lambda_i \big(g_i(\mathbf{x}) \pm \mu_i^2 - b_i\big) \tag{4.5}$$

Remember, the sign with μ_i is $+$ for $\leq$ constraints and $-$ for $\geq$ constraints.

Analogously to the method Lagrange multipliers, the computational procedure based on the KTC requires that all the partials of the Lagrangian function (4.5) equal zero. All these partials being equal to zero forms the *necessary conditions* for the solution of (4.4) to exist.

Theorem. Necessary Conditions for an Optimal Solution.
If $\mathbf{x}^*$ is an optimum for the NLP (4.4), then

$$\frac{\partial L}{\partial x_j} = 0 \text{ for } j = 1, 2, \ldots, n \tag{4.6a}$$

$$\frac{\partial L}{\partial \lambda_i} = 0 \text{ for } i = 1, 2, \ldots, m \tag{4.6b}$$

$$\frac{\partial L}{\partial \mu_i} = 2\mu_i\lambda_i = 0 \text{ for } i = 1, 2, \ldots, m \tag{4.6c}$$

Condition (4.6c) is called *the complementary slackness condition.*

The following theorem provides *sufficient conditions* for $\mathbf{x}^*$ to be an optimal solution to the NLP given in (4.4).

Theorem. Sufficient Conditions for an Optimal Solution.
Suppose each $g_i(\mathbf{x})$ is a convex function.

Maximum: If $f(\mathbf{x})$ is concave, then any point $\mathbf{x}^*$ that satisfies the necessary conditions is a maximal solution to (4.4). Further, each $\lambda_i \leq 0$.

Minimum: If $f(\mathbf{x})$ is convex, then any point $\mathbf{x}^*$ that satisfies the necessary conditions is a minimal solution to (4.4). Further, each $\lambda_i \geq 0$.

If the necessary conditions are satisfied, but the sufficient conditions are not completely satisfied, then we may use a bordered Hessian matrix to check the nature of a potential stationary or regular point. The bordered Hessian can be written in general as the block matrix

$$BdH := \left[\begin{array}{c|c} 0 & \nabla g \\ \hline (\nabla g)^T & \dfrac{\partial^2 L}{\partial x_i \partial x_j} \end{array}\right]$$

We can classify, if possible, the stationary points as maxima or minima according to the bordered Hessian's definiteness. If the bordered Hessian is indefinite, then a different classification method must be used.

The Complementary Slackness Condition

The KTC computational solution process solves the 2^m possible cases for λ_i and μ_i, where m equals the number of constraints, then applies the necessary conditions to find optimal points. The 2 comes from the number of possibilities for each λ_i: either $\lambda_i = 0$ or $\lambda_i \neq 0$. There is actually more to this process: it really involves the complementary slackness condition imbedded in the necessary condition (4.6c), $2\mu_i\lambda_i = 0$. If μ_i equals zero, then λ_i, the shadow price, can be nonzero and the ith constraint is *binding*—the optimal point lies on the constraint boundary. If μ_i is not equal to zero, then λ_i, the shadow price, must be zero and the ith constraint is *nonbinding*—there is *slack* ($\leq$ constraint) or *surplus* ($\geq$ constraint), represented by μ_i. Ensuring the complementary slackness conditions are satisfied reduces the work involved in solving the other necessary conditions from Equations (4.6a) and (4.6b).

Based on this analysis, the complementary slackness necessary conditions (4.6c) lead to the solution process that we focus on for our computational and geometric interpretation. We have defined μ_i^2 as a slack or surplus variable. Therefore, if μ_i^2 equals zero, then our optimal point lies on the ith constraint, and if μ_i^2 is greater than zero, the optimal point is interior to the ith constraint boundary. However, if the value of μ_i is undefined because μ_i^2 equals a negative number, then the point of concern is *infeasible*. Figure 4.18 illustrates these conditions.

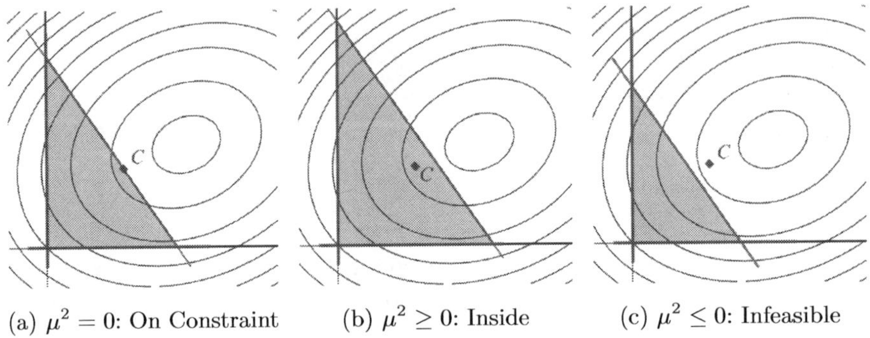

(a) $\mu^2 = 0$: On Constraint (b) $\mu^2 \geq 0$: Inside (c) $\mu^2 \leq 0$: Infeasible

FIGURE 4.18: Complementary Slackness Geometrically

Computational KTC with Maple

First, we'll step through a solution using Maple for the computations.

Example 4.23. Two Variable-Two Constraint Linear Problem.

$$\begin{aligned}&\text{Maximize } z = 3x + 2y\\&\text{subject to}\\&\qquad 2x + y \leq 100\\&\qquad x + y \leq 80\end{aligned}$$

Define the generalized Lagrangian (4.5) for this problem.

$$L(\mathbf{x}, \boldsymbol{\lambda}, \boldsymbol{\mu}) = (3x + 2y) + \lambda_1(2x + y + \mu_1^2 - 100) + \lambda_2(x + y + \mu_2^2 - 80)$$

The six *Necessary Conditions* are:

$$\begin{aligned}
&1.\quad \partial L/\partial x = 0 \implies 3 + 2\lambda_1 + \lambda_2 = 0 &\text{(4.7a)}\\
&2.\quad \partial L/\partial y = 0 \implies 2 + \lambda_1 + \lambda_2 = 0 &\text{(4.7b)}\\
&3.\quad \partial L/\partial \lambda_1 = 0 \implies 2x + y + \mu_1^2 - 100 = 0 &\text{(4.7c)}\\
&4.\quad \partial L/\partial \lambda_2 = 0 \implies x + y + \mu_2^2 - 80 = 0 &\text{(4.7d)}\\
&5.\quad \partial L/\partial \mu_1 = 0 \implies 2\mu_1\lambda_1 = 0 &\text{(4.7e)}\\
&6.\quad \partial L/\partial \mu_2 = 0 \implies 2\mu_2\lambda_2 = 0 &\text{(4.7f)}
\end{aligned}$$

Recognize that, since there are two constraints, there are four (2^2) cases required to solve for the optimal solution. These 4 cases stem from necessary conditions $2\mu_1\lambda_1 = 0$ and $2\mu_2\lambda_2 = 0$, the complementary slackness conditions.

The cases are collected in Table 4.7.

TABLE 4.7: The Four Cases for Complementary Slackness

Case	Condition Imposed	Condition Inferred
I.	$\lambda_1 = 0,\ \lambda_2 = 0$	$\mu_1^2 \neq 0,\ \mu_2^2 \neq 0$
II.	$\lambda_1 = 0,\ \lambda_2 \neq 0$	$\mu_1^2 \neq 0,\ \mu_2^2 = 0$
III.	$\lambda_1 \neq 0,\ \lambda_2 = 0$	$\mu_1^2 = 0,\ \mu_2^2 \neq 0$
IV.	$\lambda_1 \neq 0,\ \lambda_2 \neq 0$	$\mu_1^2 = 0,\ \mu_2^2 = 0$

For simplicity, we have arbitrarily made both x and $y \geq 0$ for this maximization problem. Figure 4.19 shows a graphical representation.

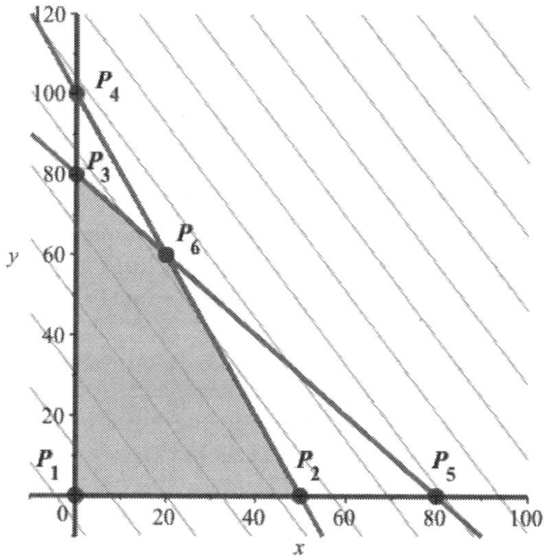

FIGURE 4.19: The Region of Feasible Solutions

Returning to Cases I-IV, we observe that:

CASE I. There is slack in both the first and second constraints as both $\mu_i^2 > 0$. Therefore, we do not fall exactly on either of the constraint boundaries. This corresponds to the intersection point labeled P_1 at $(0,0)$, since only intersection points can lead to linear optimization solutions. This point is feasible, but is clearly not optimal—moving away from $(0,0)$ in the first quadrant increases the objective function. This case will not yield an optimal solution.

Case II. The possible solution point is on the second constraint, but not on the first constraint. There are two possibilities, P_3 and P_5, from Figure 4.19. Point P_5 is infeasible. Point P_3 is feasible, but not an optimal solution. This case will not yield an optimal solution.

Case III. The possible solution point is on the first constraint, but not on the second constraint. There are two possible solutions, P_2 and P_4, from Figure 4.19. Point P_4 is infeasible. Point P_2 is feasible, but not optimal. Again, this case does not yield an optimal solution.

Case IV. The possible solution point lies on both constraint 1 and constraint 2 simultaneously. This corresponds to P_6 in Figure 4.19. Point P_6 is the optimal solution. It is the point of the feasible region furthest in the direction of increased value of the objective function. This case will yield the optimal solution to the problem.

Sensitivity analysis is also enhanced by a geometric approach. Figure 4.19 shows that increasing the right-hand side of either or both constraints will extend the feasible region in the direction of the objective function's increase, thus increasing the value of the objective function. We can also see this through the computational process and the solution's values of the λ_i. Computational sensitivity analysis can be derived with the value of the shadow price $-\lambda_i$.

Since the objective function and constraints are linear, convex, and concave functions, the sufficient conditions are also satisfied.

The following computational analysis will show that Case IV yields the optimal solution, confirming the graphical solution.

Case I. $\lambda_1 = \lambda_2 = 0$. This case violates Equations (4.7a), $2 \neq 0$, and (4.7b), $3 \neq 0$. This case also implies $\mu_i^2 \neq 0$ with slack in both inequalities.

Case II. $\lambda_1 = 0$, $\lambda_2 \neq 0$. This case violates either Equation (4.7a), $\lambda_2 = -3$, or (4.7b), $\lambda_2 = -2$.

Case III. $\lambda_1 \neq 0$, $\lambda_2 = 0$. This case similarly violates either Equation (4.7a), $\lambda_1 = -3/2$, or (4.7b), $\lambda_1 = -2$.

Case IV. $\lambda_1 \neq 0$, $\lambda_2 \neq 0$. This case implies that $\mu_1^2 = \mu_2^2 = 0$ which reduces (4.7) to the two sets

$$\{3 + 2\lambda_1 + \lambda_2 = 0, \quad 2 + \lambda_1 + \lambda_2 = 0\}$$

and

$$\{2x = y - 100 = 0, \quad x + y = 80 = 0\}$$

Solving these sets simultaneously yields the optimal solution $x^* = 20, y^* = 60$ giving the maximum $f(x^*, y^*) = 180$. We also have $\lambda_1 = -1$, $\lambda_2 = -1$, $\mu_1^2 = \mu_2^2 = 0$. The shadow price indicates for a small change Δ in the right-hand side value of either Constraint 1 or Constraint 2, the objective value will increase by approximately $-\lambda \cdot \Delta = \Delta$. The geometric interpretation reinforces the computational results, giving them meaning and fully showing the effect of binding constraints (constraints where $\mu_i^2 = 0$) on the solution.

The following is a short Maple session of the computations above.

> $with(Student[MultivariateCalculus]):$

> $f := (x, y) \to 3x + 2y:$
> $Cnstrnt := [2x + y - 100, x + y - 80]:$

> $L := (x, y, \lambda, \mu) \to f(x, y) + \lambda_1 \cdot (Cnstrnt_1 + \mu_1^2) + \lambda_2 \cdot (Cnstrnt[2] + \mu_2^2):$

> $grad := Gradient(L(x, y, \lambda, \mu), [x, y, \lambda_1, \lambda_2, \mu_1, \mu_2]):$
> $NecCond := [seq(G = 0, G \textbf{ in } grad)]:$
> $\langle NecCond \rangle;$

$$\begin{bmatrix} 3 + 2\lambda_1 + \lambda_2 = 0 \\ 2 + \lambda_1 + \lambda_2 = 0 \\ {\mu_1}^2 + 2x + y - 100 = 0 \\ {\mu_2}^2 + x + y - 80 = 0 \\ 2\lambda_1 \mu_1 = 0 \\ 2\lambda_2 \mu_2 = 0 \end{bmatrix}$$

> $soln := solve(NecCond, [x, y, \lambda_1, \lambda_2, \mu_1, \mu_2]);$

$$soln := [[x = 20, y = 60, \lambda_1 = -1, \lambda_2 = -1, \mu_1 = 0, \mu_2 = 0]]$$

> $f(20, 60);$
> $subs(x = 20, y = 60, Constraint);$

$$180$$
$$[0, 0]$$

How do we know we've found a maximum? Recall the rules for finding the maximum or minimum.

MAXIMUM: If $f(\mathbf{x})$ is a *concave* function and each constraint $g_i(\mathbf{x})$ is a convex function, then any point that satisfies the necessary conditions is an optimal solution that maximizes the function subject to the constraints and has each $\lambda_i \le 0$.

MINIMUM: If $f(\mathbf{x})$ is a *convex* function and each constraint $g_i(\mathbf{x})$ is a convex function, then any point that satisfies the necessary conditions is an optimal solution that minimizes the function subject to the constraints and has each $\lambda_i \ge 0$.

The objective function in linear, and so is both convex and concave, as are the constraints. Since the values of λ_i are both negative, we have found the maximum.

We can use Maple's own *LagrangeMultipliers*.

> $LagrangeMultipliers(f(x, y), Cnstrnt, [x, y], output = detailed);$

$$[x = 20, y = 60, \lambda_1 = 1, \lambda_2 = 1, 3 \cdot x + 2 \cdot y = 180]$$

Note the sign difference on Maple's λs. Explain how this occurs.

In the next example, we add one constraint, $x \leq 40$, to the previous problem. Adding one constraint causes the number of solution cases we must consider to grow from 2^2 to 2^3 or doubling to 8 cases—each additional constraint doubles the number of cases. The new problem with three constraints is shown in Figure 4.20. Again, for simplicity, we arbitrarily force both x and $y > 0$.

Example 4.24. Two Variable, Three Constraint Linear Problem.

$$\begin{aligned} &\text{Maximize } z = 3x + 2y \\ &\text{subject to} \\ &\qquad 2x + y \leq 100 \\ &\qquad x + y \leq 80 \\ &\qquad x \leq 40 \end{aligned}$$

The new Lagrangian is

$$L(\mathbf{x}, \boldsymbol{\lambda}, \boldsymbol{\mu}) = (3x+2y)+\lambda_1(2x+y+\mu_1^2-100)+\lambda_2(x+y+\mu_2^2-80)+\lambda_3(x+\mu_3^2-40)$$

Add the new constraint to the graph.

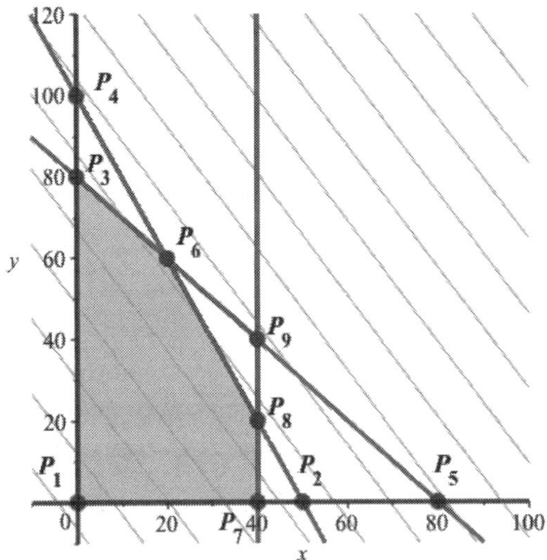

FIGURE 4.20: Contours of $f(\mathbf{x})$ with Three Constraints

A summary of the graphical interpretation is displayed in Table 4.8. The optimal solution is found using Case IV. Again, the computational solution merely looks for the point where either all the necessary conditions are met or are not violated. The geometric interpretation reinforces why the cases other than IV do not yield an optimal solution.

TABLE 4.8: The Eight Cases

Case	Condition Imposed	Point	Feasible	Optimal
I.	$\lambda_1 = 0,\ \lambda_2 = 0,\ \lambda_3 = 0$ (all constraints have slack)	P_1	yes	no
II.	$\lambda_1 \neq 0,\ \lambda_2 = 0,\ \lambda_3 = 0$ (on constraint 1, not 2 or 3)	P_2, P_4	no/no	no/no
III.	$\lambda_1 = 0,\ \lambda_2 \neq 0,\ \lambda_3 = 0$ (on constraint 2, not 1 or 3)	P_3, P_5	yes/no	no/no
IV.	$\lambda_1 = 0,\ \lambda_2 = 0,\ \lambda_3 \neq 0$ (on constraint 3, not 1 or 2)	P_7	yes	no
V.	$\lambda_1 \neq 0,\ \lambda_2 \neq 0,\ \lambda_3 = 0$ (on constraints 1 & 2, not 3)	P_6	yes	yes
VI.	$\lambda_1 \neq 0,\ \lambda_2 = 0,\ \lambda_3 \neq 0$ (on constraints 1 & 3, not 2)	P_8	yes	no
VII.	$\lambda_1 = 0,\ \lambda_2 \neq 0,\ \lambda_3 \neq 0$ (on constraints 2 & 3, not 1)	P_9	no	no
VIII.	$\lambda_1 \neq 0,\ \lambda_2 \neq 0,\ \lambda_3 \neq 0$ (on constraint 1, 2, and 3)	—	—	—

The optimal solution will be found only in Case V which geometrically shows that the solution is binding on Constraints 1 and 2, and not binding on Constraint 3 (slack still exists). The optimal solution found computationally using Case IV (as done in the previous example) is

$$f(x^*, y^*) = f(20, 60) = 180,$$

the same as before. Constraint 3 did not alter the solution as it is nonbinding; i.e., has slack in the solution. The "detailed solution" adds λ_3 and u_3^2.

$$[x = 20, y = 60, \lambda_1 = -1, \lambda_2 = -1, \lambda_3 = 0, \mu_1^2 = 0, \mu_2^2 = 0, \mu_3^2 = 20]$$

The geometric interpretation takes the mystery out of the case-wise solutions. We can visually see why in each specific case we can achieve or not achieve optimality conditions. Whenever possible, make a quick graph, and analyze the graph to eliminate as many cases as possible prior to doing the computational solution procedures. Let's apply this procedure to another example.

Example 4.25. Geometric Constrained Nonlinear Optimization Problem.

$$\text{Maximize } z = (x-14)^2+(y-11)^2$$
$$\text{subject to}$$
$$(x-11)^2+(y-13)^2 \leq 49$$
$$x+y \leq 19$$

Use Maple to generate contour plots overlaid with the constraints to obtain the geometrical interpretation shown in the worksheet below. The optimal solution, as visually shown, is the point where the *level curve of the objective function is tangent to the constraint* $x+y = 19$ in the direction of increase for the contours of f. The solution satisfies the other constraint $(x-11)^2+(y-13)^2 \leq 49$, but there is slack in this constraint. The solution corresponds to the case where Constraint 2 is binding and Constraint 1 is nonbinding. The constraints being nonbinding and binding, respectively, are shown computationally by

$$(\lambda_1 = 0, \mu_1^2 \neq 0) \text{ and } (\lambda_2 \neq 0, \mu_2^2 = 0).$$

Finish by estimating the solution in the plot below.

$> with(plots):$

$> f := (x,y) \to (x-14)^2+(y-11)^2:$
$g_1 := (x,y) \to (x-11)^2+(y-13)^2:$
$g_2 := (x,y) \to x+y:$

$> rng := (x = 2..21, y = 2..21):$
$fillopts := (filledregions = true, coloring = [wheat, white]):$

$> fcp := contourplot(f(x,y), rng, contours = 40, color = grey);$
$fcp19 := contourplot(f(x,y), rng, contours = [19], color = black);$
$g1cp := contourplot(g_1(x,y), rng, contours = [49], color = red,$
$\quad thickness = 2, fillopts);$
$g2cp := contourplot(g_2(x,y), rng, contours = [19], thickness = 2):$

> *display*(*fcp*, *fcp19*, *g1cp*, *g2cp*);

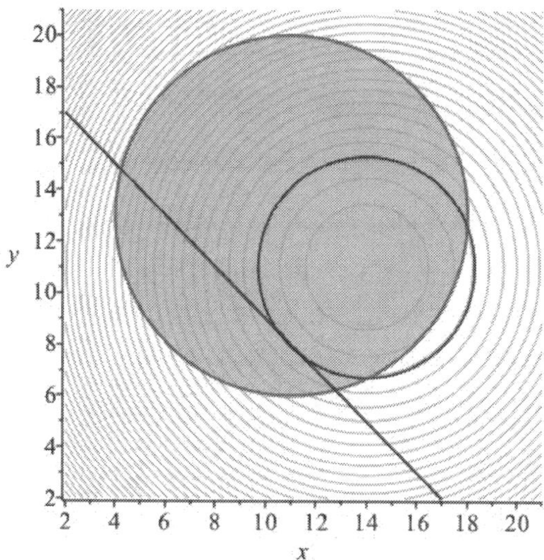

We use the fact that Constraint 1 is nonbinding and Constraint 2 is binding to directly solve this case to find the optimal solution. Graphically, we can obtain a good approximation, but we cannot obtain the shadow prices which are invaluable in sensitivity analysis. In this case, we saw that

$$(\lambda_1 = 0, \mu_1^2 \neq 0) \text{ and } (\lambda_2 \neq 0, \mu_2^2 = 0).$$

The *necessary conditions* for this case are

$$\begin{aligned}
\partial L/\partial x &\implies 2x - 28 + \lambda_2 = 0\\
\partial L/\partial y &\implies 2y - 22 + \lambda_2 = 0\\
\partial L/\partial \lambda_1 &\implies (x-11)^2 + (y-13)^2 + \mu_1^2 - 49 = 0\\
\partial L/\partial \lambda_2 &\implies x + y - 19 = 0
\end{aligned}$$

Let's use Maple to find the optimal point and the shadow prices.

We've defined f and g_i above, so start by loading the *Student*[*MultivariateCalculus*] package to make *Gradient* available.

> *with*(*Student*[*MultivariateCalculus*]) :

> $L := (x, y, \lambda, \mu) \to f(x, y) + \lambda_1(g_1(x, y) + \mu_1^2 - 49) + \lambda_2(g_2(x, y) + \mu_2^2 - 19)$:

> *grad* := *Gradient*$(L(x, y, \lambda, \mu), [x, y, \lambda_1, \lambda_2, \mu_1, \mu_2])$:

We'll add the case defining $\lambda_1 = 0$ and $\mu_2 = 0$ to the necessary conditions' system of equations.

> $\mathit{CasewNecCond} := [\lambda_1 = 0, \mu_2 = 0, \mathit{seq}(G = 0, G \textbf{ in } \mathit{grad})]$:

For solving this system, we'll use some of Maple's "cleverness:" using *RealDomain* will ignore complex solutions to our system, and we'll solve for μ_1^2, rather than μ_1.

> $\mathit{soln} := \mathit{RealDomain}\text{:-}\mathit{solve}(\mathit{CasewNecCond}, [x, y, \lambda_1, \lambda_2, \mu_1^2, \mu_2]);$

$$\mathit{soln} := [[x = 11, y = 8, \lambda_1 = 0, \lambda_2 = 6, \mu_1^2 = 24, \mu_2 = 0]]$$

> $f(11, 8);$

$$18$$

We determine the optimal solution that satisfies the conditions is: $x^* = 11$, $y^* = 8$, $\lambda_1 = 0$, $\lambda_2 = 6$, $\mu_1^2 = 24$, and $\mu_2^2 = 0$. The value of the objective function is $f(x^*, y^*) = 18$.

Interpreting the shadow prices shows that if there are more resources for Constraint 2, our objective function will decrease. If we add Δ to the right-hand side of Constraint 2, the objective function value will decrease by approximately 6Δ. If we changed the right-hand side of the constraint from 19 to 20, the optimal solution becomes $x^* = 11.5$, $y^* = 8.5$, and $f(x^*, y^*) = 12.5$ or a decrease of 5.5 units (verification is left as an exercise).

It is also possible to use the *LagrangeMultipliers* function from the *Student*[*MultivariateCalculus*] package to get extensive information. To get the extra analysis from Maple:

- write the constraints as they appear in the Lagrangian,
- write the constants in the constraints as decimals, and
- list λ_i and μ_i as variables.

> $\mathit{Soln2} := \mathit{LagrangeMultipliers}(f(x, y), [g_1(x, y) + \mu_1^2 - 49.0,$
> $g_2(x, y) + \mu_2^2 - 19.0], [x, y, \lambda_1, \lambda_2, \mu_1, \mu_2], \mathit{output} = \mathit{detailed})$:

In order to make the output more readable, normalize the decimals to four figures and increase the size of matrix displayed. (The display here is smaller and reduced to three digits to fit in the text—execute the code to see the full solution display.)

> $\mathit{interface}(\mathit{rtablesize} = 20)$:

> *Matrix(fnormal([Soln2], 4));*

$$\begin{bmatrix} x=14.0 & y=11.0 & \lambda_1=0.0 & \lambda_2=0.0 & \mu_1=6.0 & \mu_2=-2.45\,I & \dots \\ x=14.0 & y=11.0 & \lambda_1=0.0 & \lambda_2=0.0 & \mu_1=6.0 & \mu_2=2.45\,I & \dots \\ x=14.0 & y=11.0 & \lambda_1=0.0 & \lambda_2=0.0 & \mu_1=-6.0 & \mu_2=-2.45\,I & \dots \\ x=14.0 & y=11.0 & \lambda_1=0.0 & \lambda_2=0.0 & \mu_1=-6.0 & \mu_2=2.45\,I & \dots \\ x=11.0 & y=8.0 & \lambda_1=0.0 & \lambda_2=-6.0 & \mu_1=-4.90 & \mu_2=0.0 & \dots \\ x=11.0 & y=8.0 & \lambda_1=0.0 & \lambda_2=-6.0 & \mu_1=4.90 & \mu_2=0.0 & \dots \\ \vdots & \vdots & \vdots & \vdots & \vdots & \vdots & \ddots \end{bmatrix}$$

The first four lines displayed have μ_2 complex—we discard those, they are not feasible. The fifth line has μ_1 negative, this is also not feasible. The sixth line corresponds to the solution we've found. The remaining lines (*not shown here*) have λ_1 nonzero. Note, Maple's λs have an opposite signs to ours.

We close this example by once again pointing out the value of the objective function $f(x^*, y^*) = 18$ is a minimum because f is convex, the g_i are convex, and the λ_i are non-negative at (x^*, y^*). The shadow price for Constraint 2 is $\lambda_2 = 6$, and the slack in Constraint 1 is $\mu_1 = 4.9$.

Necessary and Sufficient Conditions for Computational KTC

Visual interpretation from graphs can significantly reduce the amount of work required to solve the problem. Interpreting the plot can provide the conditions involved at the optimal point, then we can solve directly for that point. However, often a graphical interpretation cannot be obtained, then we must rely on the computational method alone. When this occurs, we must solve all the cases and interpret the results.

Let's redo the previous example without any graphical analysis.

Example 4.26. Computational Constrained Nonlinear Optimization.

$$\begin{aligned} &\text{Maximize } z = (x-14)^2 + (y-11)^2 \\ &\text{subject to} \\ &\qquad (x-11)^2 + (y-13)^2 \le 49 \\ &\qquad\qquad\qquad\quad x + y \le 19 \end{aligned}$$

The Lagrangian is

$$L := (x, y, \boldsymbol{\lambda}, \boldsymbol{\mu}) \to f(x,y) + \lambda_1(g_1(x,y) + \mu_1^2 - 49) + \lambda_2(g_2(x,y) + \mu_2^2 - 19)$$

Therefore, the necessary conditions are

$$\partial L/\partial x \implies 2(x-14)+2\lambda_1(x-11)+\lambda_2 x=0 \tag{4.8a}$$
$$\partial L/\partial y \implies 2(y-11)+2\lambda_1(y-13)+\lambda_2 y=0 \tag{4.8b}$$
$$\partial L/\partial \lambda_1 \implies (x-11)^2+(y-13)^2+\mu_1^2-49=0 \tag{4.8c}$$
$$\partial L/\partial \lambda_2 \implies x+y+\mu_2^2-19=0 \tag{4.8d}$$
$$\partial L/\partial \mu_1 \implies 2\lambda_1\mu_1=0 \tag{4.8e}$$
$$\partial L/\partial \mu_2 \implies 2\lambda_2\mu_2=0 \tag{4.8f}$$

Define the functions in Maple, then investigate the cases for λ_i and μ_i being zero or nonzero. We'll use decimals for the constants to force floating-point arithmetic to simplify the calculations. Remember to load the *Student[MultivariateCalculus]* package to make the *Gradient* function available.

$$> f := (x,y) \to (x-14)^2+(y-11)^2 :$$
$$g_1 := (x,y) \to (x-11)^2+(y-13)^2 :$$
$$g_2 := (x,y) \to x+y :$$

$$> L := (x,y,\lambda,\mu) \to f(x,y)+\lambda_1\cdot(g_1(x,y)+\mu_1^2-49.0)$$
$$+\lambda_2\cdot(g_2(x,y)+\mu_2^2-19.0) :$$

$$> \mathit{grad} := \mathit{Gradient}(L(x,y,\lambda,\mu),[x,y,\lambda_1,\lambda_2,\mu_1,\mu_2]) :$$
$$\mathit{NecessaryConditions} := [\mathit{seq}(G=0, G \textbf{ in } \mathit{grad})] :$$

In each of the cases, we will solve for variables μ_i^2 rather than μ_i to reduce complexity. We'll also use *RealDomain:-solve* to avoid complex solutions. Defining the list of variables as *vars* reduces typing and simplifies the command structure making it more readable.

$$> \mathit{vars} := [x,y,\lambda_1,\lambda_2,\mu_1^2,\mu_2^2] :$$

CASE 1. $\lambda_1=0$ and $\lambda_2=0$. Then $\mu_1^2\neq 0$ and $\mu_2^2\neq 0$.

$$> \mathit{Case1} := \lambda_1=0, \lambda_2=0 :$$
$$\mathit{subs}(\mathit{Case1}, \mathit{NecessaryConditions});$$
$$s1 := \mathit{RealDomain{:-}solve}(\%, \mathit{vars}) :$$
$$\mathit{Matrix}(\mathit{subs}(\mathit{Case1}, s1));$$

$$[2x-28=0, 2y-22=0, (x-11)^2+(y-13)^2+\mu_1^2-49.0=0,$$
$$x+y+\mu_2^2-19.0=0, 0=0, 0=0]$$

$$\begin{bmatrix} x=14. & y=11. & 0=0 & 0=0 & \mu_1^2=36. & \mu_2^2=-6. \end{bmatrix}$$

This case is infeasible since $\mu_2^2 = -6$ which indicates that Condition 2, $g_2(x,y) \le 19$, is violated. It would have been easy to solve this necessary conditions system by hand. The first two necessary conditions give $x = 14$ and $y = 11$ directly, then substituting the values for x and y in the two constraints easily gives $\mu_1^2 = 36$ and $\mu_2^2 = -6$. Using Maple, however, provided a template for solving all four cases.

CASE 2. $\lambda_1 = 0$ and $\lambda_2 \neq 0$. Then $\mu_1^2 \neq 0$ and $\mu_2^2 = 0$.

```
> Case2 := λ1 = 0, μ2 = 0 :
  subs(Case2, NecessaryConditions);
  s2 := RealDomain:-solve(%, vars) :
  Matrix(subs(Case2, s2));
```

$$[2x - 28 + \lambda_2 = 0, 2y - 22 + \lambda_2 = 0, (x - 11)^2 + (y - 13)^2 + \mu_1^2 - 49.0 = 0, x + y - 19.0 = 0, 0 = 0, 0 = 0]$$

$$\begin{bmatrix} x = 11. & y = 8. & 0 = 0 & \lambda_2 = 6 & \mu_1^2 = 24. & 0 = 0. \end{bmatrix}$$

All the necessary conditions are satisfied at the point $(x^*, y^*) = (11, 8)$ where $\lambda_1 = 0$, $\lambda_2 = 6$, $\mu_1^2 = 24$, and $\mu_2^2 = 0$. (*Check this!*) The value of f at this point is 18. It is left as an exercise to show that $(11, 8)$ is an optimal point.

CASE 3. $\lambda_1 \neq 0$ and $\lambda_2 = 0$. Then $\mu_1^2 = 0$ and $\mu_2^2 \neq 0$.

```
> Case3 := λ1 = 0, μ2 = 0 :
  subs(Case3, NecessaryConditions);
  s3 := RealDomain:-solve(%, vars) :
  fnormal(Matrix(subs(Case3, s3)), 4);
```

$$[2x - 28 + \lambda_1(2x - 22) = 0, 2y - 22 + \lambda_1(2y - 26) = 0, (x - 11)^2 + (y - 13)^2 - 49.0 = 0, x + y + \mu_2^2 - 19.0 = 0, 0 = 0, 0 = 0]$$

$$\begin{bmatrix} x = 5.176 & y = 16.88 & \lambda_1 = -1.515 & 0. = 0. & 0. = 0. & \mu_2^2 = -3.059 \\ x = 16.82 & y = 9.117 & \lambda_1 = -0.4849 & 0. = 0. & 0. = 0. & \mu_2^2 = -6.941 \end{bmatrix}$$

This case is also infeasible as μ_2^2 is negative in both instances which indicates that Condition 2, g_2, is violated.

CASE 4. $\lambda_1 \neq 0$ and $\lambda_2 \neq 0$. Then $\mu_1^2 = 0$ and $\mu_2^2 = 0$.

```
> Case4 := μ1 = 0, μ2 = 0 :
  subs(Case4, NecessaryConditions);
  s4 := RealDomain:-solve(%, vars) :
  fnormal(Matrix(subs(Case4, s4)), 4);
```

$$[2x - 28 + \lambda_1(2 * x - 22) + \lambda_2 = 0, 2y - 22 + \lambda_1(2y - 26) + \lambda_2 = 0, (x - 11)^2 + (y - 13)^2 - 49.0 = 0, x + y - 19.0 = 0, 0 = 0, 0 = 0]$$

$$\begin{bmatrix} x = 12.77 & y = 6.228 & \lambda_1 = -0.4148 & \lambda_2 = 3.926 & 0. = 0. & 0. = 0. \\ x = 4.228 & y = 14.77 & \lambda_1 = -1.585 & \lambda_2 = -1.926 & 0. = 0. & 0. = 0. \end{bmatrix}$$

This case is again infeasible. The functional values are $f(12.77, 6.228) = 24.284$ and $f(4.228, 14.77) = 109.720$. These are not optimal values because they do not satisfy the sufficient conditions for λ_i for a relative minimum. Show that $f(4.228, 14.77)$ is a relative maximum. How does this happen?

Use Figure 4.21 which shows the contours of f, the constraints, and the six points found in the cases above to geometrically explain why each point appeared as a potential solution.

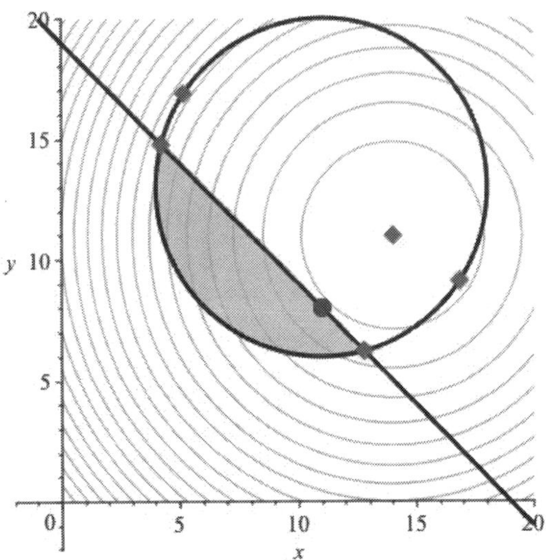

FIGURE 4.21: Infeasible Solutions and the Real Solution

Example 4.27. Computational Constrained Nonlinear Optimization Redux.

$$\begin{aligned} &\text{Maximize } z = 2x^2 - 8x + y^2 - 6y \\ &\text{subject to} \\ &\qquad x + y \le 4 \\ &\qquad\qquad y \le 2 \end{aligned}$$

We'll use the template we developed in the previous example: Choose from the four values of λ_1 and λ_2, substitute these into the necessary conditions system of equations, solve the system (using μ_i^2 as a variable), and analyze the solution.

```
> f := (x, y) → 2x^2 − 8x + y^2 − 6y :
  g1 := (x, y) → x + y :
  g2 := (x, y) → y :
> L := (x, y, λ, μ) → f(x, y) + λ1 · (g1(x, y) + μ1^2 − 4.0)
       + λ2 · (g2(x, y) + μ2^2 − 2.0) :
> grad := Gradient(L(x, y, λ, μ), [x, y, λ1, λ2, μ1, μ2]) :
> vars := [x, y, λ1, λ2, μ1^2, μ2^2] :
```

> $NecessaryConditions := [seq(G = 0, G \textbf{ in } grad)];$

$$\begin{bmatrix} 4\,x - 8 + \lambda_1 = 0 \\ 2\,y - 6 + \lambda_1 + \lambda_2 = 0 \\ x + y + {\mu_1}^2 - 4.0 = 0 \\ y + {\mu_2}^2 - 2.0 = 0 \end{bmatrix}$$

We're ready to begin working through the four cases.

> $Case1 := \lambda_1 = 0, \lambda_2 = 0 :$
> $subs(Case1, NecessaryConditions);$
> $s1 := RealDomain\text{:-}solve(\%, vars) :$
> $Matrix(subs(Case1, s1));$

$$[4x - 8 = 0, 2y - 6 = 0, x + y + \mu_1^2 - 4.0 = 0, y + \mu_2^2 - 2.0 = 0]$$
$$\begin{bmatrix} x = 2. & y = 3. & 0 = 0 & 0 = 0 & \mu_1^2 = -1. & \mu_2^2 = -1. \end{bmatrix}$$

The point $(2, 3)$ is not a solution; μ_1^2 and μ_2^2 are both negative so both constraints are violated.

> $Case2 := \lambda_1 = 0, \mu_2^2 = 0 :$
> $subs(Case2, NecessaryConditions);$
> $s2 := RealDomain\text{:-}solve(\%, vars) :$
> $Matrix(subs(Case2, s2));$

$$[4x - 8 = 0, 2y - 6 + \lambda_2 = 0, x + y + \mu_1^2 - 4.0 = 0, y - 2.0 = 0]$$
$$\begin{bmatrix} x = 2. & y = 2. & 0 = 0 & \lambda_2 = 2. & \mu_1^2 = 0. & 0 = 0. \end{bmatrix}$$

All necessary conditions are satisfied. Since the objective function is convex, the constraints are linear (and therefore convex), and λ_i for all binding constraints are positive ($\lambda_2 = 2$), the sufficient conditions are met. This solution is the only solution that satisfies all the necessary and the sufficient conditions. Thus $(2, 3)$ is the optimal solution.

> $Case3 := \lambda_1 = 0, \mu_2^2 = 0 :$
> $subs(Case3, NecessaryConditions);$
> $s3 := RealDomain\text{:-}solve(\%, vars) :$
> $fnormal(Matrix(subs(Case3, s3)), 4);$

$$[4x - 8 + \lambda_1 = 0, 2y - 6 + \lambda_1 = 0, x + y - 4.0 = 0, y + \mu_2^2 - 2.0 = 0]$$
$$\begin{bmatrix} x = 1.667 & y = 2.333 & 0 = 0 & \lambda_1 = 1.333 & 0 = 0. & \mu_2^2 = -0.333 \end{bmatrix}$$

The point $(1.667, 2.333)$ is not an optimal solution as μ_2^2 is negative violating Constraint 2.

```
> Case4 := μ1^2 = 0, μ2^2 = 0 :
  subs(Case4, NecessaryConditions);
  s4 := RealDomain:-solve(%, vars) :
  Matrix(subs(Case4, s4));
```

$$[4x - 8 + \lambda_1 = 0, 2y - 6 + \lambda_1 + \lambda_2 = 0, x + y - 4.0 = 0, y - 2.0 = 0]$$

$$\begin{bmatrix} x = 2. & y = 2. & 0 = 0 & \lambda_2 = 2. & \mu_1^2 = 0. & 0 = 0. \end{bmatrix}$$

Case IV yields the same solution as Case II.

Use Figure 4.22 which shows the contours of f, the constraints, and the points found in the cases above to geometrically explain why each point appeared as a potential solution.

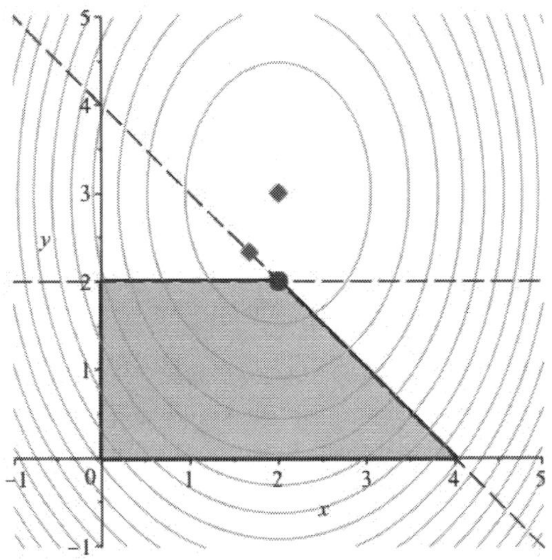

FIGURE 4.22: Infeasible Solutions and the Real Solution Redux

Applications Using the Kuhn-Tucker Conditions Method

Example 4.28. Maximizing Profit from Perfume Manufacturing.
A company manufactures perfumes. They can purchase up to 1,925 oz of the main chemical ingredient at \$10 per oz. An ounce of the chemical can produce an ounce of Perfume #1 with a processing cost of \$3 per oz. On the other hand, the chemical can produce an ounce of the higher priced Perfume #2 with a processing cost of \$5 per oz. The Analytics department used historical data to estimate that Perfume #1 will sell for \$$(30 - 0.01x)$ per ounce if x ounces are manufactured, and Perfume #2 can sell for \$$(50 - 0.02y)$ per ounce if y ounces are manufactured. The company wants to maximize profits.

MODEL FORMULATION. Let

$$x = \text{ounces of Perfume \#1 produced}$$
$$y = \text{ounces of Perfume \#2 produced}$$
$$z = \text{ounces of main chemical purchased}$$

Then

$$\text{Maximize } P(x,y,z) = \underbrace{x\cdot(30-0.01x)+y\cdot(50-0.02y)}_{Revenue} - \underbrace{(3x+5y+10z)}_{Cost}$$

subject to

$$x + y \le z$$
$$z \le 1925$$

SOLUTION. Set up the model, define the Lagrangian function L, and use the techniques from the previous examples.

$> P := (x,y,z) \to x\cdot(30-0.01x)+y\cdot(50-0.02y)-(3x+5y+10z):$
$g_1 := (x,y,z) \to x+y-z:$
$g_2 := (x,y,z) \to z:$

$> L := (x,y,z,\lambda,\mu) \to f(x,y,z)+\lambda_1\cdot(g_1(x,y,z)+\mu_1^2-0.0)$
$+\lambda_2\cdot(g_2(x,y,z)+\mu_2^2-1925.0):$

$> grad := Gradient(L(x,y,z,\lambda,\mu),[x,y,z,\lambda_1,\lambda_2]):$

$> vars := [x,y,z,\lambda_1,\lambda_2,\mu_1^2,\mu_2^2]:$

$> NecessaryConditions := [seq(G=0, G \textbf{ in } grad)]:$
$\langle NecessaryConditions \rangle;$

$$\begin{bmatrix} 27-0.02\,x+\lambda_1=0 \\ 45-0.04\,y+\lambda_1=0 \\ -10-\lambda_1+\lambda_2=0 \\ {\mu_1}^2+x+y-z=0 \\ z+{\mu_2}^2-1925.0=0 \end{bmatrix}$$

Once again, we're ready to begin working through the four cases. A summary of the results from Maple is shown in Table 4.9.

TABLE 4.9: Perfume Application's Four Cases

CASE	x	y	z	λ_1	λ_2	μ_1^2	μ_2^2
I.	–	–	–	0	0	–	–
II.	1350	1125	1925	0	10	−550	0
III.	850	875	1725	−10	0	0	200
IV.	983.33	941.67	1925	−7.333	2.667	0	0

Remarks:

CASE I. $\partial L/\partial z = 0$ becomes $-10 = 0$; this case is infeasible.

CASE II. μ_1^2 is negative violating Constraint 1.

CASE III. $\lambda_i \leq 0$ and $\mu_i \geq 0$: **Candidate Solution.**

CASE IV. λ_is have different signs so not an optimal point.

We have a concave profit function P, linear constraints, and λ_i ($\lambda_1 = -10$) is negative for all binding constraints. Thus, we have met the sufficient conditions for the point $(850, 875)$ to be the optimal solution. The optimal manufacturing strategy is to purchase $z = 1725$ ounces of the chemical and produce $x = 850$ ounces of Perfume #1 and $y = 875$ ounces of Perfume #2 yielding a profit of $P = \$22{,}537.50$.

Consider the significance of the shadow price for λ_1. How do we interpret the shadow price in terms of this scenario? If we could obtain an extra ounce ($\Delta = 1$) of the main chemical at no cost, it would improve the profit to about \$22,37.50 + \$10. What would be the largest cost for an extra ounce of the main chemical that would still yield a higher profit?

Example 4.29. Minimum Variance of Expected Investment Returns. A new company has \$5,000 to invest to generate funds for a planned project; the company needs to earn about 12% interest. A stock expert has suggested three mutual funds, A, B, and C, in which the company could invest. Based upon previous year's returns, these funds appear relatively stable. The expected return, variance on the return, and covariance between funds are shown in Table 4.10 below.

TABLE 4.10: Mutual Fund Investment Data

	A	B	C
Expected Value	0.14	0.11	0.10
Variance	0.2	0.08	0.18
	AB	AC	BC
Covariance	0.05	0.02	0.03

We use laws of expected value, variance, and covariance in our model.

MODEL FORMULATION. Let x_j be the number of dollars invested in fund j for $j = 1, 2, 3$, representing Funds A, B, and C. Our objective is to minimize the variance of the investment, so

$$\begin{aligned} V &= \text{var}(x_1 A + x_2 B + x_3 C) \\ &= x_1^2 \text{var}(A) + x_2^2 \text{var}(B) + x_3^2 \text{var}(C) + 2x_1x_2 \text{cov}(AB) + 2x_1x_3 \text{cov}(AC) \\ &\quad + 2x_2x_3 \text{cov}(BC) \\ &= 0.20x_1^2 + 0.08x_2^2 + 0.18x_3^2 + 0.10x_1x_2 + 0.04x_1x_3 + 0.06x_2x_3 \end{aligned}$$

Our constraints are: g_1, the expected return is at least 12%, and g_2, the sum of the investments is no more than \$5000. We have the NLP

$$\text{Minimize } V = 0.20x_1^2 + 0.08x_2^2 + 0.18x_3^2 + 0.10x_1x_2 + 0.04x_1x_3 + 0.06x_2x_3$$

subject to

$$\begin{aligned} 0.14x_1 + 0.11x_2 + 0.10x_3 &\geq 0.12 \cdot 5000 = 600 \\ x_1 + x_2 + x_3 &\leq 5000 \end{aligned}$$

SOLUTION. Set up the Lagrangian function L.

$$L(\mathbf{x}, \boldsymbol{\lambda}, \boldsymbol{\mu}) = f(\mathbf{x}) + \lambda_1(g_1(\mathbf{x}) - \mu_1^2 - 600) + \lambda_2(g_2(\mathbf{x}) + \mu_2^2 - 5000)$$

(*Why is μ_1^2 subtracted rather than added?*)

Define the functions for the model and calculate the necessary conditions.

```
> with(Student[MultivariateCalculus]) :
```

```
> V = x → 0.20x1^2 + 0.08x2^2 + 0.18x3^2 + 0.10x1x2 + 0.04x1x3 + 0.06x2x3 :
  g1 := x → 0.14x1 + 0.11x2 + 0.10x3 :
  g2 := x → x1 + x2 + x3 :
```

```
> grad := Gradient(L(x, λ, μ), [x1, x2, x3, λ1, λ2]) :
```

Notice that we have again left μ_1 and μ_2 out of the list of variables—the complementary slackness conditions are taken care of by considering our standard four cases ($4 = 2^{(number\ of\ constraints)}$).

> $NecessaryConditions := [seq(G = 0, G \textbf{ in } grad)]:$
> $\langle NecessaryConditions \rangle;$

$$\begin{bmatrix} 0.40\,x_1 + 0.10\,x_2 + 0.04\,x_3 + 0.14\,\lambda_1 + \lambda_2 = 0 \\ 0.16\,x_2 + 0.10\,x_1 + 0.06\,x_3 + 0.11\,\lambda_1 + \lambda_2 = 0 \\ 0.36\,x_3 + 0.04\,x_1 + 0.06\,x_2 + 0.10\,\lambda_1 + \lambda_2 = 0 \\ 0.14\,x_1 + 0.11\,x_2 + 0.10\,x_3 - {\mu_1}^2 - 600.0 = 0 \\ x_1 + x_2 + x_3 + {\mu_2}^2 - 5000.0 = 0 \end{bmatrix}$$

Let Maple do the work.

> $s := LagrangeMultipliers(V(x), [g_1(x) - \mu_1^2 - 600, g_2(x) + \mu_2^2 - 5000],$
> $[x_1, x_2, x_3, \lambda_1, \lambda_2, \mu_1, \mu_2], output = detailed):$
> $S := fnormal([s], 4):$

The output would have overflowed the page and would be very hard to read, so we'll use some "Maple Magic" to put it in an easier-to-handle form.

> $Lgnd := \langle op(map(lhs, S[1, 1..7])) \,|\, V \rangle :$
> $Sdata := map(rhs, Matrix(S)[.., [1..7, 10]]):$
> $\langle Lgnd, Sdata \rangle;$

x_1	x_2	x_3	λ_1	λ_2	μ_1	μ_2	V
1905.0	2381.0	714.3	13810.0	−904.8	0.0	0.0	1881000.0
702.2	3118.0	1180.0	0.0	639.9	−6.382 *I*	0.0	1600000.0
702.2	3118.0	1180.0	0.0	639.9	6.382 *I*	0.0	1600000.0
0.0	0.0	0.0	0.0	0.0	−24.49 *I*	−70.71	0.0
0.0	0.0	0.0	0.0	0.0	−24.49 *I*	70.71	0.0
0.0	0.0	0.0	0.0	0.0	24.49 *I*	−70.71	0.0
0.0	0.0	0.0	0.0	0.0	24.49 *I*	70.71	0.0
1250.0	2929.0	1028.0	5958.0	0.0	0.0	−14.39 *I*	1787000.0
1250.0	2929.0	1028.0	5958.0	0.0	0.0	14.39 *I*	1787000.0

Inspect the results to see that only the first row has a feasible solution; the rest have a negative μ_i^2 (imaginary μ_i). Notice that, in row 1, both $\lambda_i > 0$, that is, both constraints are binding. Then

> $s_1[1..3];$
> $Optimum := eval(V(x), s_1);$

$$[x_1 = 1904.761905, x_2 = 2380.952381, x_3 = 714.2857143]$$
$$1.880952381\,10^6$$

Check the Hessian.

> $H := Student\text{:-}VectorCalculus\text{:-}Hessian(V(x), [x_1, x_2, x_3]);$

$$\begin{bmatrix} 0.40 & 0.10 & 0.04 \\ 0.10 & 0.16 & 0.06 \\ 0.04 & 0.06 & 0.36 \end{bmatrix}$$

The Hessian matrix H has all positive leading principal minors. Therefore, since H is always positive definite, then our solution is the optimal minimum. The expected return is 12.0% found from $g_1(\mathbf{x})/5000 = (0.14 \cdot 1904.8 + 0.11 \cdot 2381. + 0.10 \cdot 714.2)/5000.$

Exercises

1. Solve the following constrained problems using the Kuhn-Tucker Conditions (KTC) approach.

   (a) Minimize $z = x^2 + y^2$ subject to $x + 2y = 4$.

   (b) Maximize $z = (x-3)^2 + (y-2)^2$ subject to $x + 2y = 4$.

   (c) Maximize $z = x^2 + 4xy + y^2$ subject to $x^2 + y^2 = 1$.

   (d) Maximize $z = x^2 + 4xy + y^2$ subject to $x^2 + y^2 = 4$ and $x + 2y = 4$.

2. Maximize $Z = 3X^2 + Y^2 + 2XY + 6X + 2Y$ subject to $2X - Y = 4$. Did you find the maximum? Explain.

3. Two manufacturing processes, f_1 and f_2, both use a resource with b units available. Maximize $f_1(x_1) + f_2(x_2)$ subject to $x_1 + x_2 = b$.

   If $f(1(x_1) = 50 - (x_1 - 2)^2$ and $f_2(x_2) = 50 - (x_2 - 2)^2$, analyze the manufacturing processes using the KTC approach to

   (a) determine the amount of x_1 and x_2 to use to maximize $f_1 + f_2$,

   (b) determine the amount b of the resource to use.

4. Use the Kuhn-Tucker Conditions to find the optimal solution to the following nonlinear problems.

   (a) Maximize $f(x, y) = -x^2 - y^2 + xy + 7x + 4y$ subject to $2x + 3y \geq 16$ and $-5x + 12y \leq 20$.

(b) Minimize $f(x, y) = 2x + xy + 3y$ subject to $x^2 + y \geq 3$ and $2.5 - 0.5x - y \leq 0$.

(c) Minimize $f(x, y) = 2x + xy + 3y$ subject to $x^2 + y \geq 3,\ x + 0.5 \geq 0$, and $y \geq 0$.

(d) Maximize $f(x, y) = -(x - 0.4)^2 - (y - 5)^2$ subject to $-x + 2y \leq 4$, $x^2 + y^2 \leq 14$, and $x, y \geq 0$.

5. Minimize $z = x^2 + y^2$ subject to

$$2x + y \leq 100$$
$$x + y \leq 80$$

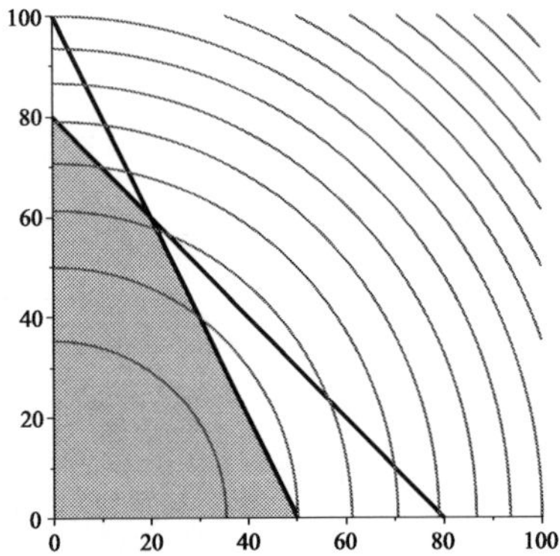

6. Maximize $z = -(x-4)^2 + xy - (y-4)^2$ subject to

$$2x + 3y \leq 18$$
$$2x + y \leq 8$$

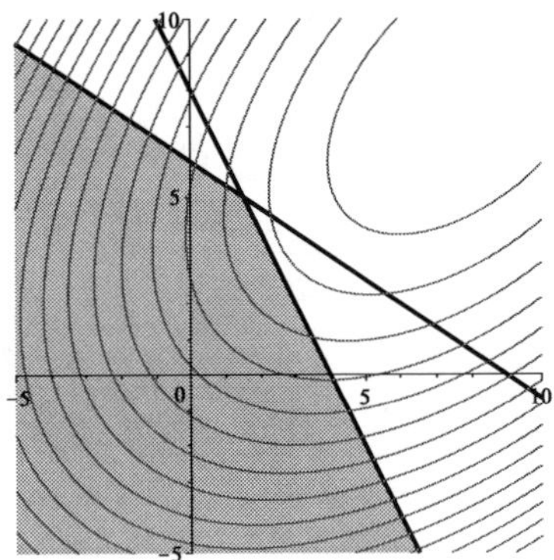

Projects

Project 4.1. A newspaper publisher must purchase three types of paper stock. The publisher must meet the demand, but desires to minimize costs in the process. An *Economic Lot Size* model is chosen to assist them in their decisions. An *Economic Order Quantity* (EOQ) model with constraints where the total cost is the sum of the individual quantity costs is $C(Q_1, Q_2, Q_3) = C(Q_1) + C(Q_2) + C(Q_3)$ where

$$C(Q_i) = a_i d_i / Q_i + h_i \cdot (Q_i/2)$$

where

$$\begin{aligned} d_i &= \text{the order rate} \\ h_i &= \text{the cost per unit time for storage (holding cost)} \\ Q_i/2 &= \text{is the average inventory/amount on hand} \\ a &= \text{is the cost of placing an order} \end{aligned}$$

The constraint is the amount of storage area available to the publisher for storing the three kinds of paper. The items cannot be stacked, but can be laid side by side. The available storage area is $S = 200$ sq ft.

Table 4.11 shows data that has been collected on the publisher's usage and costs.

TABLE 4.11: Paper Usage and Cost Data

	TYPE I.	TYPE II.	TYPE III.	*units*
d_i	32	24	20	*rolls/week*
a_i	25	18	20	*dollars*
h_i	1.00	1.50	2.00	*dollars*
s_i	4	3	2	*sq ft/roll*

REQUIRED.

(a) Find the paper quantities that give the unconstrained minimum total cost, and show that these values would not satisfy the constraint. What purpose do these values serve?

(b) Find the constrained optimal solution by using Lagrange Multipliers assuming all 200 sq feet is used.

(c) Determine and interpret the shadow prices.

Project 4.2. In the tank storage problem, Example 4.22 (pg. 175), determine whether it is better to have cylindrical storage tanks or rectangular storage tanks of 50 cubic units.

Project 4.3. Use the Cobb-Douglas function $P(L, K) = \alpha L^a K^b$ where L is labor and K is capital, to predict output in thousands, based upon amount of labor and capital used. Suppose the price of capital and labor per year are \$10,000 and \$7,000, respectively. The company estimates the values of α as 1.2, $a = 0.3$, and $b = 0.6$. Your total cost is assumed to be $T = P_L L + P_k k$, where P_L and P_k are the price of capital and labor. There are three possible funding levels: \$63,940, \$55,060, or \$71,510. Determine which budget yields the best solution for the company. Interpret the Lagrange multiplier.

References and Further Reading

[BSS2013] Mokhtar S. Bazaraa, Hanif D. Sherali, and Chitharanjan M. Shetty, *Nonlinear Programming: Theory and Algorithms*, Wiley, 2013.

[EK1988] J. Ecker and M. Kupferschmid, *Introduction to Operations Research*, John Wiley & Sons Inc., 1988.

[HL2009] Frederick Hillier and G Lieberman, *Introduction to operations research*, vol. 9, McGraw-Hill Science/Engineering/Math, 2009.

[W2002] W. L. Winston, *Introduction to mathematical programming applications and algorithms*, 4th ed., Duxbury Press, 2002.

5

Problem Solving with Linear Systems of Equations Using Linear Algebra Techniques

Objectives:

(1) Set up a system of equations to solve a problem.

(2) Recognize unique solutions, no solution and infinite solutions as results.

5.1 Introduction

In the mid-1800s, a large number of ironwork bridges were constructed as railways crossed the continents. One of the popular designs for trusses, rigid triangular support structures, was the Warren truss with vertical supports.[1] The center span of the 1926 Bridge of the Gods (Figure 5.1) over the Columbia River is a nice example.

FIGURE 5.1: Bridge of the Gods Warren Truss with Vertical Supports

[1]See Frank Griggs' "The Warren Truss," *Structure*, July, 2015, for a brief history of the Warren truss; available at https://www.structuremag.org/?p=8715.

Warren trusses supported carrying heavy loads. The civil engineering technique "Method of Joints" is used to analyze the forces acting on the truss. Individual parts of the truss are connected with rivets, rotatable pin joints, or welds that permit forces to be transferred from one member of the truss to another.

Figure 5.2 shows a truss that is fixed at the lower left endpoint p_1, and can move horizontally at the lower right endpoint p_4. The truss has pin-joints at p_1, p_2, p_3, and p_4. A load of 10 kilonewtons (kN) is placed at joint p_3; the forces on the members of the truss have magnitude f_1, f_2, f_3, f_4, and f_5 as indicated in the figure. The stationary support member has both a horizontal force F_1 and a vertical force F_2 (see Figure 5.3); the horizontally movable support member has only a vertical force F_3.[2]

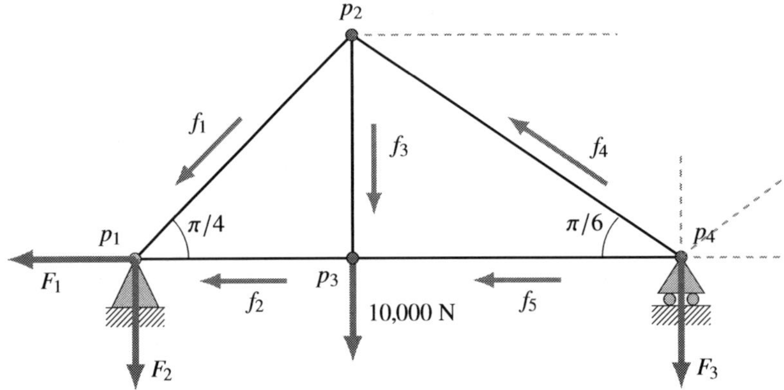

FIGURE 5.2: A Warren Truss with Vertical Supports

If the truss is in static equilibrium, the forces at each joint must sum to the zero vector. If there were net nonzero forces, the joint would move—the truss would not be in static equilibrium. Therefore, the corresponding components of the vector must also be zero; i.e., the sum of the horizontal components of the forces at each joint must be zero and the sum of the vertical components must be zero.

[2]Variants of this problem appear in many textbooks; see, e.g., Burden and Faires [BurdenFaires2005], Fox [Fox2011Maple].

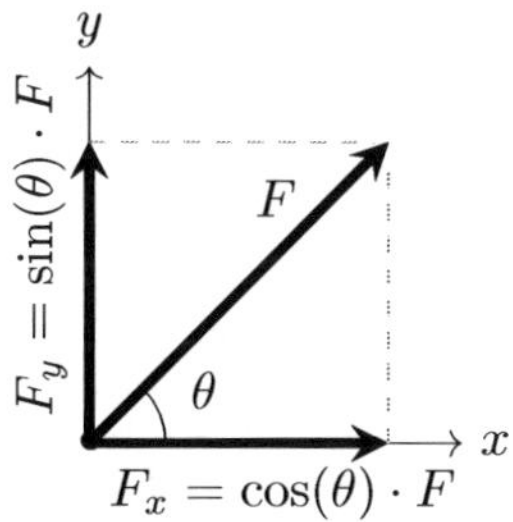

FIGURE 5.3: Force Vector Components

A system of linear equations will model the forces. We will build this model and solve its linear system later in the chapter.

5.2 Introduction to Systems of Equations

In this chapter, we illustrate the use of systems of equations to solve real-world applications from the sciences, engineering, and economics. Previously, we have seen examples of linear systems in solving discrete dynamical systems and in model fitting with least squares. Maple's *LinearAlgebra* package has a number of functions and programs that will be very useful for solving these problems.

There are exactly three possibilities for a system of linear equations: the system has a unique solution, an infinite number of solutions, or no solution. Consider the following two-dimensional system of linear equations:

$$\begin{aligned} a_1x + b_1y &= c_1 \\ a_2x + b_2y &= c_2 \end{aligned}$$

The augmented matrix for this system is

$$A = \left[\begin{array}{cc|c} a_1 & b_1 & c_1 \\ a_2 & b_2 & c_2 \end{array}\right]$$

Row reduce the augmented matrix to find one of the three possible results.

Unique Solution. Row reduction of A gives

$$\left[\begin{array}{cc|c} 1 & 0 & d \\ 0 & 1 & e \end{array}\right]$$

which yields the unique solution $(x, y) = (d, e)$.

Infinitely Many Solutions. Row reduction of A gives one of three possibilities. First,

$$\left[\begin{array}{cc|c} 1 & b_1/a_1 & c_1/a_1 \\ 0 & 0 & 0 \end{array}\right]$$

which yields solutions $(x, y) = (a_1 t, c_1 - b_1 t)$ with arbitrary t. Second, if $a_1 = 0$ and $b_1 \neq 0$, then row reduction gives

$$\left[\begin{array}{cc|c} 0 & 1 & d \\ 0 & 0 & 0 \end{array}\right],$$

and the solution is (t, d) for arbitrary t. Third, if $a_1 \neq 0$ and $b_1 = 0$, then row reduction gives

$$\left[\begin{array}{cc|c} 1 & 0 & d \\ 0 & 0 & 0 \end{array}\right],$$

and the solution is (d, t) for arbitrary t.

No Solution. Row reduction gives

$$\left[\begin{array}{cc|c} 1 & h & d \\ 0 & 0 & e \end{array}\right]$$

with $e \neq 0$. Then there is no solution as $0x + 0y = e \neq 0$ is impossible.

These alternatives are represented visually in Figure 6.4.

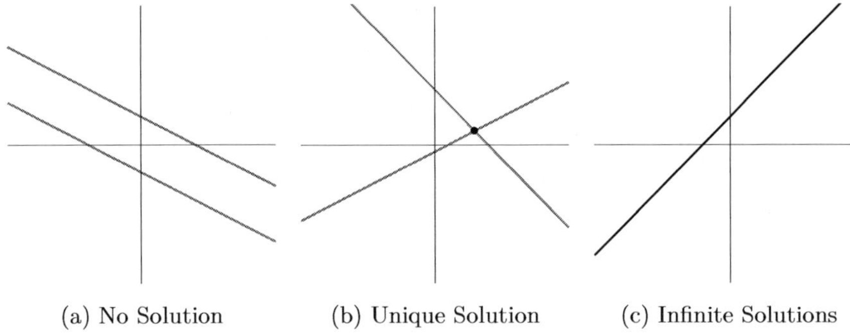

(a) No Solution (b) Unique Solution (c) Infinite Solutions

FIGURE 5.4: The Three Alternatives for Solutions of a Linear System

Describe the possibilities for three linear equations in three dimensions; that is, for the intersection of three planes in 3D.

We will rely on the matrix forms in recognizing our solutions in this chapter.

The following description is from the *Details of the LinearAlgebra Package* Help page.[3]

[3]Enter *help("LinearAlgebra, Details")* for the full description.

- Maple's **LinearAlgebra** package is an efficient and robust suite of routines for doing computational linear algebra.
- Full rectangular and sparse matrices are fully supported at the data structure level,[4] as well as upper and lower triangular matrices, unit triangular matrices, banded matrices, and a variety of others. Further, **symmetric**, **skew-symmetric**, **hermitian**, and **skew-hermitian** are known qualifiers that are used appropriately to reduce storage and select amongst algorithms.

According to V. Z. Aladjev,

> The **LinearAlgebra** module has been designed to accommodate different sets of usage scenarios: casual use and programming use. Correspondingly, there are functions and notations designed for easy casual use (*sometimes at the cost of some efficiency*), and some functions designed for maximal efficiency (*sometimes at the cost of ease-of-use*). In this way, the **LinearAlgebra** facilities scale easily from first-year classroom use to heavy industrial usage, emphasizing the different qualities that each type of use needs.[5]

The *LinearAlgebra* commands we will use most are:

GaussianElimination: performs Gaussian elimination on the matrix A and returns an upper triangular matrix U.

ReducedRowEchelonForm: performs Gaussian elimination on the matrix A and returns the reduced row echelon form of A.

(*Note for those who have studied linear algebra:* Both of these commands use *LinearAlgebra's LUDecomposition* function to determine the upper triangular factor of A.)

Recall there are many ways to enter a matrix. One of the easiest is to use Maple's *Matrix palette.* Enter *help("worksheet, matpalette")* for details.

Begin with a simple 3×3 system of linear equations.

Example 5.1. A 3×3 System.
Determine the solution of the following system of linear equations.

$$\begin{aligned} 2x + 4y + 4z &= 4 \\ x + 3y + z &= 4 \\ -x + 3y + 2z &= -1 \end{aligned}$$

---

[4]For those interested in the internal structure of matrices, etc., enter the command *help("linearalgebra,about,data")*.

[5]V. Z. Aladjev, *Computer Algebra Systems: A New Software Toolbox for Maple*, Fultus Corp., 2004, pg. 451.

Load the linear algebra package, enter the system's matrix in augmented form, row reduce the matrix using Gaussian elimination, and then interpret the results.

```
> with(|MCLinearAlgebra) :
> A := <2,1,-1>|<4,3,3>|<4,1,2>|<4,4,-1>;
```

$$A := \begin{bmatrix} 2 & 4 & 4 & 4 \\ 1 & 3 & 1 & 4 \\ -1 & 3 & 2 & -1 \end{bmatrix}$$

```
> ReducedRowEchelonForm(A);
```

$$\begin{bmatrix} 1 & 0 & 0 & 2 \\ 0 & 1 & 0 & 1 \\ 0 & 0 & 1 & -1 \end{bmatrix}$$

We interpret the results as a unique solution: $(x, y, z) = (2, 1, -1)$.

Let's increase the dimension by one.

Example 5.2. A 4×4 System.
Determine the solution of the following system of linear equations.

$$\begin{aligned} 2x + 4y + 4z + 2w &= 2 \\ 2x + 2y + z - w &= -1 \\ x + 4y - z - 2w &= 1 \\ -x + 2y + w &= 1 \end{aligned}$$

```
> B := Matrix([[2,4,4,2,2],[2,2,1,-1,-1],[1,4,-1,-2,1],[-1,2,0,1,1]]);
```

$$B := \begin{bmatrix} 2 & 4 & 4 & 2 & 2 \\ 2 & 2 & 1 & -1 & -1 \\ 1 & 4 & -1 & -2 & 1 \\ -1 & 2 & 0 & 1 & 1 \end{bmatrix}$$

```
> ReducedRowEchelonForm(B);
```

$$\begin{bmatrix} 1 & 0 & 0 & -1 & 0 \\ 0 & 1 & 0 & 0 & 0 \\ 0 & 0 & 1 & 1 & 0 \\ 0 & 0 & 0 & 0 & 1 \end{bmatrix}$$

We interpret the results: The system has no solution as the last row of the row reduced augmented matrix gives the equation $0 = 1$.

Exercises

Use Maple to solve the following system of equations and interpret the results.

1. $$\begin{aligned} x - 5y &= -154 \\ x - 3y &= -84 \end{aligned}$$

2. $$\begin{aligned} 11x - 6y &= 494 \\ x + 7y &= -23 \end{aligned}$$

3. $$\begin{aligned} 9x + y &= 56 \\ 6x - 5y &= 128 \end{aligned}$$

4. $$\begin{aligned} 6x + y &= 50 \\ 18x + 3y &= 150 \end{aligned}$$

5. $$\begin{aligned} x + 2y + 3z &= 5 \\ x - y + 6z &= 1 \\ 3x - 2y &= 4 \end{aligned}$$

6. $$\begin{aligned} x + y + 3z - 4w &= 12 \\ 3x + y - 2z - w &= 0 \end{aligned}$$

7. $$\begin{aligned} 2x + 3y + 4z &= 5 \\ x - y + 2z &= 6 \\ 3x - 5y - z &= 0 \end{aligned}$$

8. $$\begin{aligned} 2x - 3y + 2z &= 21 \\ x + 4y - z &= 1 \\ -x + 2y + z &= 17 \end{aligned}$$

9. $$\begin{aligned} 3x - 4y + 5z - 4w &= 12 \\ x - y + z - 2w &= 0 \\ 2x + y + 2z + 3w &= 52 \\ 2x - 2y + 2z - 3w &= 1 \end{aligned}$$

Projects

Project 5.1. Model a solution methodology for $A\mathbf{x} = \mathbf{b}$ using linear algebra and illustrate your method with Maple commands.

Project 5.2. Let A be a random $n \times (n+1)$ matrix representing a linear system of n linear equations in n variables. Estimate the probability that the system has a unique solution when

(a) $n = 3$.

(b) $n = 4$.

(c) $n = 5$.

(d) $n > 5$.

5.3 Models with Unique Solutions Using Systems of Linear Equations

Let's use the method of joints to analyze the Warren truss from the Introduction.

Example 5.3. Analysis of a Truss by the Method of Joints.
A segment of a Warren truss with vertical supports that is fixed at the lower left endpoint p_1, and can move horizontally at the lower right endpoint p_4 is displayed in Figure 5.5. The pin-joints of the truss are at p_1, p_2, p_3, and p_4. A load of 10 kilonewtons (kN) is placed at joint p_3; the forces on the truss with magnitudes f_1, f_2, f_3, f_4, and f_5 are indicated in the figure. The stationary support member has a horizontal force F_1 and a vertical force F_2; the horizontally movable support member has only a vertical force F_3.

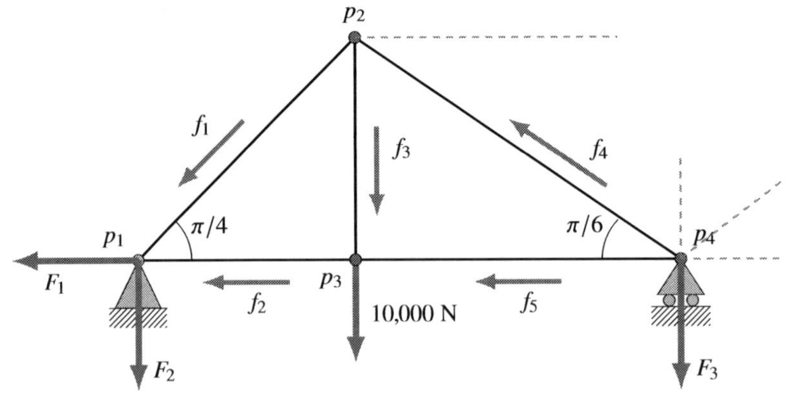

FIGURE 5.5: A Warren Truss with Vertical Supports Segment

Since the truss is in static equilibrium, Newton's Law for equilibrium of forces implies that the sum of the force vectors at each joint must be the zero vector. Therefore, the sum of the horizontal components of the forces and the sum of the vertical components must each be zero.

Begin with joint p_1, the fixed joint at the lower left of the truss. Both f_1 and f_2 are forces into p_1, while forces F_1 and F_2 pull away from p_1. Force f_1 acts at angle of $\pi/4$ which resolves into the horizontal component $(\sqrt{2}/2)f_1$ and vertical component $(\sqrt{2}/2)f_1$. These forces are at equilibrium, so that

$$-F_1 + \frac{\sqrt{2}}{2}f_1 + f_2 = 0 \qquad \text{and} \qquad -F_2 + \frac{\sqrt{2}}{2}f_1 = 0.$$

Next, we consider joint p_2 at the top of the truss segment. Both f_1 and f_3 pull away, while f_4 acts into the joint. Resolving the angular forces and summing gives

$$-\frac{\sqrt{2}}{2}f_1 + \frac{\sqrt{3}}{2}f_3 = 0 \qquad \text{and} \qquad -\frac{\sqrt{2}}{2}f_1 - f_3 + \frac{1}{2}f_4 = 0.$$

Complete the model for the forces at each joint and place the force vector equations in Table 5.1. Verify that these are correct!

TABLE 5.1: Forces on the Truss's Joints

Joint	Horizontal Component	Vertical Component
p_1	$-F_1 + \frac{\sqrt{2}}{2}f_1 + f_2 = 0$	$-F_2 + \frac{\sqrt{2}}{2}f_1 = 0$
p_2	$-\frac{\sqrt{2}}{2}f_1 + \frac{\sqrt{3}}{2}f_4 = 0$	$-\frac{\sqrt{2}}{2}f_1 - f_3 + \frac{1}{2}f_4 = 0$
p_3	$-f_2 + f_5 = 0$	$f_3 - 10000 = 0$
p_4	$-\frac{\sqrt{3}}{2}f_4 - f_5 = 0$	$-\frac{1}{2}f_4 - F_3 = 0$

Write this model as a system of equations in matrix notation $A\mathbf{x} = \mathbf{b}$ with 8 equations in 8 unknowns:

$$\begin{bmatrix} -1 & 0 & 0 & \sqrt{2}/2 & 1 & 0 & 0 & 0 \\ 0 & -1 & 0 & \sqrt{2}/2 & 0 & 0 & 0 & 0 \\ 0 & 0 & 0 & -\sqrt{2}/2 & 0 & 0 & \sqrt{3}/2 & 0 \\ 0 & 0 & 0 & -\sqrt{2}/2 & 0 & -1 & 1/2 & 0 \\ 0 & 0 & 0 & 0 & -1 & 0 & 0 & 1 \\ 0 & 0 & 0 & 0 & 0 & 1 & 0 & 0 \\ 0 & 0 & 0 & 0 & 0 & 0 & -\sqrt{3}/2 & -1 \\ 0 & 0 & -1 & 0 & 0 & 0 & -1/2 & 0 \end{bmatrix} \begin{bmatrix} F_1 \\ F_2 \\ F_3 \\ f_1 \\ f_2 \\ f_3 \\ f_4 \\ f_5 \end{bmatrix} = \begin{bmatrix} 0 \\ 0 \\ 0 \\ 0 \\ 0 \\ 10000 \\ 0 \\ 0 \end{bmatrix}$$

Enter the augmented matrix, and then reduce the matrix using *ReducedRowEchelonForm*. Interpret the results to solve the problem.

As this matrix is large (for entering by hand, that is), use the *Matrix palette* to enter it. Set the size to 8×9, click the "Insert Matrix" button, then fill in the entries.

$$> T := \begin{bmatrix} -1 & 0 & 0 & \frac{\sqrt{2}}{2} & 1 & 0 & 0 & 0 & 0 \\ 0 & -1 & 0 & \frac{\sqrt{2}}{2} & 0 & 0 & 0 & 0 & 0 \\ 0 & 0 & 0 & -\frac{\sqrt{2}}{2} & 0 & 0 & \frac{\sqrt{3}}{2} & 0 & 0 \\ 0 & 0 & 0 & -\frac{\sqrt{2}}{2} & 0 & -1 & \frac{1}{2} & 0 & 0 \\ 0 & 0 & 0 & 0 & -1 & 0 & 0 & 1 & 0 \\ 0 & 0 & 0 & 0 & 0 & 1 & 0 & 0 & 10000 \\ 0 & 0 & 0 & 0 & 0 & 0 & -\frac{\sqrt{3}}{2} & -1 & 0 \\ 0 & 0 & -1 & 0 & 0 & 0 & -\frac{1}{2} & 0 & 0 \end{bmatrix} :$$

> *evalf*(*ReducedRowEchelonForm*(T));

$$\begin{bmatrix} 1.0 & 0.0 & 0.0 & 0.0 & 0.0 & 0.0 & 0.0 & 0.0 & 0.0 \\ 0.0 & 1.0 & 0.0 & 0.0 & 0.0 & 0.0 & 0.0 & 0.0 & -23660.25405 \\ 0.0 & 0.0 & 1.0 & 0.0 & 0.0 & 0.0 & 0.0 & 0.0 & 13660.25405 \\ 0.0 & 0.0 & 0.0 & 1.0 & 0.0 & 0.0 & 0.0 & 0.0 & -33460.65216 \\ 0.0 & 0.0 & 0.0 & 0.0 & 1.0 & 0.0 & 0.0 & 0.0 & 23660.25405 \\ 0.0 & 0.0 & 0.0 & 0.0 & 0.0 & 1.0 & 0.0 & 0.0 & 10000.0 \\ 0.0 & 0.0 & 0.0 & 0.0 & 0.0 & 0.0 & 1.0 & 0.0 & -27320.50810 \\ 0.0 & 0.0 & 0.0 & 0.0 & 0.0 & 0.0 & 0.0 & 1.0 & 23660.25405 \end{bmatrix}$$

Note that we used *evalf* to force decimal arithmetic.

We interpret the solution of the linear system as the magnitude of the forces: $F_1 = 0$, $F_2 = -23660.25$, $F_3 = -13660.25$, $f_1 = -33460.65$, $f_2 = 23660.25$, $f_3 = 10000$, $f_4 = -27320.51$, and $f_5 = 23660.25$.

This problem is continued as an exercise at the end of the section.

Wassily Leontief, recipient of the 1973 Nobel Prize in Economics, explained his input-output model showing the interdependencies of the U.S. economy in the April 1965 issue of *Scientific American*.[6] He organized the 1958 American economy into an 81×81 matrix. The 81 sectors of the economy, such as steel, agriculture, manufacturing, transportation, and utilities, each represented resources that rely on input from the output of other resources. For example, the production of clothing requires an input from manufacturing, transportation, agriculture, and other sectors. The following is a brief example of a Leontief model and its solution.

[6]Wassily W. Leontief, "The Structure of the U.S. Economy," *Scientific American*, April, 1968, pp. 25–35.

Example 5.4. A Leontief Input-Output Economic Model.
Consider an open production model, one that doesn't consume all of its production, where to produce 1 unit of output (units are in millions of dollars):

- the petroleum sector requires 0.1 units of itself, 0.2 units of transportation, and 0.4 units of chemicals;
- the textiles sector requires 0.4 units of petroleum, 0.1 units of itself, 0.15 units of transportation, 0.3 units of chemicals, and 0.35 units of manufacturing;
- the transportation sector requires 0.6 units of petroleum, 0.1 units of itself, and 0.25 units of chemicals;
- the chemicals sector requires 0.2 units of petroleum, 0.1 units of textiles, 0.3 units of transportation, 0.2 units of itself, and 0.1 units of manufacturing;
- the manufacturing sector requires 0.1 units of petroleum, 0.3 units of transportation, and 0.2 units of itself.

Table 5.2 shows the technology matrix representing this model.

TABLE 5.2: Leontief Input-Output Table for Our Five-Sector Economy

		Input Consumed per Unit of Output				
		Petrol.	Textiles	Transport.	Chemicals	Manufact.
Purchased From	Petrol.	0.1	0.4	0.6	0.2	0.1
	Textiles	0.0	0.1	0.0	0.1	0.0
	Transport.	0.2	0.15	0.1	0.3	0.3
	Chemicals	0.4	0.3	0.25	0.2	0.0
	Manufact.	0.0	0.35	0.0	0.1	0.2

The entries of Table 5.2 form the *consumption matrix* C for this economy. We use C to answer Leontief's question, "Is there a production level that will balance the total demand?" Let the vector $\mathbf{x}$ be the total production of the economy, and let the vector $\mathbf{d}$ be the *final demand*, the demand external to production. Then $C\mathbf{x}$ is the *intermediate demand*, the amount of production consumed internally by the individual sectors of the economy. The Leontief Exchange Input-Output model, or Production Equation, is

$$\mathbf{x} = C\mathbf{x} + \mathbf{d}.$$

Let I be the identity matrix. We can solve for $\mathbf{x}$ as follows.

$$\begin{aligned}\mathbf{x} &= C\mathbf{x} + \mathbf{d} \\ \mathbf{x} - C\mathbf{x} &= \mathbf{d} \\ (I - C)\mathbf{x} &= \mathbf{d} \\ \mathbf{x} &= (I - C)^{-1}\mathbf{d}\end{aligned}$$

If the economy produces 900 million dollars of petroleum, 300 million dollars of textiles, 850 million dollars of transportation, 800 million dollars of chemicals, and 750 million dollars of manufacturing, how much of this production is internally consumed by the economy?

Use Maple to enter the augmented matrix $L = [C \mid \mathbf{d}]$ and row reduce it. Remember to load *LinearAlgebra* first.

$> C := \langle\langle 0.9, 0, -0.2, -0.4, 0\rangle \mid \langle -0.4, 0.9, -0.15, -0.3, -0.35\rangle$
$\mid \langle -0.6, 0, 0.9, -0.25, 0\rangle \mid \langle -0.2, -0.1, -0.3, 0.8, -0.1\rangle$
$\mid \langle -0.1, 0, -0.3, 0, 0.8\rangle\rangle :$
$d := \langle 900, 300, 850, 800, 750\rangle :$
$L := \langle C \mid d\rangle;$

$$L := \begin{bmatrix} 0.9 & -0.4 & -0.6 & -0.2 & -0.1 & 900 \\ 0 & 0.9 & 0 & -0.1 & 0 & 300 \\ -0.2 & -0.15 & 0.9 & -0.3 & -0.3 & 850 \\ -0.4 & -0.3 & -0.25 & 0.8 & 0 & 800 \\ 0 & -0.35 & 0 & -0.1 & 0.8 & 750 \end{bmatrix}$$

$> \mathit{evalf} \sim (\mathit{ReducedRowEchelonForm}(L), 5);$

$$\begin{bmatrix} 1.0 & 0.0 & 0.0 & 0.0 & 0.0 & 6944.2 \\ 0.0 & 1.0 & 0.0 & 0.0 & 0.0 & 1070.0 \\ 0.0 & 0.0 & 1.0 & 0.0 & 0.0 & 5620.6 \\ 0.0 & 0.0 & 0.0 & 1.0 & 0.0 & 6629.8 \\ 0.0 & 0.0 & 0.0 & 0.0 & 1.0 & 2234.3 \end{bmatrix}$$

We interpret the unique solution as the amounts the sectors need to produce to meet the total demand:

$$\begin{aligned}\text{petroleum: } x_1 &= 6944.2 \\ \text{textiles: } x_2 &= 1070.0 \\ \text{transportation: } x_3 &= 5620.6 \\ \text{chemicals: } x_4 &= 6629.8 \\ \text{manufacturing: } x_5 &= 2234.3\end{aligned}$$

We studied the method of least squares in Chapter 5 of Volume 1. We used multivariable calculus to minimize the sum of squares of the errors leading to least squares in Chapter 4, Volume 2. Now, we will consider least squares curve fitting as solving a system of linear equations.

The sum of squares of the errors from fitting a parabola $f(x) = ax^2+bx+c$ is given by

$$S = \sum_{i=1}^{n} \left(y_i - (ax_i^2 + bx_i + c)\right)^2.$$

To minimize S, first set the first partial derivatives equal to 0 and then solve the resulting system. After a little algebra, we have the normal equations

$$\begin{aligned}
\frac{\partial S}{\partial a} &= 0 = \sum y_i x_i^2 - a \cdot \sum x_i^4 - b \cdot \sum x_i^3 - c \cdot \sum x_i^2 \\
\frac{\partial S}{\partial b} &= 0 = \sum y_i x_i - a \cdot \sum x_i^3 - b \cdot \sum x_i^2 - c \cdot \sum x_i \\
\frac{\partial S}{\partial c} &= 0 = \sum y_i - a \cdot \sum x_i^2 - b \cdot \sum x_i - c \cdot \sum 1
\end{aligned}$$

Rewrite these equations as a system of linear equations.

$$\begin{aligned}
a \cdot \sum x_i^4 + b \cdot \sum x_i^3 + c \cdot \sum x_i^2 &= \sum y_i x_i^2 \\
a \cdot \sum x_i^3 + b \cdot \sum x_i^2 + c \cdot \sum x_i &= \sum y_i x_i \\
a \cdot \sum x_i^2 + b \cdot \sum x_i + c \cdot \sum 1 &= \sum y_i
\end{aligned} \tag{5.1}$$

Observe the pattern in the equations; what would the system be for a cubic fit?

Example 5.5. A Least Squares Model as a System of Equations. Fit a quadratic model $f(x) = ax^2 + bx + c$ to the following data:

x	0	1	2	3	4
y	62	50	39	18	2

Substitute the following into (5.1).

$$\sum 1 = 5, \quad \sum x_i = 10, \quad \sum x_i^2 = 30, \quad \sum x_i^3 = 96,$$

$$\sum y_i = 171, \qquad \sum y_i x_i = 190, \qquad \sum y_i x_i^2 = 400$$

We have the linear system

$$\begin{aligned}
354a + 96b + 30c &= 400 \\
96a + 30b + 10c &= 190 \\
30a + 10b + 5c &= 171
\end{aligned}$$

Enter the augmented coefficient matrix for the system into Maple and row reduce to find the solution.

```
> A := Matrix([[354, 96, 30, 400], [96, 30, 10, 190], [30, 10, 5, 171]]) :
```

```
> ReducedRowEchelonForm(A);
```

$$\begin{bmatrix} 1 & 0 & 0 & -\frac{197}{111} \\ 0 & 1 & 0 & -\frac{326}{37} \\ 0 & 0 & 1 & \frac{11557}{185} \end{bmatrix}$$

The output yields $a = -197/111$, $b = -326/36$, and $c = 11557/185$. Therefore, the quadratic model for our data is

$$f(x) = -\frac{197}{111}x^2 - \frac{326}{37}x + \frac{11557}{185}.$$

Check the fit.

```
> with(plots) :
> f := x -> -197/111 x^2 - 326/37 x + 11557/185 :
> Pts := zip((x, y) -> [x, y], [0..4], []) :
> DataPlot := pointplot(Pts, symbol = solidcircle, symbolsize = 18) :
  fPlot := plot(f, -1..5, -10..70, thickness = 2) :
> display(DataPlot, fPlot);
```

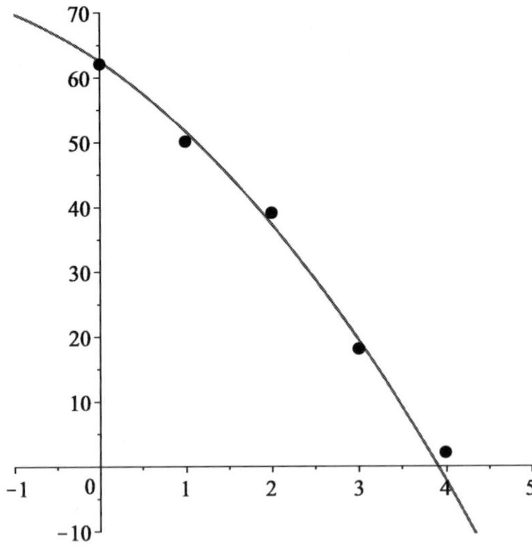

The plot shows visually that the fit captures the trend of the data.

Our next example of an application of systems of linear equations is a different method of fitting a curve to data: this time, smoothly connecting data points via cubic polynomials to form a piecewise curve. These curves are called "cubic splines," cubic for the third degree and spline after the flexible draftsman's ruler originally used to draw the curves. The data points are called "knots." Pierre Bézier, a French design engineer at Renault, introduced using splines for designing auto bodies in the 1960s. Splines are ubiquitous today, appearing in numerous applications. TrueType, developed by Apple in the 1990s, uses two-dimensional quadratic splines to form the characters on your phone or computer screen, and PostScript, developed by Adobe in the 1980s, uses two-dimensional cubic splines for printed characters.

Example 5.6. Natural Cubic Splines as a System of Equations. Fit a *natural cubic spline*[7] to the data

x	7	14	21	28	35	42
y	125	275	800	1200	1700	1650

We will need 5 third order equations.

$$\begin{aligned}
\text{for } x \in [7,14]&: S_1(x) = a_{1,3}x^3 + a_{1,2}x^2 + a_{1,1}x + a_{1,0}\\
\text{for } x \in [14,21]&: S_2(x) = a_{2,3}x^3 + a_{2,2}x^2 + a_{2,1}x + a_{2,0}\\
\text{for } x \in [21,28]&: S_3(x) = a_{3,3}x^3 + a_{3,2}x^2 + a_{3,1}x + a_{3,0}\\
\text{for } x \in [28,35]&: S_4(x) = a_{4,3}x^3 + a_{4,2}x^2 + a_{4,1}x + a_{4,0}\\
\text{for } x \in [35,42]&: S_5(x) = a_{5,3}x^3 + a_{5,2}x^2 + a_{5,1}x + a_{5,0}
\end{aligned}$$

There are 20 unknowns: $a_{i,j}$, so we need 20 equations to uniquely solve for these unknowns. The (x, y) data pairs give 10 equations, $S_1(7) = 125$, $S_1(14) = 275$, $S_2(14) = 275$, $S_2(21) = 800$, and so forth. We need 10 more equations.

To make the spline smooth, we require the curve to have both first and second derivatives equal, i.e., match slope and concavity, at the knots (data points). So, for $i = 1..4$, make

$$S_i'(x_{i+1}) = S_{i+1}'(x_{i+1}) \quad \text{and} \quad S_i''(x_{i+1}) = S_{i+1}''(x_{i+1}).$$

These requirements give us 8 more equations. For the last two equations, use the condition for "natural" splines:

$$S_1'(x_1) = 0 \quad \text{and} \quad S_5'(x_6) = 0$$

We have the requisite 20 equations. Let's use Maple to build the equations, the augmented coefficient matrix, and find the spline.

[7]Natural cubic splines mimic the original engineer's flexible ruler by requiring that the second derivative of the spline at the two end data points be 0. "Clamped" cubic splines use specified values for the first derivative at the two end data points.

First, we load the needed packages and enter the data.

```
> with(LinearAlgebra) :
  with(plots) :
> N := 6 :
  X := [(7k) $ k = 1..6] :
  Y := [125, 275, 800, 1200, 1700, 1650] :
  Pts := zip((x, y) -> [x, y], X, Y) :
      Pts := [[7, 125], [14, 275], [21, 800], [28, 1200], [35, 1700], [42, 1650]]
```

We define the individual cubic functions using an "indexed procname" to make the code more general.

```
> S := proc(x)
    local i;
    i := op(procname);
    return(a[i,3] · x^3 + a[i,2] · x^2 + a[i,1] · x + a[i,0];
    end proc :
```

Now define the natural cubic spline function for our 6 point data set.

```
> Spline := proc(x)
    piecewise(x ≤ X[2], S[1](x), x ≤ X[3], S[2](x), x ≤ X[4], S[3](x),
      x ≤ X[5], S[4](x), S[5](x));
    end proc :
```

In order to avoid some convoluted statements, we'll define the cubic segment's derivative functions directly.

```
> dS := proc(x)
    locali;
    i := op(procname);
    return(3 · a[i,3] · x^2 + 2a[i,2] · x + a[i,1]);
    end proc :
  d2S := proc(x)locali;
    i := op(procname);
    return(6 · a[i,3] · x + 2a[i,2]);
    end proc :
```

Let's calculate the 3 sets of equations that make up the system of 20 linear equations.

```
> KnotEqs := seq('S[i](X[i]) = Y[i], S[i](X[i+1]) = Y[i+1]', i = 1..N − 1) :
  SmoothEqs := seq('dS[i](X[i+1]) = dS[i+1](X[i + 1])', i = 1..N − 2),
    seq('d2S[i](X[i+1]) = d2S[i+1](X[i+1])', i = 1..N − 2) :
  NaturalEqs := d2S[1](X[1]) = 0, d2S[5](X[6]) = 0 :
```

Displaying the sets as vectors make them much easier to read.

> ⟨*KnotEqs*⟩;

$$\begin{bmatrix} 343\,a_{1,3} + 49\,a_{1,2} + 7\,a_{1,1} + a_{1,0} = 125 \\ 2744\,a_{1,3} + 196\,a_{1,2} + 14\,a_{1,1} + a_{1,0} = 275 \\ 2744\,a_{2,3} + 196\,a_{2,2} + 14\,a_{2,1} + a_{2,0} = 275 \\ 9261\,a_{2,3} + 441\,a_{2,2} + 21\,a_{2,1} + a_{2,0} = 800 \\ 9261\,a_{3,3} + 441\,a_{3,2} + 21\,a_{3,1} + a_{3,0} = 800 \\ 21952\,a_{3,3} + 784\,a_{3,2} + 28\,a_{3,1} + a_{3,0} = 1200 \\ 21952\,a_{4,3} + 784\,a_{4,2} + 28\,a_{4,1} + a_{4,0} = 1200 \\ 42875\,a_{4,3} + 1225\,a_{4,2} + 35\,a_{4,1} + a_{4,0} = 1700 \\ 42875\,a_{5,3} + 1225\,a_{5,2} + 35\,a_{5,1} + a_{5,0} = 1700 \\ 74088\,a_{5,3} + 1764\,a_{5,2} + 42\,a_{5,1} + a_{5,0} = 1650 \end{bmatrix}$$

> ⟨*SmoothEqs*⟩;

$$\begin{bmatrix} 588\,a_{1,3} + 28\,a_{1,2} + a_{1,1} = 588\,a_{2,3} + 28\,a_{2,2} + a_{2,1} \\ 1323\,a_{2,3} + 42\,a_{2,2} + a_{2,1} = 1323\,a_{3,3} + 42\,a_{3,2} + a_{3,1} \\ 2352\,a_{3,3} + 56\,a_{3,2} + a_{3,1} = 2352\,a_{4,3} + 56\,a_{4,2} + a_{4,1} \\ 3675\,a_{4,3} + 70\,a_{4,2} + a_{4,1} = 3675\,a_{5,3} + 70\,a_{5,2} + a_{5,1} \\ 84\,a_{1,3} + 2\,a_{1,2} = 84\,a_{2,3} + 2\,a_{2,2} \\ 126\,a_{2,3} + 2\,a_{2,2} = 126\,a_{3,3} + 2\,a_{3,2} \\ 168\,a_{3,3} + 2\,a_{3,2} = 168\,a_{4,3} + 2\,a_{4,2} \\ 210\,a_{4,3} + 2\,a_{4,2} = 210\,a_{5,3} + 2\,a_{5,2} \end{bmatrix}$$

> ⟨*NaturalEqs*⟩;

$$\begin{bmatrix} 42\,a_{1,3} + 2\,a_{1,2} = 0 \\ 252\,a_{5,3} + 2\,a_{5,2} = 0 \end{bmatrix}$$

Get a list of the variables of the system: the coefficients of the equations.

> *indets*([*KnotEqs*]) :
vars := *convert*(%, *list*);

$$vars := [a_{1,0}, a_{1,1}, a_{1,2}, a_{1,3}, a_{2,0}, a_{2,1}, a_{2,2}, a_{2,3}, a_{3,0}, a_{3,1}, a_{3,2}, a_{3,3}, a_{4,0}, a_{4,1}, a_{4,2}, a_{4,3}, a_{5,0}, a_{5,1}, a_{5,2}, a_{5,3}]$$

Use *LinearAlgebra*'s *GenerateMatrix* to create the augmented matrix for the system. (To see the full matrices, execute *interface*(*rtablesize*=50) before executing the next Maple command.)

```
> A := GenerateMatrix([KnotEqs, SmoothEqs, NaturalEqs], vars,
    augmented = true);
```

$$A := \begin{bmatrix} 1 & 7 & 49 & 343 & 0 & 0 & 0 & 0 & 0 & 0 & \cdots \\ 1 & 14 & 196 & 2744 & 0 & 0 & 0 & 0 & 0 & 0 & \cdots \\ 0 & 0 & 0 & 0 & 1 & 14 & 196 & 2744 & 0 & 0 & \cdots \\ 0 & 0 & 0 & 0 & 1 & 21 & 441 & 9261 & 0 & 0 & \cdots \\ 0 & 0 & 0 & 0 & 0 & 0 & 0 & 0 & 1 & 21 & \cdots \\ 0 & 0 & 0 & 0 & 0 & 0 & 0 & 0 & 1 & 28 & \cdots \\ 0 & 0 & 0 & 0 & 0 & 0 & 0 & 0 & 0 & 0 & \cdots \\ 0 & 0 & 0 & 0 & 0 & 0 & 0 & 0 & 0 & 0 & \cdots \\ 0 & 0 & 0 & 0 & 0 & 0 & 0 & 0 & 0 & 0 & \cdots \\ 0 & 0 & 0 & 0 & 0 & 0 & 0 & 0 & 0 & 0 & \cdots \\ \vdots & \vdots & \vdots & \vdots & \vdots & \vdots & \vdots & \vdots & \vdots & \vdots & \end{bmatrix}$$

20 × 21 Matrix

Compute the reduced row echelon form of A.

```
> RA := ReducedRowEchelonForm(A);
```

$$A := \begin{bmatrix} 1 & 0 & 0 & 0 & 0 & 0 & 0 & 0 & 0 & 0 & \cdots \\ 0 & 1 & 0 & 0 & 0 & 0 & 0 & 0 & 0 & 0 & \cdots \\ 0 & 0 & 1 & 0 & 0 & 0 & 0 & 0 & 0 & 0 & \cdots \\ 0 & 0 & 0 & 1 & 0 & 0 & 0 & 0 & 0 & 0 & \cdots \\ 0 & 0 & 0 & 0 & 1 & 0 & 0 & 0 & 0 & 0 & \cdots \\ 0 & 0 & 0 & 0 & 0 & 1 & 0 & 0 & 0 & 0 & \cdots \\ 0 & 0 & 0 & 0 & 0 & 0 & 1 & 0 & 0 & 0 & \cdots \\ 0 & 0 & 0 & 0 & 0 & 0 & 0 & 1 & 0 & 0 & \cdots \\ 0 & 0 & 0 & 0 & 0 & 0 & 0 & 0 & 1 & 0 & \cdots \\ 0 & 0 & 0 & 0 & 0 & 0 & 0 & 0 & 0 & 1 & \cdots \\ \vdots & \vdots & \vdots & \vdots & \vdots & \vdots & \vdots & \vdots & \vdots & \vdots & \end{bmatrix}$$

20 × 21 Matrix

Use *LinearAlgebra*'s *GenerateEquations* to capture the values of the coefficients.

> *theFit* := *GenerateEquations*(*RA*, *vars*);

$$theFit := [a_{1,0} = -25, a_{1,1} = \frac{79000}{1463}, a_{1,2} = -\frac{71475}{10241}, a_{1,3} = \frac{23825}{71687},$$
$$a_{2,0} = \frac{511375}{209}, a_{2,1} = -\frac{695900}{1463}, a_{2,2} = \frac{28725}{931}, a_{2,3} = -\frac{40750}{71687}, a_{3,0} = -\frac{1525100}{209},$$
$$a_{3,1} = \frac{1340575}{1463}, a_{3,2} = -\frac{362850}{10241}, a_{3,3} = \frac{1825}{3773}, a_{4,0} = \frac{3953300}{209}, a_{4,1} = -\frac{2768225}{1463},$$
$$a_{4,2} = \frac{664350}{10241}, a_{4,3} = -\frac{7275}{10241}, a_{5,0} = -\frac{6559200}{209}, a_{5,1} = \frac{3539275}{1463}, a_{5,2} = -\frac{597150}{10241},$$
$$a_{5,3} = \frac{33175}{71687}]$$

Set the values of the coefficients.

> *assign*(*theFit*) :

And plot the results.

> *SplinePlot* := *plot*(*Spline*(*x*), $x = X_1..X_N$) :
DataPlot := *pointplot*(*Pts*, *symbol* = *solidcircle*, *symbolsize* = 18) :
display(*SplinePlot*, *DataPlot*);

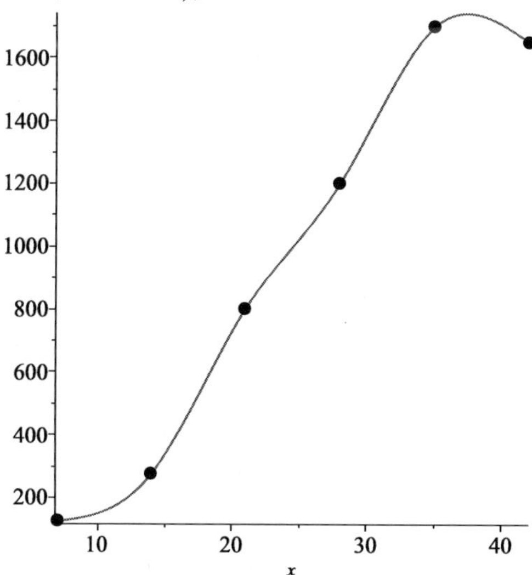

The natural cubic spline fits the data very well. Caution: do not use the spline outside the data! Plot over a larger range to see why.

Use Maple's

$$CurveFitting{:}\text{-}Spline(Pts, x, 3, endpoints = \text{'}natural\text{'});$$

to verify that we have found the curve.

In general, to connect N knots (data points), we need $N-1$ cubic equations. To set up the system of equations, we will need:

1. $2(N-2)$ equations matching the endpoints of the $N-1$ connecting cubic curves,

2. $(N-1)$ equations matching the first derivatives at the $N-1$ connecting points of the cubic curves,
3. $(N-1)$ equations matching the second derivatives at the $N-1$ connecting points of the cubic curves,
4. • either 2 equations for natural cubic splines setting the second derivative of the left end of the first cubic and the right end of the last cubic to 0, or
   • 2 equations for clamped cubic splines setting the first derivative of the left end of the first cubic and the right end of the last cubic to given values.

Combining these equations gives a total of $4n-4$ equations for our system with $(N-1) \times 4$ variables.

Exercises

1. A Bridge Too Far. Revisit the truss of Example 5.3. Set up and solve a model of the truss:
   (a) if the angles at joint p_1 and joint p_4 are both $\pi/3$.
   (b) if the angles at joint p_1 and joint p_4 are both $\pi/4$.
   (c) if the angles at joint p_1 and joint p_4 are both $\pi/6$.
2. Consider the Leontief Input-Output model of Example 5.4.
   (a) Determine the solution for the following technology matrix:

	Petrol.	Textiles	Transport.	Chem.	Manufact.
Petrol.	0.2	0.3	0.6	0.2	0.1
Textiles	0.0	0.2	0.0	0.1	0.0
Transport.	0.2	0.15	0.2	0.3	0.3
Chem.	0.4	0.35	0.25	0.2	0.0
Manufact.	0.0	0.35	0.0	0.1	0.2

   (b) If the economy now produces 1,000 million dollars of petroleum, 400 million dollars of textiles, 950 million dollars of transportation, 750 million dollars of chemicals, and 950 million dollars of manufacturing, how much of this production is internally consumed by the economy?
3. Use the least squares technique to fit the model $f(x) = kx^2$ to the following data:

x	0.5	1.0	1.5	2.0	2.5
y	0.7	3.4	7.2	12.4	20.1

4. Use the least squares technique to fit the model $W = kL^3$ to the following data:

Length L	12.5	12.625	12.625	14.125	14.5	14.5	17.27	17.75
Weight W	17	16	17	23	26	27	43	49

5. Use cubic spline interpolation to fit the following sets of data points.

(a)

x	1	2	3
y	9	27	48

(b)

x	0.5	1.0	1.5	2.0	2.5
y	0.7	3.4	7.2	12.4	20.1

Projects

Project 1. One of the main decisions that manufacturers have is to choose how much of a product to produce. Considering that an economy consists of producers of many things, this problem gets very complex very quickly. Leontief received the Nobel Prize in Economics in 1973 "for the development of the input-output method and for its application to important economic problems."[8] What we will discuss now comes from his work, hence the name *Leontief Model*. An example of such a model follows.

First, divide an economy into certain sectors. In reality, there are hundreds of sectors, but for the sake of simplicity, we will posit three sectors: manufacturing (m), electronics (e), and agriculture (a). We must decide how many units of each sector to produce, expressing the amounts as a *production vector*

$$\mathbf{x} = \begin{bmatrix} m \\ e \\ a \end{bmatrix}$$

Now, let's say that the public wants 100 units of manufacturing, 200 units of electronics, and 300 units of agriculture. Put this data into an *(external) demand vector*

$$\mathbf{d} = \begin{bmatrix} 100 \\ 200 \\ 300 \end{bmatrix}$$

Now, one might say that if this is what the economy wants, then this is what should be produced; i.e., $\mathbf{x} = \mathbf{d}$. The problem, however, is that the

[8]See https://www.nobelprize.org/prizes/economic-sciences/1973/leontief/biographical/.

production of certain resources actually *consumes* resources as well. In other words, it takes stuff to make stuff. How much is consumed by production can be expressed by using an *input-output matrix.*

$$T = \begin{array}{c} \\ \rotatebox{90}{\textit{inputs}} \end{array} \begin{array}{c} \\ m \\ e \\ a \end{array} \overset{\textit{outputs}}{\begin{array}{c} \begin{array}{ccc} m & e & a \end{array} \\ \begin{bmatrix} 0.1 & 0.2 & 0.3 \\ 0.2 & 0.1 & 0.3 \\ 0.1 & 0.1 & 0.2 \end{bmatrix} \end{array}}$$

The matrix T indicates that the production of 1 unit of manufacturing uses up 0.1 units of manufacturing, 0.2 units of electronics, and 0.1 units of agriculture. The production of 1 unit of electronics uses up 0.2 units of manufacturing, 0.1 units of electronics, and 0.1 units of agriculture. Finally, the production of 1 unit of agriculture uses up 0.3 units of manufacturing, 0.3 units of electronics, and 0.1 units of agriculture. So, not only must we account for what the people want, but we must also make up for what is used up in the process of producing what the people want. The amount consumed by production is called *internal demand*, it is given by the product $T\mathbf{x}$. Hence, what we produce needs to satisfy both internal demand and external demand. That is,

$$\mathbf{x} = T\mathbf{x} + \mathbf{d}$$

Using the matrix T above, determine the production vector $\mathbf{x}$, also called the *total demand*, and the internal demand $T\mathbf{x}$ for our three-sector economy.

Project 2. Wassily Leontief (1906-1999), the Russian-born, Nobel Prize winning American economist who, aside from developing highly sophisticated economic theories, also enjoyed trout fishing, ballet and fine wines. In this project, we will look at a very simple special case of his work called a *closed exchange model.*

Long, long ago, far, far away in the land of Eigenbazistan, in a small country town called Matrixville, there lived a Farmer, a Tailor, a Carpenter, a Coal Miner and Slacker Bob. The Farmer produced food; the Tailor, clothes; the Carpenter, housing; the Coal Miner supplied energy; and Slacker Bob made High Quality 100-Proof Moonshine, half of which he drank himself. Let's make the following assumptions:

- Everyone buys from and sells to the central pool (i.e., there is no outside supply or demand).
- Everything produced is consumed.

This type of economy is called a *closed exchange model.* Table 5.3 specifies what fraction of each of the goods is consumed by each person in our town.

TABLE 5.3: Matrixville's Closed Exchange Model

	Food	Clothes	Housing	Energy	Moonshine
Farmer	0.25	0.15	0.25	0.18	0.20
Tailor	0.15	0.28	0.18	0.17	0.05
Carpenter	0.22	0.19	0.22	0.22	0.10
Coal Miner	0.20	0.15	0.20	0.28	0.15
Slacker Bob	0.18	0.23	0.15	0.15	0.50

So for example, the Carpenter consumes 22% of all food, 19% of all clothes, 22% of all housing, 22% of all energy and 10% of all High Quality 100 Proof Moonshine.

If the matrix $I - T$ is invertible, the total demand equation, $\mathbf{x} = T\mathbf{x} + \mathbf{d}$ can be solved for $\mathbf{x}$.

(a) Determine $C = I - T$ using Table 5.3.

(b) If possible, determine C^{-1}.

(c) If feasible, compute the total demand $\mathbf{x}$ for this closed exchange model.

(d) For those who have studied linear algebra: what is the relation between the total demand $\mathbf{x}$ and the eigenvalues and eigenvectors of T, if any?

5.4 Stoichiometric Chemical Balancing and Infinitely Many Solutions

During your life you have witnessed numerous chemical reactions. How would you describe them to someone else? How could you obtain quantitative information about the reaction? Chemists use *stoichiometric chemical equations*[9] to answer these questions.

By definition, a chemical equation is a written representation of a chemical reaction, showing the reactants and products, their physical states, and the direction in which the reaction proceeds. In addition, many chemical equations designate the conditions necessary (such as high temperature) for the reaction to occur. A chemical equation provides stoichiometric information about a chemical reaction, only if the equation is balanced.

For a chemical equation to be balanced, the same number of each kind of atom must be present on both sides of the chemical equation. The French

[9]See LibreTexts' *Chemistry Library* "Stoichiometry and Balancing Equations".

chemist Antoine-Laurent de Lavoisier[10] (1743–1794) introduced the law of *conservation of matter* during the latter half of the eighteenth century. The conservation law states that matter can neither be created nor destroyed. de Lavoisier's principles of naming chemical substances are still used today.

John Dalton[11] (1766–1844), developer of the first useful atomic theory of matter in the early 1800s, was the first to associate the ancient idea of atoms with stoichiometry. Dalton concluded, while studying meteorology, that evaporated water exists in air as an independent gas. Solid bodies cannot occupy the same space at the same time, but obviously water and air could. If the water and air were composed of discrete particles, evaporation might be viewed as mixing their separate particles. He performed a series of experiments on mixtures of gases to determine what effect properties of the individual gases had on the properties of the mixture as a whole. While trying to explain the results of those experiments, Dalton hypothesized that the sizes of the particles making up different gases must be different. Dalton wrote

> it became an object to determine the relative sizes and weights, together with the relative numbers of atoms entering into such combinations... Thus a train of investigation was laid for determining the number and weight of all chemical elementary particles which enter into any sort of combination one with another.[12]

According to Dalton's *Atomic Theory of Matter* of 1803, all substances are composed of atoms. During a chemical reaction atoms may be combined, separated, or rearranged, but not created or destroyed. The postulates include:

1. All matter is composed of atoms, indivisible and indestructible objects, which are the ultimate chemical particles.

2. All atoms of a given element are identical, both in mass and in properties. Atoms of different elements have different masses and different properties.

3. Compounds are formed by combination of two or more different kinds of atoms. Atoms combine in the ratio of small whole numbers.

4. Atoms are the units of chemical change. A chemical reaction involves only the combination, separation, or rearrangement of atoms.

Let's examine the meaning of chemical equations and compare them to mathematical equations to gain some insights. A chemical equation identifies the starting and finishing chemicals as reactants and products: reactants $\to$ products.

[10]See https://www.britannica.com/biography/Antoine-Lavoisier.

[11]See https://www.britannica.com/biography/John-Dalton.

[12]The quote is from Dalton's notes for a lecture to the Royal Institution in London in 1810.

Chemical Equations vs. Mathematical Equations

We usually think of an equation like $x+2x = 3x$ as purely mathematical, even if x represents a physical quantity like distance or mass. A chemical equation may look like a mathematical equation, but it describes experimental observations: the quantities and kinds of reactants and products for a particular chemical reaction. Reactants appear on the left side of a chemical equation; products on the right. For example, see Table 5.4. The products, which are the result of combining the reactants, are known from experimental observations – they are not derived mathematically.

TABLE 5.4: Chemical Equation vs. Mathematical Equation

Chemical Equation	Mathematical Equation
Combustion of Propane	Linear Equation
$C_3H_8 + 5O_2 \longrightarrow 3CO_2 + 4H_2O$	$x = 2x + 3$
Left: reactants; right: products	No standard left/right order
Balanced when it reflects the conservation of matter	Solved when all values that give a true statement are found

In fact, combining the same reactants at different concentrations or temperatures can often produce different products from the same reactants. First-year chemistry students cannot predict these effects, and are generally not asked to predict them. On the other hand, a chemical equation is similar to a mathematical equation in that there are certain restrictions on what may appear on the left and right sides. These mathematical rules represent the effects of conservation of matter on the reaction—conservation of matter says that no atoms are destroyed or created during a chemical reaction.

How a Chemist Approaches Balancing an Equation

Many chemical equations, in the view of the chemists, can be balanced by inspection, that is, by the process of "trial and error." The objectives of a chemist are:

- Recognize a balanced equation.
- Recognize an unbalanced equation.
- Balance by inspecting chemical equations with given reactants and products.
- Write the unbalanced equation when given compound names for reactants and products.

According to chemistry textbooks, the step-wise procedure to balance equations is:

STEP 1. Determine what reaction is occurring: know the reactants, the products, and the physical states.

STEP 2. Write the unbalanced equation that summarizes the reaction described in Step 1.

STEP 3. Balance the equation by inspection, starting with the most complicated molecules. Do not change the identities of any reactant or product.

Thus, balancing the equation is done by inspection, a "trial and error" process that some students catch on to quickly, but others struggle with in frustration.

Balancing Equations with Systems of Equations

Balancing chemical equations can be an application of solving a system of linear equations. Placing variables as the multipliers for each compound and making equations for each type of atom in the reaction results in a system of linear equations. This system is usually under-determined, there are more variables than equations. An under-determined system has infinitely many solutions. Our goal is to provide a procedure to find an integer solution from among the infinitely many solutions. This method is best grasped through an example which also lays the foundation for balancing more complicated oxidation-reduction equations.

Example 5.7. A Sulfur Dioxide Reaction.
Balance the chemical equation

$$\mathrm{S_6} + \mathrm{O_2} \longrightarrow \mathrm{SO_2}$$

STEP 1. Introduce multipliers for each compound. There are three compounds, so we identify three multipliers: $\{x_1, x_2, x_3\}$. The reaction equation becomes

$$x_1\mathrm{S_6} + x_2\mathrm{O_2} \longrightarrow x_3\mathrm{SO_2}$$

STEP 2. Identify all elements: S (sulfur) and O (oxygen). Set up an equation for each element involving the amounts (multipliers) for that chemical on the reactant side equaling the amount on the product side.

$$\begin{aligned} \mathrm{S:}\;\; 6x_1 &= x_3 \\ \mathrm{O:}\;\; 2x_2 &= 2x_3 \end{aligned}$$

STEP 3. Create a homogeneous system; that is, put all variables on the left side of the equation and zero on the right.

$$\begin{aligned} \mathrm{S:}\;\; 6x_1 - x_3 &= 0 \\ \mathrm{O:}\;\; 2x_2 - 2x_3 &= 0 \end{aligned}$$

STEP 4. Write the system in matrix notation $A\mathbf{x} = \mathbf{b}$ where A is the coefficient matrix and $\mathbf{b}$ is the column vector of zeros. Then form the augmented matrix for the system.

$$[A \mid \mathbf{b}] = \left[\begin{array}{ccc|c} 6 & 0 & -1 & 0 \\ 0 & 2 & -2 & 0 \end{array}\right]$$

STEP 5. Apply Gaussian elimination (row reduction) to the augmented matrix $[A \mid \mathbf{b}]$ to obtain $[H \mid \mathbf{c}]$, where H is in reduced row echelon form.

$$[H \mid \mathbf{c}] = \left[\begin{array}{ccc|c} 1 & 0 & -1/6 & 0 \\ 0 & 1 & -1 & 0 \end{array}\right]$$

Read the system's solution: $x_1 - x_3/6 = 0$ and $x_2 - x_3 = 0$.

Balancing the chemical equation means finding the *smallest whole numbers* x_1, x_2, and x_3 solving the system. Since there are more equations than unknowns, there is an infinite number of solutions. Note that x_3 is part of every equation, so x_3 can equal anything. The fraction $1/6$ is a coefficient of x_3 in one of the equations, so choose x_3 to be the smallest whole number that eliminates the fraction; i.e., choose $x_3 = 6$. Then $x_1 = 1$ and $x_2 = 6$.

The balanced reaction equation is

$$S_6 + 6\,O_2 \longrightarrow 6\,SO_2.$$

We are presented with the following more complicated unbalanced equation that would be difficult to solve by inspection.

Example 5.8. Ethylenediamine Mixed with Dinitrogen Tetroxide. Balance the reaction

$$C_2H_8N_2 + N_2O_4 \longrightarrow N_2 + CO_2 + H_2O$$

STEP 1. Introduce five multipliers x_1 through x_5, one for each of the 5 terms.

$$x_1\,C_2H_8N_2 + x_2\,N_2O_4 \longrightarrow x_3\,N_2 + x_4\,CO_2 + x_5\,H_2O$$

STEP 2. List the equation for each element.

$$\begin{aligned} &\text{C: } 2x_1 = x_4 \\ &\text{H: } 8x_1 = 2x_5 \\ &\text{N: } 2x_1 + 2x_2 = 2x_3 \\ &\text{O: } 4x_2 = 2x_4 + x_5 \end{aligned}$$

STEP 3. Write the augmented matrix for the homogeneous system.

$$[A \mid \mathbf{b}] = \left[\begin{array}{ccccc|c} 2 & 0 & 0 & -1 & 0 & 0 \\ 8 & 0 & 0 & 0 & -2 & 0 \\ 2 & 2 & -2 & 0 & 0 & 0 \\ 0 & 4 & 0 & -2 & -1 & 0 \end{array}\right]$$

STEP 4. Maple's *ReducedRowEchelonForm* yields

> $B := \langle\langle 2,8,2,0\rangle \mid \langle 0,0,2,4\rangle \mid \langle 0,0,-2,0\rangle \mid \langle -1,0,0,-2\rangle \mid \langle 0,-2,0,-1\rangle$
> $\mid \langle 0,0,0,0\rangle\rangle;$

> *LinearAlgebra:-ReducedRowEchelonForm*(B);

$$\left[\begin{array}{ccccc|c} 1 & 0 & 0 & 0 & -\frac{1}{4} & 0 \\ 0 & 1 & 0 & 0 & -\frac{1}{2} & 0 \\ 0 & 0 & 1 & 0 & -\frac{3}{4} & 0 \\ 0 & 0 & 0 & 1 & -\frac{1}{2} & 0 \end{array}\right]$$

STEP 5. The least common multiple of the denominators for x_5's coefficients is 4; choose $x_5 = 4$. Then $x_1 = 1$, $x_2 = 2$, $x_3 = 3$, and $x_4 = 2$.

The balanced equation is

$$C_2H_8N_2 + 2\,N_2O_4 \longrightarrow 3\,N_2 + 2\,CO_2 + 4\,H_2O$$

Check the result—count the elements in the equations to make sure each balances.

Exercises

In Exercises 1 to 4 balance the basic chemical reactions.

1. Copper plus silver nitrate displacement/redux reaction:

$$Cu + AgNO_3 \longrightarrow Ag + Cu(NO_3)_2$$

(Elements: copper (Cu), silver (Ag), nitrogen (N) and oxygen (O).)

2. Zinc and hydrochloric acid replacement reaction:

$$Zn + HCl \longrightarrow ZnCl_2 + H_2$$

(Elements: zinc (Zn), hydrogen (H), and chlorine (Cl).)

3. Ferrous oxide to ferric oxide reaction:

$$FeO + O_2 \longrightarrow Fe_2O_3$$

(Elements: iron (Fe) and oxygen (O).)

4. Calcium hydroxide and phosphoric acid neutralization reaction:

$$Ca(OH)_2 + H_3PO_4 \longrightarrow Ca_3(OH_4)_2 + H_2O$$

(Elements: calcium (Ca), oxygen (O), hydrogen (H), and phosphorus (P).)

Projects

Project 1. Explain why we need to use the smallest integer solution for the balanced chemical reactions. Provide examples.

Project 2. The reduction of kerosene takes place in three steps

$$C_{10}H_{16} + O_2 \longrightarrow CO + H_2$$
$$CO + O_2 \longrightarrow CO_2$$
$$H_2 + O_2 \longrightarrow H_2O$$

Balance the three reactions simultaneously.

References and Further Reading

[BKWK2009] Robert T. Balmer, William D. Keat, George Wise, Philip Kosky, *Exploring Engineering: an Introduction to Engineering and Design*, Elsevier Science & Tech., 2009.

[FVC2007a] William P. Fox, K. Varanzo, and J. Croteau, "Mathematical Modeling of Chemical Stoichiometry", PRIMUS, XVII(4), 301–315, 2007.

[FVC2007b] William P. Fox, K. Varanzo, and J. Croteau, "Oxidation-Reduction Chemical Equation Balancing with MAPLE", Computers in Education Journal (COED), VOL XII (2), April-June 2007, pages 50–57.

[LLM2016] D. Lay, S. Lay, amd J. McDonald, *Linear Algebra and its Applications*, Pearson Education, London, U.K., 2016.

[MB1985] Ronald E. Miller and Peter D. Blair, *Input-Output Analysis: Foundations and Extensions*, Prentice Hall, 1985.

[Tro2016] Nivaldo J. Tro, *Chemistry: A Molecular Approach*, 4th ed, Pearson, 2016.

6

Review of Regression Models and Advanced Regression Models

Objectives:

(1) **Understand the concepts of correlation and linearity.**

(2) **Build and interpret nonlinear regression models.**

(3) **Build and interpret logistic regression models.**

(4) **Build and interpret Poisson regression models.**

(5) **Understand when to use each type of regression model.**

(6) **Understand the use of technology in regression.**

The Philippine National Statistics Coordination Board (NSCB) has collected data on acts of violence committed by terrorists and insurgents over the past decade. Over the same period, the NSCB has collected data on the population such as education levels (literacy), employment, government satisfaction, ethnicity, and so forth. The government is looking to see which areas it can target for improvements that might reduce the number of violent acts committed. What should the administration do to improve the situation?

In this chapter, we will discuss several regression techniques as background information and keys to finding adequate models. We do not try to comprehensively cover regression, but we will use real examples to illustrate the techniques used to gain insights, predict, explain, and answer scenario-related questions such as the Philippines question above.

The forms we will study are simple linear regression, multiple regression, exponential and sine nonlinear regression, binary logistic regression, and Poisson regression.[1]

[1]This chapter is adapted from Fox and Hammond, "Advanced Regression Models: Least Squares, Nonlinear, Poisson and Binary Logistics Regression Using R" in Márquez and Lev, *Data Science and Digital Business*, Springer, 2019.

6.1 Re-Introduction to Regression

Simple linear regression[2] builds the model $f(x) = ax + b$ by minimizing the sum of errors squared.[3] As an optimization problem, this is

$$\text{Minimize } SSE = \sum_{i=1}^{n} \left(y_i - f(x_i)\right)^2 \tag{6.1}$$

Most programs will do the computations for (6.1) easily; a calculator can handle small data sets. Excel, JMP, R, SAS, Stata, and SPSS are among the most commonly used programs. We will continue to use Maple.

For the ordinary least squares method of linear regression, we are most concerned, in this chapter, with the mathematical modeling and the use of the model for *explaining* or *predicting* the phenomena being studied or analyzed. Other standard uses are *variable screening* and *parameter estimation.* We will make use of the basic diagnostic measures, percent relative error and residual plots, for analysis. Percent relative error is calculated as

$$\%RE = \frac{y_i - f(x_i)}{y_i} \cdot 100. \tag{6.2}$$

A rule of thumb for the magnitude of percent relative error is that most should be less than 20%, and those near where a prediction is needed should be less than 10%.

A plot of residuals versus the model provides a visualization of the goodness of a model's fit. Examine the plot for trends or patterns such as those shown in Figure 6.1. If a pattern is apparent, the model is not adequate, even though we might have to use it. If there is no visual pattern, then we may take that as evidence that the model is adequate.

When should we use simple linear regression and when should we use something more advanced? Let's begin an answer with the concept of *correlation.*

Correlation

Many decision makers have misconceptions about correlation in linear regression that is often engendered by non-technical, common-usage definitions of the term. *The Oxford English Dictionary* defines

corrolation: *noun*

1. A mutual relationship or connection between two or more things.

[2]Linear regression was introduced in Chapter 5, Model Fitting and Linear Regression, of Volume 1.

[3]"Least squares" was introduced by Gauss in the early 1800s; Legendre was the first to publish the technique in 1805. Galton introduced the term "regression" in his classic 1885 study of children's size relating to their parents' size.

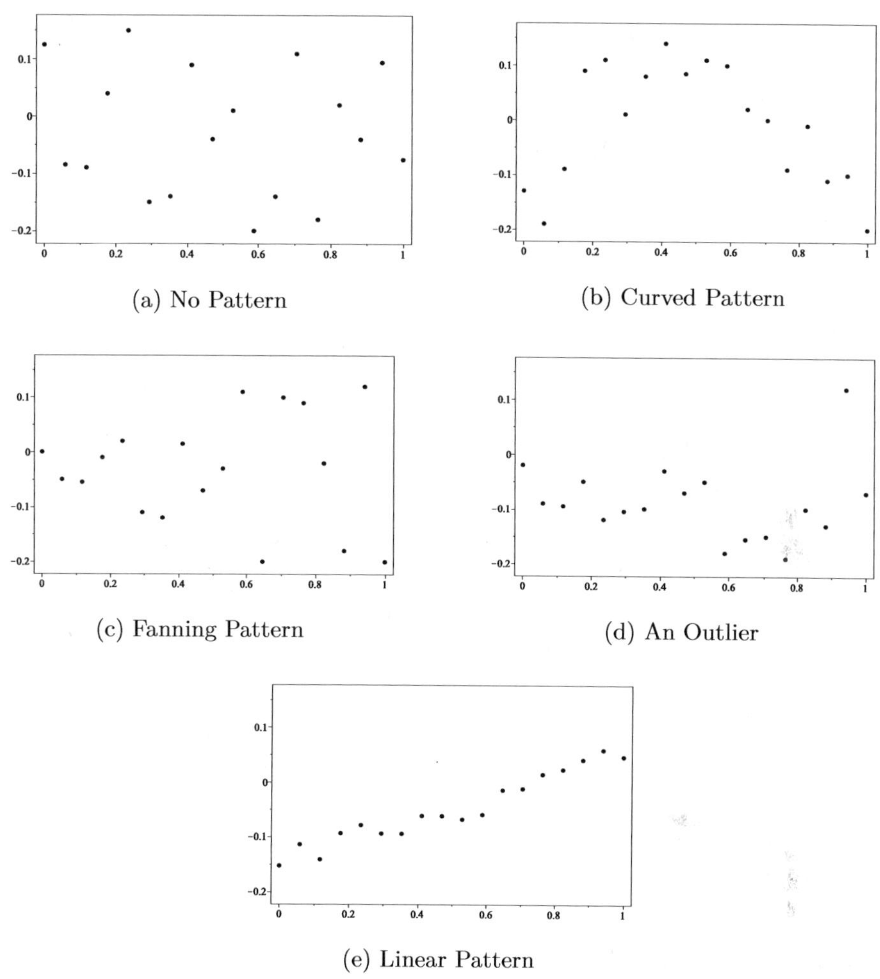

(a) No Pattern

(b) Curved Pattern

(c) Fanning Pattern

(d) An Outlier

(e) Linear Pattern

FIGURE 6.1: Patterns in the Residual Plot

A *statistical* definition in the context of linear regression is given in terms of *Pearson's product-moment correlation coefficient*, often reduced to *correlation coefficient*, or even just *correlation*,

> correlation is a measure of the strength of the linear relationship between two variables.[4]

Linear relationship is a key term in the definition. Some definitions for correlation merely state it is a measure of the relationship between two variables,

[4]See, e.g., Lane et al, *Introduction to Statistics*, Online ed., pg. 170.

and do not mention linearity. Definitions like the one from the help menu of a popular spreadsheet

> CORREL: Returns the correlation coefficient of the Array1 and Array2 cell ranges. Use the correlation coefficient to determine the relationship between two properties. For example, you can examine the relationship between a location's average temperature and the use of air conditioners.

help fuel misconceptions.

It is no wonder decision makers have misinterpretations, thinking that correlation completely measures the relationship between variables. The most common misunderstandings that decision makers have expressed include:

- Correlation implies causation.
- Correlation measures everything.
- When the correlation value is large and it looks linear (visually), then the relation must be linear.
- Model diagnostics are not needed when the correlation value is large.
- The regression package that was taught in class is what must be used.

The next section begins with examples that illustrate some of the misconceptions surrounding correlation and shows possible corrections that we use in our mathematical modeling courses. One diagnostic test that we now cover is the "common-sense test"—does the model *answer the question* and does it provide *realistic results*?

6.2 Modeling, Correlation, and Regression

Return to the Philippines scenario introduced at the beginning of the chapter. The researcher actually began with linear regression. For example, in 2008, an analysis of the literacy index per region versus the number of significant acts of violence (by terrorists or insurgents) produced the linear model

$$f(x) = -1.5463x + 146.54$$

with correlation coefficient -0.3742 or $R^2 = 0.14$ (R^2 is the square of the correlation coefficient). The model is plotted in Figure 6.2.

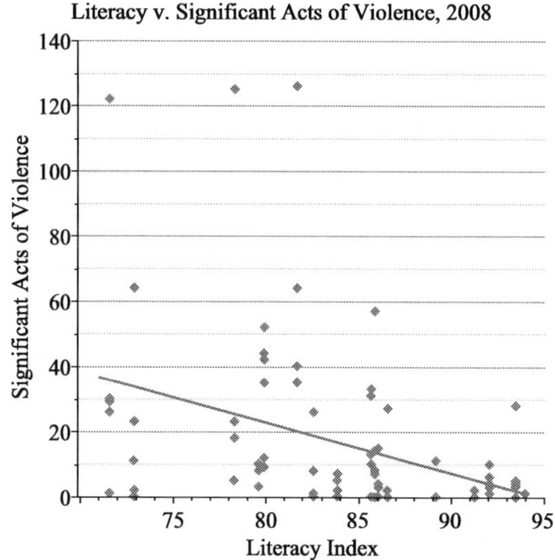

FIGURE 6.2: Philippines: Literacy v. Significant Acts of Violence, 2008

Within all the data, Figure 6.2 represents one of the "best" resulting overall models—note that R^2 is only 0.14 which represents a really poor linear fit or a very weak linear relationship. We will return to this scenario again later in the chapter, but suffice it to say, linear regression is not useful to explain, predict, or answer the questions the Philippine government had posed.

6.3 Linear, Nonlinear, and Multiple Regression

Many applications are not linear. For example, a pendulum is modeled by the differential equation $\theta'' + \omega^2 \sin(\theta) = 0$; this equation can be "linearized" by replacing $\sin(\theta)$ with θ, a good linear approximation for small values of θ. Two widely used rules of thumb for relating correlation to the likelihood of a linear relationship come from Devore [D2012] for math, science, and engineering data, shown in Table 6.1, and a more liberal interpretation from Johnson [J2012] for non-math, non-science, and non-engineering data, given in Table 6.2.

TABLE 6.1: Linearity Likelihood for Math, Science, and Engineering Data; Devore

Correlation ρ	Relationship
$0.8 < \lvert\rho\rvert \leq 1.0$	Strong linear relationship
$0.5 < \lvert\rho\rvert \leq 0.8$	Moderate linear relationship
$0.0 \leq \lvert\rho\rvert \leq 0.5$	Weak linear relationship

TABLE 6.2: Linearity Likelihood for Non-Math, Non-Science, and Non-Engineering Data; Johnson

Correlation ρ	Relationship
$0.5 < \lvert\rho\rvert \leq 1.0$	Strong linear relationship
$0.3 < \lvert\rho\rvert \leq 0.5$	Moderate linear relationship
$0.1 \leq \lvert\rho\rvert \leq 0.3$	Weak linear relationship
$0.0 \leq \lvert\rho\rvert \leq 0.1$	No linear relationship

In our modeling efforts, we emphasize the interpretation of $\rho \approx 0$. A very small correlation coefficient can be interpreted as indicating either no linear relationship or the existence of a nonlinear relationship. Many neophyte analysts fail to pick up on the importance of potential nonlinear relationships in their interpretation.

Let's look at an example where we attempt to fit a linear model to data that may not be linear.

Example 6.1. Exponential Decay.
Build a mathematical model to predict the degree of recovery after discharge for orthopedic surgical patients. There are two variables: time t in days in the hospital, and a medical prognostic index for recovery y with larger values indicating a better prognosis. The data in Table 6.3 comes from Neter et al. [NKNW1996, p. 469].

TABLE 6.3: Hospital Stay vs. Recovery Index Data

t	2	5	7	10	14	19	26	31	34	38	45	52	53	60	65
y	54	50	45	37	35	25	20	16	18	13	8	11	8	4	6

First, a scatterplot of the data.

$> t := [2, 5, 7, 10, 14, 19, 26, 31, 34, 38, 45, 52, 53, 60, 65]:$
$y := [54, 50, 45, 37, 35, 25, 20, 16, 18, 13, 8, 11, 8, 4, 6]:$
$Pts := zip((x, y) \to [x, y], t, y):$

$> pointplot(Pts, labels = [days, \text{`}Recovery\ index\text{`}], labeldirections = [horizontal, vertical], title = \text{"Hospital Stay vs. Recovery Index"});$

Hospital Stay vs. Recovery Index

The scatterplot shows a clear negative trend.

Check the correlation ρ.

$> \rho := Statistics{:}\text{-}Correlation(t, y);$
$\text{`}R^2\text{`} = \rho^2;$

$$\rho := -0.941052825413336$$
$$R^2 = 0.885580420218423$$

Using either rule of thumb, the correlation coefficient $\rho = -0.94$ indicates a strong linear relation. With this value in hand, looking at the scatterplot leads us to think we will have an excellent linear regression model.

Let's determine the linear model, calling it *RI*.

$> Statistics{:}\text{-}Fit(a \cdot x + b, t, y, x);$
$RI := unapply(evalf(\%, 6), x);$

$$RI := x \mapsto -0.752508\,x + 46.4604$$

It's time for diagnostics. Calculate the residuals and percent relative errors, then plot the residuals.

$> Residuals := [seq(y_i - RI(t_i), i = 1..15)]:$

$> RelErr := \left[seq\left(\frac{Residuals_i}{y_i} \cdot 100, i = 1..15\right)\right];$

Look at the worst percent relative error values.

```
> SRE := sort(RelErr);
  SRE[1..2], SRE[-2.. - 1];
```

$$[-57.46925000, -44.57907500], [67.25200000, 140.8770000]$$

```
> pointplot(ResidualPts, symbol = soliddiamond, symbolsize = 16,
    title = "Residuals");
```

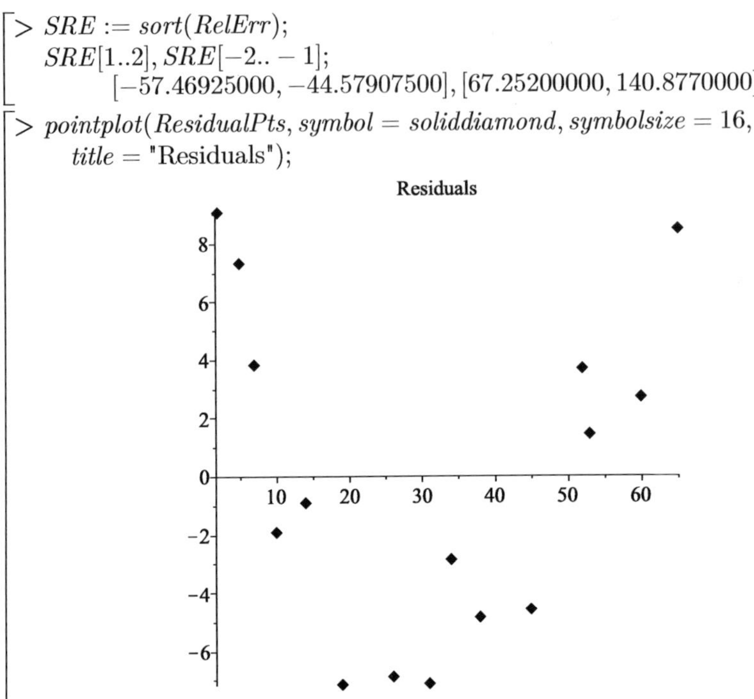

We've obtained the model, $y = f(t) = 49.4601 - 0.75251t$. The sum of squared error is 451.1945, the correlation is -0.94, and R^2, the coefficient of determination which is the correlation coefficient squared, is 0.886. These are all indicators of a "good" linear model.

But...

Although some percent relative errors are small, others are quite large, with eight of the fifteen over 20%. The largest two positive errors are $\approx 67\%$ and 140%, and the largest two negative errors are $\approx -45\%$ and -57%. How much confidence would you have in making predictions with this model? The residual plot clearly shows a curved pattern. The residuals and percent relative errors show that we do not have an adequate model. Advanced courses in statistical regression will show how to attempt to correct this inadequacy.

Further, suppose we need to predict the index when time was 100 days.

```
> RI(100);
```

$$-28.790400$$

A negative value is clearly unacceptable and makes no sense in the context of our problem since the index is always positive. This model does not pass the common sense test.

With a strong correlation of -0.94, what went wrong? The residual plot diagnostic shows a curved pattern. In many regression analysis books, the suggested first-attempt cure for a curved residual is adding a nonlinear term that is missing from the model. Since we only have (t, y) data points, we'll add a nonlinear term a_2x^2 to the model.

Example 6.2. Multiple Linear Regression.
Fit a parabolic model $f(x) = a_0 + a_1x + a_2x^2$ to the Hospital Stay vs. Recovery Index Data of Table 6.3.

This model, although a parabola, is linear in the "variables" a_0, a_1, and a_2.

Let's find the fit with Maple.

$> with(Statistics):$

$> Fit(a_0 + a_1 \cdot x + a_2 \cdot x^2, t, y, x):$
$QF := unapply(evalf(\%, 4), x);$

$$QF := x \mapsto 55.82 - 1.710x + 0.01481x^2$$

Use the formula $RS = 1 - SSE/SST$ to compute R^2, the coefficient of determination, for the quadratic fit.

$> SSE := sum((y_i - QF(t_i))^2, i = 1..15);$
$\mu := Mean(y):$
$SST := sum((y_i - \mu)^2, i = 1..15);$

$$SSE := 72.34557123$$
$$SST := 3943.333333$$

$> RSq := 1 - \frac{SSE}{SST};$

$$RSq := 0.9816537013$$

These diagnostics look quite good. The sum of the squares of the errors is much smaller. The coefficient of determination R^2 is larger, and the correlation has increased to 0.99.

Let's plot the new residuals to see if there is still a curved pattern.

$> QResidualPts := [seq([t_i, y_i - QF(t_i)], i = 1..15)]:$

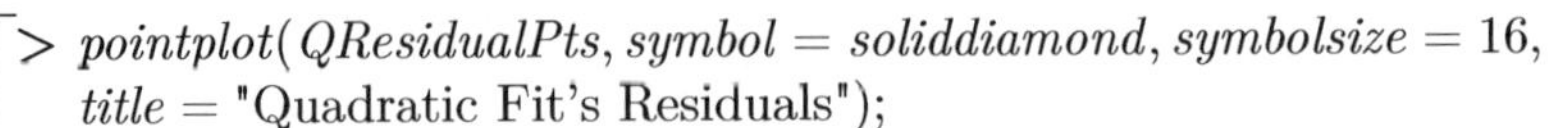

> pointplot(QResidualPts, symbol = soliddiamond, symbolsize = 16,
title = "Quadratic Fit's Residuals");

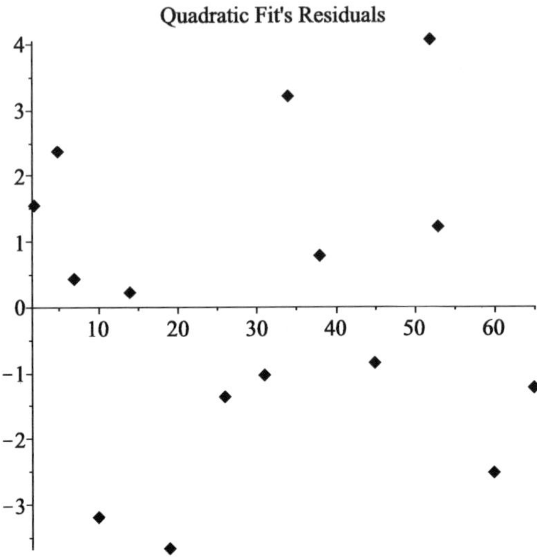

Use the model to predict the index at 100 days. We find $QF(100) = 32.92$. The answer is now positive, but again does not pass the common-sense test. The quadratic function is now curving upwards toward positive infinity. *Graph it!* An upward curve is an unacceptable outcome; an increasing curve for the index indicates that staying in the hospital for an extended time leads to a better prognosis. We certainly cannot use the model to predict the values of the index for t beyond 60.

We must try a different model. Looking at the original scatterplot leads us to attempt fitting an exponential decay curve.

Example 6.3. Nonlinear Regression: Exponential Decay.
Fit an exponential decay model $f(x) = a_0 e^{a_1 x}$ to the Hospital Stay vs. Recovery Index Data of Table 6.3.[5]

Maple's *NonlinearFit* command from the *Statistics* package does not do well unless we specify reasonable estimates for the parameters of our model, here a_0 and a_1.

> $NonlinearFit(a_0 \cdot exp(a_1 \cdot x), t, y, x)$:
> $NEDF := unapply(evalf(\%, 5), x);$

$$NEDF := x \mapsto 5.6790\, 10^{-29}\, e^{1.0282x}$$

Test the fit at $t = 100$.

> $NEDF(100)$

$$2.561123364\, 10^{16}$$

[5]Adapted from Neter et al. [NKNW1996] and Fox [Fox2012].

A ridiculous value!

Care must be taken with the selection of the initial values for the unknown parameters (see, e.g., Fox [Fox2012]). Maple yields good models when the calculations are based upon good input parameters. How can we get reasonable approximations to a_0 and a_1? Let's use a standard technique: transform the model to make it linear. Apply logs to our model.

$$\begin{aligned} y &= a_0 e^{a_1 x} \\ \ln(y) &= \ln\left(a_0 e^{a_1 x}\right) \\ &= \ln(a_0) + a_1 x \end{aligned}$$

Relabel $\ln(y) = Y$ and $\ln(a_0) = A_0$ to obtain

$$Y = A_0 + a_1 x$$

Now use Maple to fit the transformed model.

> $LinearFit(A_0 + a_1 \cdot x, t, ln\sim(y), x);$

$$4.03715886613379 - 0.0379741808112946 \cdot x$$

Good estimates to use for the parameters are

$$a_0 = e^{A_0} = 56.667 \quad \text{and} \quad a_1 = -0.03797.$$

> $NonlinearFit(a_0 \cdot \exp(a_1 \cdot x), t, y, x,$
> $\quad initialvalues = [a_0 = 56.667, a_1 = -0.03797]);$
> $NEDF := unapply(evalf(\%, 5), x);$
> $NEDF(100);$

$$NEDF := x \mapsto 58.607\, e^{-0.039586x}$$
$$1.118797159$$

This prediction is much closer to what we expect. A common technique used to test the result is to recompute the model with different, but close, initial values. *Try it!*

Note how close the nonlinear-fitted parameter values are to the estimates. The difference comes from where least squares is applied in the computation: to the logs of data points, rather than to the points. Write the general least squares formulas for the model and the transformed model to see the difference. Also, compare the model from Maple's *ExponentialFit* to the one we have found.

For a nonlinear model, linear correlation and R^2 have no meaning. Computing the sum of squared error gives $SSE = 45.495$. This value is substantially smaller than $SSE = 451.1945$ obtained by the linear model. The nonlinear model appears reasonable. Now check the nonlinear model's residual plot for patterns.

> $EDResidualPts := [seq([t_i, y_i - NEDF(t_i)], i = 1..15)]:$

> *pointplot*(*EDResidualPts*, *symbol* = *soliddiamond*, *symbolsize* = 16, *title* = "Exponential Decay Fit's Residuals");

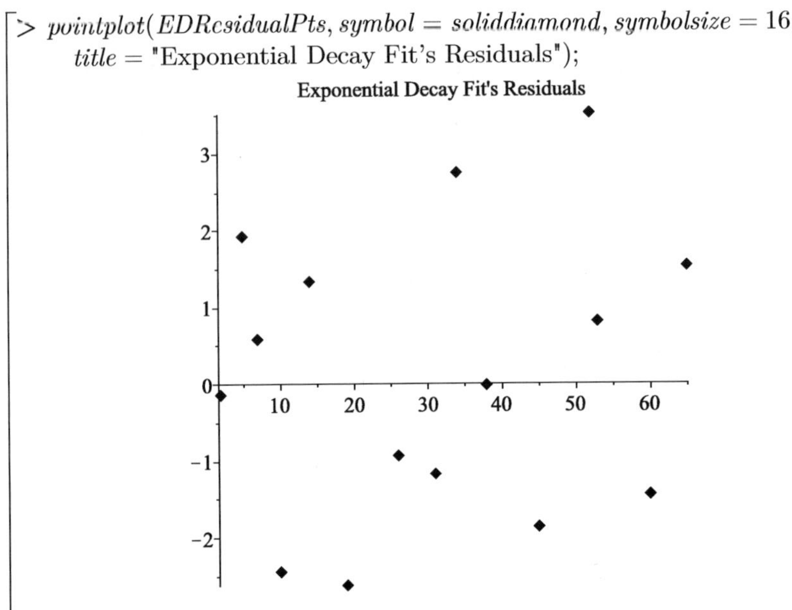

Figure 6.3 shows the exponential decay model overlaid on the data

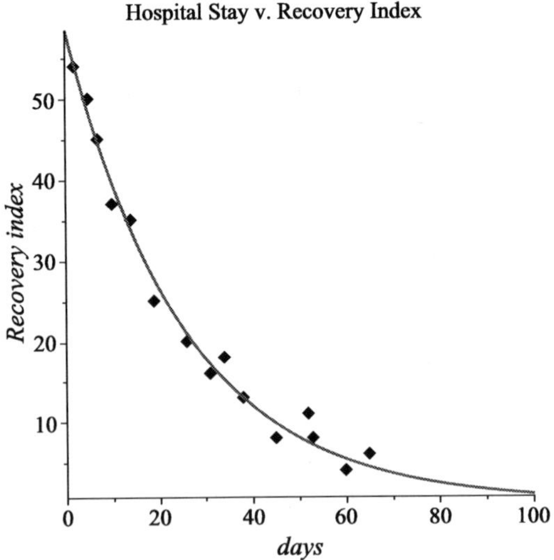

FIGURE 6.3: Exponential Decay Fit to Hospital Recovery Data

The percent relative errors are also much improved—only four are larger than 20%, and none are beyond 36.3%. *Verify this!*

Using this model to predict the index when $t = 100$ gave $y = 1.12$. This new result passes the common-sense test. This model would be our recommendation for the hospital's data.

The text's *PSMv2* library contains *NonlinearRegression*, a program we wrote for nonlinear regression in Maple that prints a diagnostic report, and then returns a fit of the specified model.

$> with(PSMv2):$

$> Describe(NonlinearRegression);$

```
# Usage: NonlinearRegression(X_data, Y_data, model, param_estimates)
NonlinearRegression( )
```

$> model := a_0 \cdot \exp(a_1 \cdot x);$
$paramestimates := [a_0 = 50, a_1 = -0.04];$
$NonlinearRegression(t, y, model, paramestimates)$

	Coefficient	*Standard Error*	*T-Statistic*	*P-value*
a_0	58.607	1.4722	39.810	0.0
a_1	−0.039586	0.0017113	−23.132	0.0

$$\textit{Model } m(x) = 58.607\, e^{-0.039586x}$$

We see that both coefficients are statistically significant (*P-values* < 0.05).

Just for illustration, change the model to $a + b \cdot \exp(c \cdot x)$ and rerun *NonlinearRegression.*

$> model2 := a + b \cdot \exp(c \cdot x);$
$paramestimates2 := [a = 2.5, b = 57, c = -0.04];$
$NonlinearRegression(t, y, model2, paramestimates2)$

	Coefficient	*Standard Error*	*T-Statistic*	*P-value*
a	2.4302	1.9655	1.2364	0.23995
b	57.332	1.8284	31.356	0.0
c	−0.044604	0.0048777	−9.1445	0.0

$$\textit{Model } m(x) = 2.4302 + 57.332\, e^{-0.044604x}$$

Select the model with:

$> rhs(\%);$

$$2.4302 + 57.332\, e^{-0.044604x}$$

The new constant term a is not statistically significant (*P-value* $\gg 0.05$).

Exercises

In Exercises 1 through 4 compute the correlation coefficient and then fit the following data with the given models using linear regression.

1.

x	1	2	3	4	5
y	1	1	2	2	4

   (a) $y = a + bx$

   (b) $y = ax^2$

2. Data from stretching a spring:

$x\,(\times 10^{-3})$	5	10	20	30	40	50	60	70	80	90	100
$y\,(\times 10^{5})$	0	19	57	94	134	173	216	256	297	343	390

   (a) $y = ax$

   (b) $y = a + bx$

   (c) $y = ax^2$

3. Data for the ponderosa pine:

x	17	19	20	22	23	25	28	31	32	33	36	37	39	42
y	19	25	32	51	57	71	113	140	153	187	192	205	250	260

   (a) $y = ax$

   (b) $y = a + bx$

   (c) $y = ax^2$

   (d) $y = ax^3$

   (e) $y = a_3x^3 + a_2x^2 + a_1x + a_0$

4. Fit the model $y = ax^{3/2}$ to Kepler's planetary data:

Body	*Period*	*Distance from sun* (m)
Mercury	7.60×10^6	5.79×10^{10}
Venus	1.94×10^7	1.08×10^{11}
Earth	3.16×10^7	1.5×10^{11}
Mars	5.94×10^7	2.28×10^{11}
Jupiter	3.74×10^8	7.79×10^{11}
Saturn	9.35×10^8	1.43×10^{12}
Uranus	2.64×10^9	2.87×10^{12}
Neptune	5.22×10^9	4.5×10^{12}

5. Fit the following data, Experience vs. Flight time, with a nonlinear exponential model $y = a\,e^{bx}$.

Crew Experience (months)	Total Flight Time	Crew Experience (months)	Total Flight Time
91.5	23.79812835	116.1	33.85966964
84	22.32110752	100.6	25.21871384
76.5	18.59206246	85	25.63021418
69	15.06492527	69.4	18.52772164
61.5	14.9960869	53.9	12.88023798
80	24.63492506	112.3	31.42422451
72.5	18.87937085	96.7	26.299381
65	18.60416326	81.1	19.6670982
57.5	13.6173644	65.6	15.21024716
50	15.62588379	50	11.977759
103	28.46073261	120	36.33722835
95.5	30.39886433	104.4	29.35965197
88	25.06898139	88.9	21.78073392
80.5	21.30460092	73.7	21.72030963
73	21.98724383	57.8	16.19643014

Projects

Project 6.1. Write a program without using statistical programs/software to

(a) compute the correlation between the data, and

(b) find the least squares fit for a general proportionality model y is proportional to x^n.

Project 6.2. Write a program without using statistical programs/software to

(a) compute the correlation between the data, and

(b) find the least squares fit for a general polynomial model $y = a_0 + a_1 x + a_2 x^2 + \cdots + a_n x^n$.

6.4 Advanced Regression Techniques with Examples

In this section, we will consider sine regression, one-predictor logistics regression, and one-predictor Poisson regression. First, consider data that has an oscillating component.

Nonlinear Regression

Example 6.4. Model Shipping by Month.
Management is asking for a model that explains the behavior of tons of material shipped over time so that predictions might be made concerning future allocation of resources. Table 6.4 shows logistical supply train information collected over 20 months.

TABLE 6.4: Total Shipping Weight vs. Month

Month	*Shipped* (tons)	*Month*	*Shipped* (tons)
1	20	11	19
2	15	12	25
3	10	13	32
4	18	14	26
5	28	15	21
6	18	16	29
7	13	17	35
8	21	18	28
9	28	19	22
10	22	20	32

First, we find the correlation coefficient.

```
> with(plots) :
  with(Statistics) :
  with(PSMv2) :
> Month := [$1..20)] :
  Tons := [20, 15, 10, 18, 28, 18, 13, 21, 28, 22, 19, 25, 32, 26, 21, 29, 35, 28, 22, 32] :
> Correlation(Month, Tons);
                              0.672564359308119
```

According to our rules of thumb, 0.67 is a moderate to strong value for linear correlation. So is the model to use linear? Plot the data, looking for trends and patterns. Figure 6.4a shows the data as a scatterplot, while 6.4b "connects the dots."

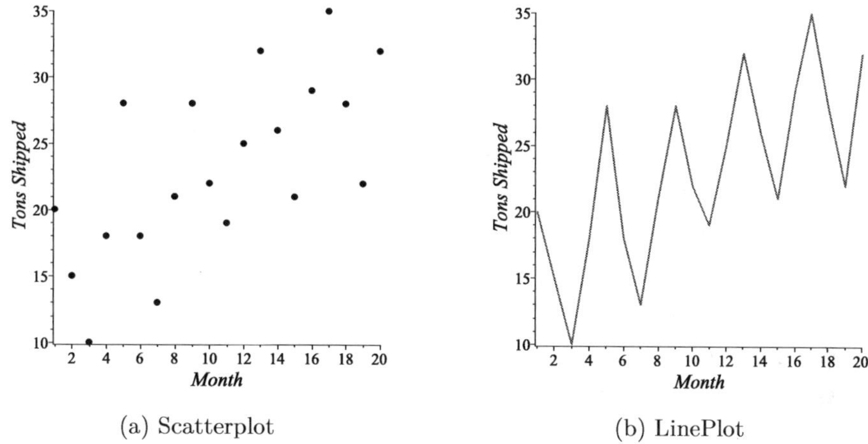

(a) Scatterplot (b) LinePlot

FIGURE 6.4: Shipping Data Graphs

Although linear regression can be used here, it will not capture the seasonal trends. There appears to be an oscillating pattern with a linear upward trend.

For purposes of comparison, find a linear model.

> $Fit(a + b \cdot x, Month, Tons, x, summarize = embed);$

$$15.1263157894737 + 0.759398496240602\,x$$

Summary

Model: $15.126316 + 0.75939850\,x$				
Coefficients	Estimate	Standard Error	t-value	P(>|t|)
a	15.1263	2.35928	6.41140	$\mathbf{4.90702 \cdot 10^{-6}}$
b	0.759398	0.196949	3.85581	**0.00115802**

R-squared: 0.452343

Adjusted R-squared: 0.421917

▸ **Residuals**

The R^2 value of 0.45 does not indicate a strong fit of the data as we expected. Since we need to represent oscillations with a slight linear upward trend, we'll try a sine model with a linear component

$$f(x) = a_0 + a_1 x + a_2 \sin(a_3 x + a_4).$$

As noted before, good estimates of the parameters a_i are necessary for obtaining a good fit. *Check Maple's fit with default parameter values!* Use the linear fit from above for estimating a_0 and a_1; use your knowledge of trigonometry to estimate the other parameters.

$$a_0 = 15, \quad a_1 = 0.8, \qquad a_2 = 6, \quad a_3 = 1.6, \quad a_4 = 1.$$

Use *NonlinearRegression* from the *PSMv2* package.

$$> sinemodel := a_1 + a_1 \cdot x + a_2 \cdot sin(a_3 \cdot x + a_4) :$$
$$paramestimates := [a_0 = 15, a_1 = 0.8, a_2 = 6, a_3 = 1.6, a_4 = 1.0] :$$

$$> NonlinearRegression(Month, Tons, sinemodel, paramestimates);$$

	Coefficient	*Standard Error*	*T-Statistic*	*P-value*
a_0	14.187	0.56742	25.002	0.
a_1	0.84795	0.047473	17.862	0.
a_2	6.6892	0.38351	17.442	0.
a_3	1.5735	0.010024	156.98	0.
a_4	0.082625	0.12415	0.66555	0.51580

Model: $m(x) = 14.187 + 0.84795\,x + 6.6892\sin(1.5735\,x + 0.082625)$

The coefficient's p-values look very good, save the phase shift a_4. Plot the model with the data.

$$> m := rhs(\%) :$$

$$> display(plot(m, x = 0..21, thickness = 2),$$
$$pointplot(Pts, symbol = solidcircle, symbolsize = 14),$$
$$labels = [month, \text{`Tons Shipped`}],$$
$$labeldirections = [horizontal, vertical]) :$$

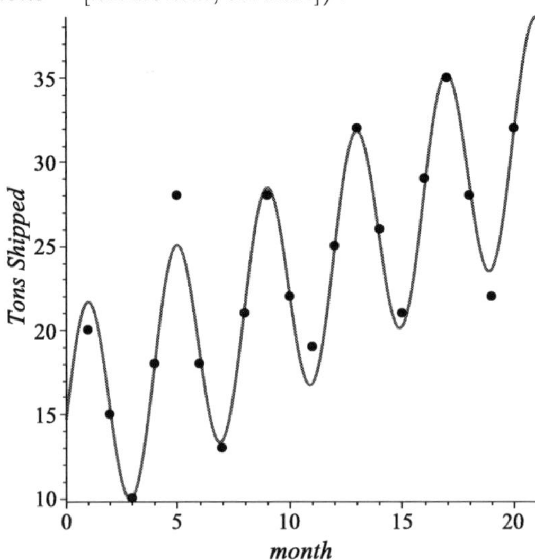

This model captures the oscillations and upward trend nicely. The sum of squared error is only $SSE = 21.8$. The new SSE is quite a bit smaller than that of the linear model. Clearly the model based on sine+linear regression does a much better job in predicting the trends than just using a simple linear regression.

Example 6.5. Modeling Casualties in Afghanistan.
In a January 2010 news report, General Barry McCaffrey, USA, Retired, stated that the situation in Afghanistan would be getting much worse.[6] General McCaffrey claimed casualties would double over the next year. The problem is to analyze the data to determine whether it supports his assertion.

The data that Gen. McCaffrey used for his analysis was the available 2001–2009 figures shown in Table 6.5. The table also shows casualties for 2010 and part of 2011 that were not available at the time.

TABLE 6.5: Casualties in Afghanistan by Month

Month	2001	2002	2003	2004	2005	2006	2007	2008	2009	2010	2011
1		12	10	25	6	7	21	19	83	199	308
2		13	9	17	5	17	39	18	52	247	245
3		53	14	12	16	7	26	53	78	346	345
4		8	13	11	29	13	61	37	60	307	411
5		2	8	31	34	39	87	117	156	443	
6		3	4	34	60	68	100	167	213	583	
7		6	10	25	38	59	100	151	394	667	
8		2	13	22	72	56	103	167	493	631	
9		5	19	34	47	70	88	122	390	674	
10	5	8	5	38	27	68	131	90	348	631	
11	10	4	27	18	12	51	75	37	214	605	
12	28	6	12	9	20	23	46	50	168	359	

First, do a quick "reasonability model." Sum the numbers across the years that Gen. McCaffrey had data for (Table 6.6) and graph a scatterplot.

TABLE 6.6: Casualties in Afghanistan by Year

2002	2003	2004	2005	2006	2007	2008	2009
122	144	276	366	478	877	1028	2649

[6]See www.newser.com/story/77563/general-brace-for-thousands-of-gi-casualties.html.

The scatterplot's shape suggests that we use a parabola as our "reasonability model."

```
> with(plots) :
  with(Statistics) :
> CasYr := [122, 144, 276, 366, 478, 877, 1028, 2649] :
  Yr := [$1..8];
  Pts := zip((x, y) -> [x, y], Yr, CasYr) :
> Fit(a + b·x + c·x^2, Yr, CasYr, x) :
  RM := unapply(evalf(%, 5), x);
  RM(9);
```

$$RM := x \mapsto 606.39 - 404.54x + 76.726x^2$$
$$3180.336$$

The model's prediction, while much smaller than the actual 2010 value, is not a doubling. However, the model does suggest further analysis is required.

We will focus on the four years before 2010, that is 2006 to 2009, and ask, do we expect the casualties in Afghanistan to double over the next year, 2010, based on those casualty figures?

```
> Cas := [7, 17, 7, 13, 39, 68, 59, 56, 70, 68, 51, 23, 21, 39, 26, 61, 87, 100, 100,
    103, 88, 131, 75, 46, 19, 18, 53, 37, 117, 167, 151, 167, 122, 90, 37, 50, 83, 52,
    78, 60, 156, 213, 394, 493, 390, 348, 214, 168] :
  Mon := [$1..nops(Cas)] :
  Pts := zip((x, y) -> [x, y], Mon, Cas) :
```

In the same fashion as before, plot both a scatterplot and a line plot of the data available to Gen. McCaffrey over that period. See Figure 6.5. The line plot may better show trends in the data, such as an upward tendency or oscillations that are not apparent in the scatterplot. However, a line plot can be very difficult to read or interpret when there are a large number of data points connected. Good graphing is always a balancing act. After modeling the data from 2006 to 2009, we can use the 2010 values to test our model for goodness of prediction. There are two trends apparent from the graphs. First, the data oscillates seasonally. This time, however, the oscillations grow in magnitude. We will try to capture that with an $x \cdot \sin(x)$ term. Second, the data appears to have an overall upward trend. We will attempt to capture that feature with a linear component. The nonlinear model we choose is

$$m(x) = a_0 + a_1 \cdot x + a_2 \cdot x \cdot \sin(a_3 \cdot x + a_4).$$

Using the techniques described in the previous example, we fit the nonlinear model: a growing-amplitude sine plus a linear trend. We estimate the parameters from the scatterplot:

$$[a_0 = 10, a_1 = 2, a_2 = 47, a_3 = 0.5, a_4 = 0.3].$$

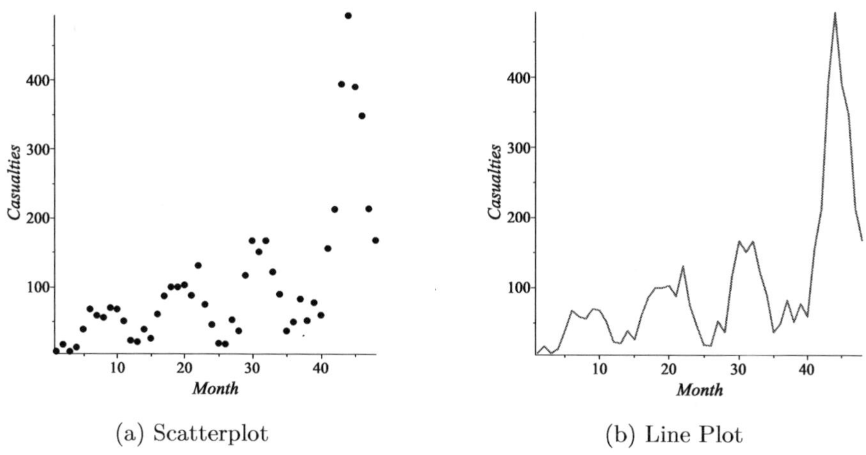

(a) Scatterplot (b) Line Plot

FIGURE 6.5: Afghanistan Casualties Graphs

Use our *NonlinearRegression* program.

$$> model := a_0 + a_1 \cdot x + a_2 \cdot x \cdot \sin(a_3 \cdot x + a_4) :$$
$$paramest := [a_0 = 10, a_1 = 2, a_2 = 47, a_3 = 0.5, a_4 = 0.3] :$$

$$> NonlinearRegression(Mon, Cas, model, paramest) :$$
$$m := unapply(rhs(\%), x);$$

	Coefficient	*Standard Error*	*T-Statistic*	*P-value*
a_0	−7.1709	14.589	−0.49152	0.62556
a_1	4.3320	0.52772	8.2089	0.0
a_2	−3.4228	0.37369	−9.1596	0.0
a_3	0.50094	0.010784	46.454	0.0
a_4	1.4436	0.41046	3.5171	0.0010433

$$m := x \mapsto -7.1709 + 4.3320\, x - 3.4228\, x \sin(0.50094\, x + 1.4436)$$

The p=values for all the parameters, except the constant term, are quite good. Plot a graph to see the model capturing the oscillations and linear growth fairly well. Does the model also show the increase in amplitude as well?

Considering the residuals will be our next diagnostic.

$$> Residuals := [seq([Mon_i, Cas_i - m(Mon_i)], i = 1..nops(Cas))] :$$

Now graph the residuals looking for patterns and warning signs.

> $pointplot(Residuals, symbol = solidcircle, symbolsize = 14);$

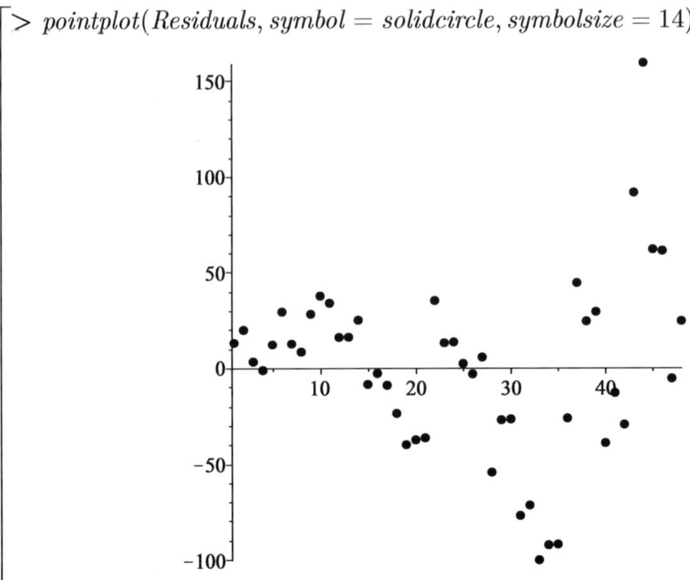

The residual plot shows no clear pattern suggesting the model appears to be adequate. Although we note that the model did not "keep up" with the change in amplitude of the oscillations.

What does the model predict for 2010 in relation to 2009?

> $Yr2009 := sum(Cas_k, k = 37..48);$
$Yr2010 := sum(m(k), k = 49..60);$
$\frac{Yr2010}{Yr2009};$

$$Yr2009 := 2649$$
$$Yr2010 := 2869.751675$$
$$1.083333966$$

This model does not show a doubling effect from year four. Thus, the model does not support General McCaffrey's hypothesis.

Consider the ratios of casualties for each month of 2009 to 2008 and then 2010 to 2009. How would this information affect your conclusions?

Logistic Regression and Poisson Regression

Often the dependent variable has special characteristics. Here we examine two notable cases: (a) logistic regression, also known as a logit model, where the dependent variable is binary, and (b) Poisson regression where the dependent variable measures integer counts that follow a Poisson distribution.

One-Predictor Logistic Regression

We begin with three one-predictor logistic regression model examples in which the dependent variable is binary, i.e., $\{0, 1\}$. The logistic regression model form that we will use is

$$f(x) = \frac{e^{b_0+b_1 x}}{1 + e^{b_0+b_1 x}}.$$

The logistic function, approximating a unit step function, gave the name logistic regression. The most general form handles dependent variables with a finite number of states.

Example 6.6. Damages versus Flight Time.
After a number of hours of flight time, equipment is either damaged or not. Let the dependent variable y be a binary variable with

$$y = \begin{cases} 1 & \text{there is damage} \\ 0 & \text{there is no damage} \end{cases},$$

and let t be the flight time in hours.

Over a reporting period, the data of Table 6.7 has been collected.

TABLE 6.7: Damage vs. Flight Time

t	4	2	4	3	9	6	2	11	6	7	3	2	5	3	3	8
y	1	1	0	1	0	0	0	0	1	0	1	1	0	0	0	0

t	10	5	13	7	3	4	2	3	2	5	6	6	3	4	10
y	0	1	0	0	1	0	1	1	0	0	0	1	0	1	0

Calculate a logistic regression for damage.

```
> with(Statistics) :
```

```
> t := [4, 2, 4, 3, 9, 6, 2, 11, 6, 7, 3, 2, 5, 3, 3, 8,
    10, 5, 13, 7, 3, 4, 2, 3, 2, 5, 6, 6, 3, 4, 10] :
  y := [1, 1, 0, 1, 0, 0, 0, 0, 1, 0, 1, 1, 0, 0, 0, 0,
    0, 1, 0, 0, 1, 0, 1, 1, 0, 0, 0, 1, 0, 1, 0] :
  Pts := zip((u, v) -> [u, v], t, y) :
```

$$> LM := \frac{\exp(a + b \cdot x)}{1 + \exp(a + b \cdot x)} :$$

Now, the fit.

```
> NonlinearFit(LM, t, y, x, initialvalues = [a = 1.5, b = -0.5]) :
  Damage := unapply(evalf(%, 5), x);
```

$$Damage := x \mapsto \frac{e^{-0.39190\,x+1.4432}}{1. + e^{-0.39190\,x+1.4432}}$$

> $plot(Damage, 0..1 + max(t), thickness = 2);$

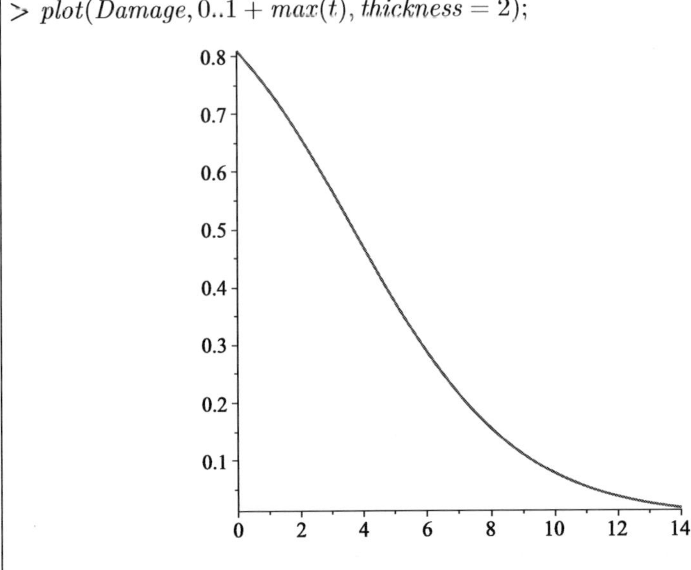

The analyst must decide over what intervals of x we call the y probability a 1 or a 0 using the logistic S-curve shown from the fit.

We switch from times to time differentials in the next example.

Example 6.7. Damages vs. Time Differentials.
Replace the times in the previous example with *time differentials* given in Table 6.8.

TABLE 6.8: Damage vs. Time Differentials (TD)

TD	19.2	24.1	−7.1	3.9	4.5	10.6	−3	16.2
y	1	1	0	1	0	0	0	0

TD	72.8	28.7	11.5	56.3	−0.5	−1.3	12.9	34.1
y	1	0	0	1	0	0	1	1

TD	6.6	−2.5	24.2	2.3	36.9	−11.7	2.1	10.4
y	0	0	0	0	1	0	1	1

TD	9.1	2	12.6	18	1.5	27.3	−8.4
y	0	0	0	1	0	1	0

Repeat the procedure of the previous example.

$> TD := [19.2, 24.1, -7.1, 3.9, 4.5, 10.6, -3, 16.2, 72.8, 28.7, 11.5, 56.3, -0.5, -1.3, 12.9, 34.1, 6.6, -2.5, 24.2, 2.3, 36.9, -11.7, 2.1, 10.4, 9.1, 2, 12.6, 18, 1.5, 27.3, -8.4]:$

$> NonlinearFit(LM, TD, y, x, initialvalues = [a = 0.5, b = 1]):$
$TDDamage := unapply(evalf(\%, 5), x);$

$$TDDamage := x \mapsto \frac{e^{0.054553\,x - 1.1997}}{1. + e^{0.054553\,x - 1.1997}}$$

$> plot(TDDamage, min(TD) - 1..1 + max(TD), thickness = 2);$

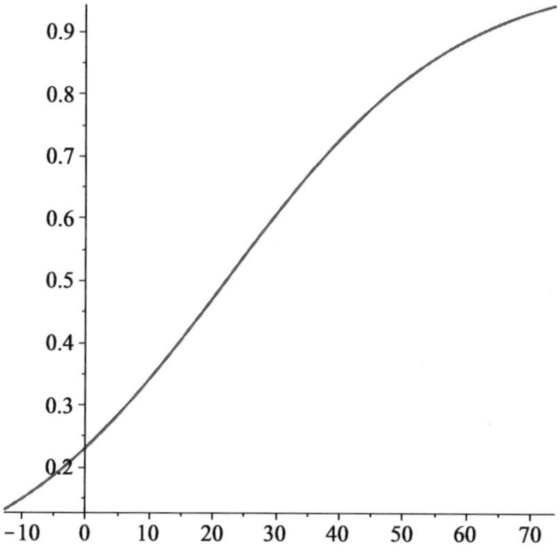

Once again, the analyst must decide over what intervals of x we call the y probability a 1 or a 0 using the logistic S-curve shown above.

Dehumanization is not a new phenomenon in human conflict. Societies have dehumanized their adversaries since the beginnings of civilization in order to allow them to seize, coerce, maim, or ultimately to kill while avoiding the pain of conscience for committing these extreme, violent actions. By taking away the human traits of these opponents, adversaries are made to be objects deserving of wrath and meriting the violence as justice.[7] Dehumanization still occurs today in both developed and underdeveloped societies. The next example analyzes the impact that dehumanization has in its various forms on the outcome of a state's ability to win a conflict.

Example 6.8. Conflict and Dehumanization.
To examine dehumanization as a quantitative statistic, we combine a data set of 25 conflicts from Erik Melander, Magnus Oberg, and Jonathan Hall's

[7]See David L. Smith, *Less Than Human: Why We Demean, Enslave, and Exterminate Others.*

"Uppsala Peace and Conflict," (Table 1, pg. 25)[8] with Joakim Kreutz's "How and When Armed Conflicts End: Introducing the UCDP Conflict Termination Dataset"[9] to have a designated binary "win-lose" assessment for each conflict. We will use civilian casualties as a proxy indicator of the degree of dehumanization during the conflict. The conflicts in Table 6.9 run the gamut from high- to low-intensity in the spectrum, and include both inter- and intra-state hostilities. Therefore, the data is a reasonably general representation.

TABLE 6.9: Top 25 Worst Conflicts Estimated by War-Related Deaths

Year	*Side A*	*Side B*	*Side A: Win= 1 Lose= 0*	*Civilian (1,000s)*	*Military (1,000s)*	*Percentage Civilian Deaths*
1946-48	India	CPI	1	800	0	100.0
1949-62	Columbia	Mil. Junta	1	200	100	66.67
1950-51	China	Taiwan	1	1,000	*	100.0
1950-53	Korea	South Korea	0	1,000	1,889	34.60
1954-62	Algeria/France	FLN	0	82	18	82.00
1956-59	China	Tibet	1	60	40	60.00
1956-65	Rwanda/Tutsi	Hutu	0	102	3	97.14
1961-70	Iraq	KDP	1	100	5	95.24
1963-72	Sudan	Anya Nya	1	250	250	50.00
1965-66	Indonesia	OPM	1	500	*	100.0
1965-75	N. Vietnam	S. Vietnam	1	1,000	1,058	48.59
1966-87	Guatemala	FAR	1	100	38	72.46
1967-70	Nigeria	Rep. Biafra	1	1,000	1,000	50.00
1967-70	Egypt	Israel	0	50	25	66.67
1971-71	Bangladesh	JSS/SB	1	1,000	500	66.67
1971-78	Uganda	Military Fact.	1	300	0	100.0
1972-72	Burundi	Military Fact.	1	80	20	80.00
1974-87	Ethiopia	OLF	1	500	46	91.58
1975-90	Lebanon	LNM	1	76	25	75.25
1975-78	Cambodia	Khmer Rouge	0	1,500	500	75.00
1975-87	Angola	FNLA	1	200	13	93.90
1978-87	Afghanistan	USSR	1	50	50	50.00
1979-87	El Salvador	FMLN	1	50	15	76.92
1981-87	Uganda	Kikosi Maalum	1	100	2	98.04
1981-87	Mozambique	Renamo	1	350	51	87.28

'*' denotes missing values.

[8] E. Melander, M. Öberg, and J. Hall, "The 'New Wars' Debate Revisited: An Empirical Evaluation of the Atrociousness of 'New Wars'," Uppsala Univ. Press, Uppsala, 2006. Available at www.pcr.uu.se/digitalAssets/654/c_654444-l_1-k_uprp_no_9.pdf.

[9] J. Kreutz, "How and When Armed Conflicts End: Introducing the UCDP Conflict Termination Dataset," J. Peace Research, 47(2), 2010, 243–250.

By including the ratio of civilian casualties to total casualties in Table 6.9, we are able to determine what percentage of casualties in each conflict is civilian. This ratio provides a quantifiable variable to analyze.

Binary logistic regression analysis is the first method to choose to analyze the interrelation of dehumanization's effects (shown by proxy through higher percentages of civilian casualties) on the outcome of conflict as a win (1) or a loss (0). This type of regression model will allow us to infer whether or not the independent variable, civilian casualties percentage, has a statistically significant impact on the conflict's outcome, win or lose. Using the data from Table 6.9, we assign the civilian casualty percentages to be the independent variable and Side A's win/loss outcome of the conflict to be the binary dependent variable, then develop a binary logistic regression model. Use Maple to derive the logistic regression statistics from the model as follows.

$> with(Statistics):$

$> CivilianCasualtyPercent := 0.01 \cdot [100, 66.67, 100, 34.6, 82, 60, 97.14, 95.24, 50, 100, 48.59, 72.46, 50, 66.67, 66.67, 100, 80, 91.58, 75.25, 75, 93.9, 50, 76.92, 98.04, 87.28]:$

$> WinLose := [1, 1, 1, 0, 0, 1, 0, 1, 1, 1, 1, 1, 1, 0, 1, 1, 1, 1, 1, 0, 1, 1, 1, 1, 1]:$

$$> model := \frac{\exp(a \cdot x + b)}{1 + \exp(a \cdot x + b)}:$$

We derive estimates of the parameters from the data. (See, e.g., Bauldry [B1997] for simple methods.) Take $a = -1.9$ and $b = 0.05$ initially.

$> NonlinearFit(model, CivilianCasualtyPercent, WinLose, t, initialvalues = [a = -1.9, b = 0.05]);$

$$0.800002645508260$$

This result does not pass the common sense test. Ask Maple for more information by increasing *infolevel.*

$> infolevel[Statistics] := 2:$
$infolevel[Optimization] := 1:$

$> NonlinearFit(model, CivilianCasualtyPercent, WinLose, t, initialvalues = [a = -1.9, b = 0.05]);$

```
In NonlinearFit (algebraic form)
LSSolve: calling nonlinear LS solver
LSSolve: using method=modifiednewton
LSSolve: number of problem variables 3
LSSolve: number of residuals 25
attemptsolution: conditions for a minimum are not all satisfied, but
a better point could not be found
```

$$0.800002645508260$$

Maple's *NonlinearFit* could not optimize the regression. Let's try our *NonlinearRegression*.

```
> with(PSMv2) :
> NonlinearRegression(CivilianCasualtyPercent, WinLose, model,
    [a = -1.9, b = 0.05]);
In Fit
In NonlinearFit (algebraic form)
LSSolve: calling nonlinear LS solver
LSSolve: using method=modifiednewton
LSSolve: number of problem variables 2
LSSolve: number of residuals 25
attemptsolution: conditions for a minimum are not all satisfied, but
a better point could not be found
```

	Coefficient	*Standard Error*	*T-Statistic*	*P-value*
a	1.9305	2.6384	0.73170	0.47174
b	−0.044361	1.8723	−0.023693	0.98130

$$\text{Model: } m(x) = \frac{e^{1.9305\,x-0.044361}}{1+e^{1.9305\,x-0.044361}}$$

This logistic model result appears much better at first look. However, the coefficients' P-values tell us to have no confidence in the model. *Graph the model with the data!*

Analysis Interpretation: The conclusion from our analysis is that the civilian casualty percentages are not significantly correlated with whether the conflict leads to a win or a loss for Side A. Therefore, from this initial study, we can loosely conclude that dehumanization does not have a significant effect on the outcome of a state's ability to win or lose a conflict. Further investigation will be necessary.

One-Predictor Poisson Regression

According to Devore [D2012], the *simple linear regression model* is defined by:

> There exists parameters β_0, β_1, and σ^2, such that for any fixed input value of x, the dependent variable is a random variable related to x through the *model equation* $Y = \beta_0 + \beta_1 x + \varepsilon$. The quantity ε in the model equation is the "error"—a random variable assumed to be normally distributed with mean 0 and variance σ^2.

We expand this definition to when the response variable y is assumed to have a normal distribution with mean μ_y and variance σ^2. We found that the mean could be modeled as a function of our multiple predictor variables, $x_1, x_2, \ldots, x_n$, using the linear function $Y = \beta_0 + \beta_1 x_1 + \beta_2 x_2 + \cdots + \beta_k x_k$.

The key assumptions for least squares are

- the relationship between dependent and independent variables is linear,
- errors are independent and normally distributed, and
- homoscedasticity[10] of the errors.

If any assumption is not satisfied, the model's adequacy is questioned. In first courses, patterns seen or not seen in residual plots are used to gain information about a model's adequacy. (See [AA1979], [D2012]).

Normality Assumption Lost
In logistic and Poisson regression, the response variable's probability lies between 0 and 1. According to Neter [NKNW1996], this constraint loses both the normality and the constant variance assumptions listed above. Without these assumptions, the F and t tests cannot be used for analyzing the regression model. When this happens, transform the model and the data with a logistic transformation of the probability p, called logit p, to map the interval $[0,1]$ to $(-\infty,+\infty)$, eliminating the 0-1 constraint:

$$\ln\left(\frac{p}{1-p}\right) = \beta_0 + \beta_1 x_1 + \beta_2 x_2 + \cdots + \beta_n x_n$$

The βs can now be interpreted as increasing or decreasing the "log odds" of an event, and $\exp(\beta)$ (the "odds multiplier") can be used as the odds ratio for a unit increase or decrease in the associated explanatory variable.

When the response variable is in the form of a *count*, we face a yet different constraint. Counts are all positive integers corresponding to rare events. Thus, a Poisson distribution (rather than a normal distribution) is more appropriate since the Poisson has a mean greater than 0, and the counts are all positive integers. Recall that the Poisson distribution gives the probability of y events occurring in time period t as

$$P(y;\mu) = \frac{\exp(-\mu t)\cdot(\mu t)^y}{y!}.$$

Then the logarithm of the response variable is linked to a linear function of explanatory variables.

$$\ln(Y) = \beta_0 + \beta_1 x_1 + \beta_2 x_2 + \cdots + \beta_n x_n$$

Thus

$$Y = \left(e^{\beta_0}\right)\left(e^{\beta_1 x_1}\right)\left(e^{\beta_2 x_2}\right)\cdots\left(e^{\beta_n x_n}\right).$$

In other words, a Poisson regression model expresses the "log outcome rate" as a linear function of the predictors, sometimes called "exposure variables."

[10]Homoscedasticity: All random variables have the same finite variance.

Assumptions in Poisson Regression

There are several key assumptions in Poisson regression that are different from those in the simple linear regression model. These assumptions include that the logarithm of the dependent variable changes linearly with equal incremental increases in the exposure variable; i.e., the relationship between the *logarithm of the dependent variable* and the independent variables is linear. For example, if we measure risk in exposure per unit time with one group as counts per month, while another is counts per years, we can convert all exposures to strictly counts. We find that changes in the rate from combined effects of different exposures are multiplicative; i.e., changes in the log of the rate from combined effects of different exposures are additive. We find for each level of the covariates, the number of cases has variance equal to the mean, making it follow a Poisson distribution. Further, we assume the observations are independent.

Here, too, we use diagnostic methods to identify violations of the assumptions. To determine whether variances are too large or too small, plot residuals versus the *mean* at different levels of the predictor variables. Recall that in simple linear regression, one diagnostic of the model used plots of residuals against fits (fitted values). We will look for patterns in the residual or deviation plots as our main diagnostic tool for Poisson regression.

Poisson Regression Model

The basic model for Poisson regression is

$$Y_i = E[Y_i] + \varepsilon_i \quad \text{for } i = 1, 2, \ldots, n$$

The ith case mean response is denoted by u_i, where u_i can be one of many defined functions (Neter [NKNW1996]). We will only use the form

$$u_i = u(\mathbf{x}_i, \mathbf{b}) = \exp(\mathbf{x}_i^T \mathbf{b}) \quad \text{where } u_i > 0.$$

We assume that the Y_i are independent Poisson random variables with expected value u_i.

In order to apply regression techniques, we will use the likelihood function L (see [AA1979, D2012]) given by

$$L = \prod_{i=1}^{n} f_i(Y_i) = \prod_{i=1}^{n} \frac{(u(\mathbf{x}_i, \mathbf{b}))^{Y_i} \cdot \exp(-u(\mathbf{x}_i, \mathbf{b}))}{Y_i!} \tag{6.3}$$

Maximizing this function is intrinsically quite difficult. Instead, maximize the logarithm of the likelihood function shown below.

$$\ln(L) = \sum_{i=1}^{n} Y_i \ln(u_i) - \sum_{i=1}^{n} u_i - \sum_{i=1}^{n} \ln(Y_i!) \tag{6.4}$$

Numerical techniques are used to maximize $\ln(L)$ to obtain the best estimates for the coefficients of the model. Often, "good" starting points are required to obtain convergence to the maximum ([Fox2012]).

The deviations or residuals will be used to analyze the model. In Poisson regression, the deviance is given by

$$Dev = 2\left[\sum_{i=1}^{n} Y_i \cdot \ln\left(\frac{Y_i}{u_i}\right) - \sum_{i=1}^{n}(Y_i - u_i)\right] \tag{6.5}$$

where u_i is the fitted model; whenever $Y_i = 0$, we set $Y_i \cdot \ln(Y_i/u_i) = 0$.

Diagnostic testing of the coefficients is carried out in the same fashion as for logistic regression. To estimate the variance-covariance matrix, use the Hessian matrix $H(\mathbf{X})$, the matrix of second partial derivatives of the log-likelihood function $\ln(L)$ of (6.4). Then the approximated variance-covariance matrix is $VC(\mathbf{X}, \mathbf{B}) = -H(\mathbf{X})^{-1}$ evaluated at $\mathbf{B}$, the final estimates of the coefficients. The main diagonal elements of VC are estimates for the variance; the estimated standard deviations se_B are the square roots of the main diagonal elements. Then perform hypothesis tests on the coefficients using t-tests.

Two examples using the Hessian follow.

Example 6.9. Hessian-based Modeling.
Consider the model $y_i = \exp(b_0 + b_1 x_i)$ for $i = 1, 2, \ldots, n$.

Put this model into (6.4) to obtain

$$\begin{aligned}\ln(L) &= \sum_{i=1}^{n} y_i \cdot \ln\big(\exp(b_0 + b_1 x_i)\big) - \sum_{i=1}^{n} \exp(b_0 + b_1 x_i) - \sum_{i=1}^{n} y_i! \\ &= \sum_{i=1}^{n} y_i \cdot (b_0 + b_1 x_i) - \sum_{i=1}^{n} \exp(b_0 + b_1 x_i) - \sum_{i=1}^{n} y_i!\end{aligned}$$

The Hessian $H = [h_{ij}]$ comes from

$$h_{ij} = \frac{\partial^2 \ln(L)}{\partial b_i\, \partial b_{ij}} \quad \text{for all } i \text{ and } j,$$

which gives the estimate of the variance-covariance matrix $VC = -H^{-1}_{\mathbf{B}=\mathbf{b}}$. For the two-parameter model (b_0 and b_1), the Hessian is

$$H(\mathbf{X}) = \begin{bmatrix} -\sum_{i=1}^{n} y_i & -\sum_{i=1}^{n} x_i\, y_i \\ -\sum_{i=1}^{n} x_i\, y_i & -\sum_{i=1}^{n} x_i^2\, y_i \end{bmatrix}.$$

Change the model slightly adding a second independent variable with a third parameter. The model becomes $y_i = \exp(b_0 + b_1 x_{1i} + b_2 x_{2i})$ for $i = 1, 2, \ldots, n$.

Compute the new Hessian and carefully note the similarities.

$$H(\mathbf{X}) = -\begin{bmatrix} \sum_{i=1}^{n} y_i & \sum_{i=1}^{n} x_{1i}\, y_i & \sum_{i=1}^{n} x_{2i}\, y_i \\ \sum_{i=1}^{n} x_{1i}\, y_i & \sum_{i=1}^{n} x_{1i}^2\, y_i & \sum_{i=1}^{n} x_{1i}x_{2i}\, y_i \\ \sum_{i=1}^{n} x_{2i}\, y_i & \sum_{i=1}^{n} x_{1i}x_{2i}\, y_i & \sum_{i=1}^{n} x_{2i}^2\, y_i \end{bmatrix}.$$

The pattern in the matrix is easily extended to obtain the Hessian for a model with n independent variables.

Let $y_i = \exp(b_0 + b_1x_{1i} + b_2x_{2i} + \cdots + b_nx_{ni})$. The general Poisson model Hessian is

$$H(\mathbf{X}) = -\begin{bmatrix} \sum_{i=1}^{n} y_i & \sum_{i=1}^{n} x_{1i}\, y_i & \sum_{i=1}^{n} x_{2i}\, y_i & \cdots & \sum_{i=1}^{n} x_{ni}\, y_i \\ \sum_{i=1}^{n} x_{1i}\, y_i & \sum_{i=1}^{n} x_{1i}^2\, y_i & \sum_{i=1}^{n} x_{1i}x_{2i}\, y_i & \cdots & \sum_{i=1}^{n} x_{1i}x_{ni}\, y_i \\ \sum_{i=1}^{n} x_{2i}\, y_i & \sum_{i=1}^{n} x_{1i}x_{2i}\, y_i & \sum_{i=1}^{n} x_{2i}^2\, y_i & \cdots & \sum_{i=1}^{n} x_{2i}x_{ni}\, y_i \\ \vdots & \vdots & \vdots & \ddots & \vdots \\ \sum_{i=1}^{n} x_{ni}\, y_i & \sum_{i=1}^{n} x_{1i}x_{ni}\, y_i & \sum_{i=1}^{n} x_{2i}x_{ni}\, y_i & \cdots & \sum_{i=1}^{n} x_{ni}^2\, y_i \end{bmatrix}.$$

Replace the formulas with numerical values from the data. The resulting symmetric square matrix should be non-singular. Compute the inverse of the negative of the Hessian matrix to find the variance-covariance matrix VC. The main diagonal entries of VC are the (approximate) variances of the estimated coefficients b_i. The square roots of the entries on the main diagonal are the estimates of $se(b_i)$, the standard error for b_i, to be used in the hypothesis testing with $t^* = b_i/se(b_i)$.

We now have all the information we need to build the tables for a Poisson regression that are similar to a regression program's output.

Estimating the Regression Coefficients: Summary

The number of predictor variables plus one (for the constant term) gives the number of coefficients in the model $y_i = \exp(b_0 + b_1x_{1i} + b_2x_{2i} + \cdots + b_nx_{ni})$.

Estimates of the b_i are the final values from the numerical search method (if it converged) used to maximize the log-likelihood function $\ln(L)$ of (6.4). The values of $se(b_i)$, the standard error estimate for b_i, are the square roots of the main diagonal of the variance-covariance matrix $VC = -H(\mathbf{X})^{-1}_{\mathbf{B}=\mathbf{b}}$. The values of $t^* = b_i/se(b_i)$ and the p-value, the probability $P(T > |t^*|)$. In the summary table of Poisson regression analysis below, let m be the number of variables in the model, and let k be the number of data elements of y, the dependent variable. A summary appears in Table 6.10.

TABLE 6.10: Poisson Regression Variables Summary

	Degrees of Freedom (df)	Deviance	Mean Deviance ($MDev$)	Ratio
Regression	m	$D_{reg} = D_t - D_{res}$	$MDev(reg) = D_{reg}/m$	$\lvert MDev(reg)\rvert$
Residual	$k-1-m$	D_{res} = result from the full model with m predictors	$MDev(reg) = \frac{D_{reg}}{k-1-m}$	
Total	$k-1$	D_t = result from reduced model $y = e^{b_0}$	$MDev(t) = \frac{D_{reg}}{k-1}$	

Note that a prerequisite for using Poisson regression is that the dependent variable Y must be discrete counts with large numbers being a rare event.

We have chosen two data sets that have published solutions to be our basic examples. First, an outline of the procedure:

STEP 0. Enter the data for X and Y.

STEP 1. For Y:

(a) generate a histogram, and

(b) perform a chi-squared goodness-of-fit test for a Poisson distribution.[11]

If Y follows a Poisson distribution, then continue. If Y is "count data," use Poisson regression regardless of the chi-squared test.

STEP 2. Compute the value of b_0 in the constant model $y = \exp(b_0)$ that minimizes (6.5); i.e., minimize two times the deviations.

[11]See, e.g., stattrek.com/chi-square-test/goodness-of-fit.aspx for an introduction.

STEP 3. Compute the values of b_0 and b_1 in the model $y = \exp(b_0 + b_1 x)$ that minimize the deviation (6.5).

STEP 4. Interpret the results and the odds ratio.

We'll step through an example following the outline above.

Example 6.10. Hospital Surgeries.
A group of hospitals has collected data on the numbers of Caesarean surgeries vs. the total number of births (see Table 6.11).[12]

TABLE 6.11: Total Births vs. Caesarean Surgeries

Total	3246	2750	2507	2371	1904	1501	1272	1080	1027	970
Special	26	24	21	21	21	20	19	18	18	17

Total	739	679	502	236	357	309	192	138	100	95
Special	17	16	16	16	16	15	14	14	13	13

Use the hospitals' data set to perform a Poisson regression following the steps listed above.

STEP 0. Enter the data.

```
> xhc := [3246, 2750, 2507, 2371, 1904, 1501, 1272, 1080, 1027, 970, 739,
    679, 502, 236, 357, 309, 192, 138, 100, 95] :
  yhc := [26, 24, 21, 21, 21, 20, 19, 18, 18, 17, 17, 16, 16, 16, 16, 15, 14, 14,
    13, 13] :
  N := nops(yhc);
                              N := 20
```

STEP 1. Plot a histogram, and then perform a Chi-square Goodness-of-fit test on *yhc*, if appropriate.
(Note: Maple's *Histogram* function is in the *Statistics* package. There are a large number of options for binning the data; we will use *frequencyscale* = *absolute* to have the heights of the bars equal to the frequency of entries in the associated bin. Collect the bin counts with *TallyInto*.)

```
> with(Statistics) :
  Dirac(0.) := 1.0 :    # needed for the Poisson distribution
> Bins := [$(min(yhc)..max(yhc))];
          Bins := [13, 14, 15, 16, 17, 18, 19, 20, 21, 22, 23, 24, 25, 26]
```

[12] Adapted from "Research Methods II: Multivariate Analysis," J. Trop. Pediatrics, Online Feature, (2009), pp. 136–143. Originally at: www.oxfordjournals.org/our_journals/tropej/online/ma_chap13.pdf.

> $Histogram(yhc, frequencyscale = absolute);$

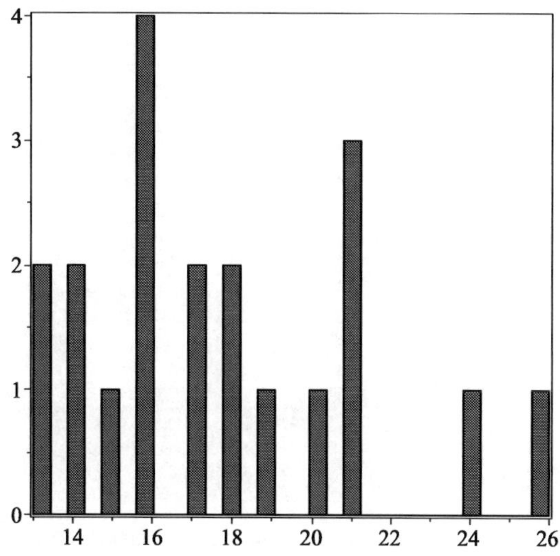

> $TallyInto(yhc, default, bins = max(yhc) - min(yhc) + 1):$
> $ObsFreq := rhs \sim (\%);$

$$ObsFreq := [2, 2, 1, 4, 2, 2, 1, 1, 3, 0, 0, 1, 0, 1]$$

Now for the chi-squared test. First, generate the predicted values from an estimated Poisson distribution.

> $\lambda_{est} := Mean(yhc);$

$$\lambda_{est} := 17.7500000000000$$

> $P := Distribution(Poisson(\lambda_{est})):$

> $PredFreq := [seq(N \cdot PDF(P, t), t \textbf{ in } Bins)];$

$$PredFreq := [1.090438738, 1.382520543, 1.635982643, 1.814918244, 1.894988167, 1.868668887, 1.745730144, 1.549335503, 1.309557389, 1.056574712, 0.8154000494, 0.6030562866, 0.4281699634, 0.2923083404]$$

We are ready to use Maple's chi-squared test, *ChiSquareGoodnessOfFitTest*, with a significance level of 0.05. Use the *summarize* = *embed* option, as it produces the most readable output. The command is terminated with a colon: "embedding the output" makes it unnecessary to return a result.

> $ChiSquareGoodnessOfFitTest(ObsFreq, PredFreq, level = 0.05,$
> $summarize = embed)$:

Chi-Square Test for Goodness-of-Fit

Null Hypothesis:	Observed sample does not differ from expected sample
Alternative Hypothesis:	Observed sample differs from expected sample

Categories	Distribution	Computed Statistic	Computed p-value	Critical Value
14.	*ChiSquare*(13)	10.8977	0.619387	22.3620

Result:	**Accepted:** This statistical test does not provide enough evidence to conclude that the null hypothesis is false.

The chi-squared test indicates that a Poisson distribution is reasonable.

STEP 2. Find the best constant model $y = \exp(b_0)$.
Let's use Maple's *LinearFit* on the function $Y = \ln(y) = b_0$.

> $LinearFit(b_0, xhc, \ln\sim(yhc), x, summarize = embed)$:

Summary

Model: 2.8581739

Coefficients	Estimate	Standard Error	t-value	P(>|t|)
Parameter 1	2.85817	0.0434058	65.8478	**0.**

R-squared: $-1.77636\,10^{-15}$

Adjusted R-squared: $-1.77636\,10^{-15}$

▼ **Residuals**

Residual Sum of Squares	Residual Mean Square	Residual Standard Error	Degrees of Freedom
0.715944	0.0376813	0.194117	19

Five Point Summary

Minimum	First Quartile	Median	Third Quartile	Maximum
0.293225	−0.123233	−0.0249605	0.166019	0.399923

STEP 3. Find the best exponential model $y = \exp(b_0 + b_1 x)$.
Let's use Maple's *ExponentialFit* to find the model.

> $ExponentialFit(xhc, yhc, x, summarize = embed)$:

Summary

Model: $14.136875\, e^{0.000190568858x}$

Coefficients	Estimate	Standard Error	t-value	P(>|t|)
Parameter 1	2.64879	0.0192928	137.294	**0.**
Parameter 2	0.000190569	0.0000132697	14.3612	**$2.66\,10^{-11}$**

R-squared: 0.999650

Adjusted R-squared: 0.999611

▼ **Residuals**

Residual Sum of Squares	Residual Mean Square	Residual Standard Error	Degrees of Freedom
0.0574687	0.00319270	0.0565040	18

Five Point Summary

Minimum	First Quartile	Median	Third Quartile	Maximum
−0.102894	−0.0420307	0.00279072	0.0449234	0.0788279

STEP 4. Conclude by calculating the odds-ratio.
Use the odds-multiplier $\exp(\beta_1)$ as the approximate odds-ratio, often called risk-ratio for Poisson regression.

> $OR := \exp(0.000190569);$

$$OR := 1.000190587$$

OR represents the potential increase resulting from one unit increase in x. (*How does this concept relate to "opportunity cost" in linear programming and "marginal revenue" in economics?*)

Return to the Philippines example relating literacy and violence described in the opening of this chapter.

Example 6.11. Violence in the Philippines.
The number of significant acts of violence, *SigActs* in Table 6.12, are integer counts.[13]

[13]Data sources: National Statistics Office (Manila, Philipppines) and the *Archives of the Armed Forces of the Philippines.*

TABLE 6.12: Literacy Rate (*Lit*) vs. Significant Acts of Violence (*SigActs*), Philippines, 2008.

Province	Lit	SigActs	Province	Lit	SigActs
Basnlan	71.6	29	Drnagat Istands	85.7	0
Larseao del Sur	71.6	30	Sungapdel Norte	85.7	10
Maguindanso	71.6	122	Sungapdel Sur	85.7	31
Suu	71.6	26	Bukidnon	85.9	14
Tawi-Tawi	71.6	1	Camigum	85.9	0
Bihran	72.9	0	Laraodel Norte	85.9	57
Eastern Samar	72.9	11	Misamis Occidental	85.9	8
Leyte	72.9	2	Misamis Onental	85.9	7
Northern Samar	72.9	23	Batanes	86.1	0
Southern Leyte	72.9	0	Cagayan	86.1	15
Western Samar	72.9	64	Isabela	86.1	4
North Cotabato	78.3	125	Nueva Vizcaya	86.1	3
Sarangani	78.3	23	Quirmo	86.1	0
South Cotabato	78.3	5	Bokal	86.6	2
Suan Ku:iarat	78.3	18	Cebu	86.6	0
Zamboanga del Norte	79.6	8	Negros Onertal	86.6	27
Zamboarga del Sur	79.6	10	Siquyjor	86.6	0
Zamboanga Sibugay	79.6	3	Abra	89.2	11
Albey	79.9	35	Apayap	89.2	0
Camarines Norte	79.9	12	Benguet	89.2	0
Camarines Sur	79.9	44	Ifugao	89.2	0
Caanduancs	79.9	9	Kahinga	89.2	11
Masbate	79.9	42	Mountain Province	89.2	0
Sorsogon	79.9	52	Veces Norte	91.3	0
Compostela Valtey	81.7	126	Lvees Sur	91.3	2
Davaodel Norte	81.7	35	La Unon	91.3	0
Davaedel Sur	81.7	64	Pangasman	91.3	0
Davao Orental	81.7	40	Aurora	92.1	10
Aklan	82.6	0	Bataan	92.1	1
Artque	82.6	1	Bulacan	92.1	6
Capuz	82.6	8	Nueva Ecya	92.1	4
Guimaras	82.6	0	Pampenga	92.1	3
lloilo	82.6	8	Tarlac	92.1	4
Negros Occidental	82.6	26	Zambales	92.1	6
Marinduque	83.9	0	Batangas	93.5	5
Occedemta Mindoro	83.9	5	Cavric	93.5	0
Onental Mindoro	83.9	7	Laguna	93.5	4
Palawan	83.9	2	Quezon	93.5	28
Romblon	83.9	0	Rizal	93.5	3
Agusandel Norte	85.7	13	Metropolzian Manila	94	1
Aguxandel Sur	85.7	33			

The literacy data has been defined as L, the SigActs as V. Examine the histogram in Figure 6.6 to see that the data appears to follow a Poisson distribution. A goodness-of-fit test (*left as an exercise*) confirms the data follows a Poisson distribution.

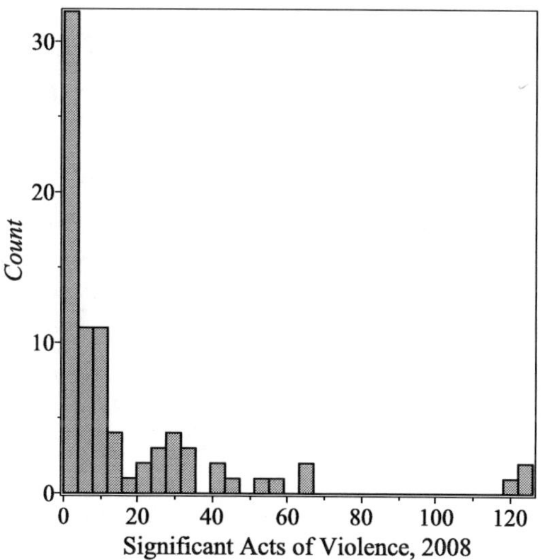

FIGURE 6.6: Histogram of *SigActs* Data

Use Maple to fit the data. First, remove the three outlier data points with values well over 100, as there are other much more significant generators of violence beyond literacy levels in those regions. We cannot use Maple's *ExponentialFit*, as it attempts a log-transformation of *SigActs* which fails due to 0 values.

> $model := \exp(b_0 + b_1 x)$:

> $outops := [leastsquaresfunction, degreesoffreedom, residualmeansquare,$
$residualstandarddeviation, residualsumofsquares]$:
$theLabels := [$*'Degrees of Freedom'*, *'Residual Mean Square'*,
'Residual Standard Deviation', *'Residual Sum of Squares'*$]$:

> $m := NonlinearFit(model, L, V, x, output = outops)$:

> $f := unapply(fnormal(m_1, 5), x);$

$$f := x \mapsto e^{-0.055437\,x+7.1439}$$

> $Matrix([theLabels, m_{2..5}])$:
$LinearAlgebra\text{:-}Transpose(\%)$;

$$\begin{bmatrix} \text{'Degrees of Freedom'} & 76 \\ \text{'Residual Mean Square'} & 235.8330320 \\ \text{'Residual Standard Deviation'} & 15.35685619 \\ \text{'Residual Sum of Squares'} & 17923.31043 \end{bmatrix}$$

Plot the fit.

```
> display(
    pointplot([L, V], symbol = solidcircle, symbolsize = 14),
    plot(f, 70..95, 0..70, thickness = 2),
    title = "Violence v. Literacy", titlefont = [TIMES, 14],
    labels = [SigActs, Count], labeldirections = [horizontal, vertical],
      labelfont = [TIMES, 14]
  );
```


We accept that the fit looks pretty good.

The odds multiplier, e^{b_1}, for our fit is $e^{-0.055437} \approx 0.946$ which means that for every 1 unit increase in literacy we expect violence to go down $\approx 5.4\%$. This value suggests improving literacy will help ameliorate the violence.

Poisson Regression with Multiple Predictor Variables in Maple

Often, there are many variables that influence the outcome under study. We'll add a second predictor to the Hospital Births problem.

Example 6.12. Hospital Births Redux.
Revisit Example 6.10 with an additional predictor: the type of hospital, rural (0) or urban (1). the new data appears in Table 6.13.

TABLE 6.13: Total Births vs. Caesarean Surgeries and Hospital Type

Total	3246	2750	2507	2371	1904	1501	1272	1080	1027	970
Special	26	24	21	21	21	20	19	18	18	17
Type	1	1	1	1	1	1	1	1	1	1

Total	739	679	502	236	357	309	192	138	100	95
Special	17	16	16	16	16	15	14	14	13	13
Type	1	1	1	1	1	0	1	0	0	0

The data has been entered as B: Total, C: Special, and T: Type. After loading the *Statistics* package, define the model.

$> model := \exp(a + b \cdot x + c \cdot y);$

$$model := e^{a+b \cdot x+c \cdot y}$$

Collect the data and use *NonlinearFit* to fit the model.

$> data := \langle\langle C\rangle \mid \langle T\rangle \mid \langle B\rangle\rangle :$

$> outops :=$ [*leastsquaresfunction, degreesoffreedom, residualmeansquare, residualstandarddeviation, residualsumofsquares*] :
theLabels := [*'Degrees of Freedom', 'Residual Mean Square', 'Residual Standard Deviation', 'Residual Sum of Squares'*] :

$> m := NonlinearFit(model, data, [x, y], output = outops) :$

$> f := unapply(fnormal(m_1, 5), x);$

$$f := x \mapsto e^{0.15397\,x+1.2047\,y+3.0022}$$

$> Matrix([theLabels, m_{2..5}]) :$
LinearAlgebra:-Transpose(%);

$$\begin{bmatrix} \text{'Degrees of Freedom'} & 17 \\ \text{'Residual Mean Square'} & 124410. \\ \text{'Residual Standard Deviation'} & 352.72 \\ \text{'Residual Sum of Squares'} & 2.1150\,10^6 \end{bmatrix}$$

Finishing the statistical analysis of the model is left as an exercise.

Exercises

1. Adjust the nonlinear model for Afghanistan casualties, Example 6.5, to increase the amplitude of the sine term more quickly. How does the conclusion change, if at all?

2. Investigate the action of parameters in the logistic function by executing the Maple statements below using the *Explore* command to make an interactive graph.

```
> LP := (a, b) → plot(1/(1 + exp(a · x + b)), x = −4..4, y = −0.1..1.1) :
> Explore(LP(a, b), a = −10.0..10.0, b = −10.0..10.0);
```

3. For the data in Table 6.14 (a) plot the data and (b) state the type of regression that should be used to model the data.

TABLE 6.14: Tire Tread Data

Number	Hours	Tread (cm)
1	2	5.4
2	5	5.0
3	7	4.5
4	10	3.7
5	14	3.5
6	19	2.5
7	26	2.0
8	31	1.6
9	34	1.8
10	38	1.3
11	45	0.8
12	52	1.1
13	53	0.8
14	60	0.4
15	65	0.6

4. Assume the suspected nonlinear model for the data of Table 6.15 is

$$Z = a \cdot \frac{x^b}{y^c}.$$

If we use a *log-log* transformation, we obtain

$$\ln(Z) = \ln(a) + b\ln(x) - c\ln(y).$$

Use regression techniques to estimate the parameters a, b, and c, and statistically analyze the resulting coefficients.

TABLE 6.15: Nonlinear Data

x	y	Z
101	15	0.788
73	3	304.149
122	5	98.245
56	20	0.051
107	20	0.270
77	5	30.485
140	15	1.653
66	16	0.192
109	5	159.918
103	14	1.109
93	3	699.447
98	4	281.184
76	14	0.476
83	5	54.468
113	12	2.810
167	6	144.923
82	5	79.733
85	6	21.821
103	20	0.223
86	11	1.899
67	8	5.180
104	13	1.334
114	5	110.378
118	21	0.274
94	5	81.304

5. Using the basic linear model $y = \beta_0 + \beta_1 x$, fit the following data sets. Provide the model, the analysis of variance information, the value of R^2, and a residual plot.

   (a)

x	100	125	125	150	150	200	200
y	150	140	180	210	190	320	280

x	250	250	300	300	350	400	400
y	400	430	440	390	600	610	670

   (b) The following data represents change in growth where x is body weight and y is normalized metabolic rate for 13 animals.

x	110	115	120	230	235	240	360
y	198	173	174	149	124	115	130

x	362	363	500	505	510	515
y	102	95	122	112	98	96

6. Use an appropriate multivariable-model for the following ten observations of college acceptances to graduate school of GRE score, high school GPA, highly selective college, and whether the student was admitted. 1 indicates "Yes" and 0 indicates "No."

GPA	GRE	Selective	Admitted
3.61	380	0	1
3.67	660	1	0
4.00	800	1	0
3.19	640	0	0
2.93	520	0	1
3.00	760	0	0
2.98	560	0	0
3.08	400	0	1
3.39	540	0	0
3.92	700	1	1

7. The data set for lung cancer in relation to cigarette smoking in Table 6.16 is from Frome, Biometrics 39, 1983, pg. 665–674. The number of person years in parentheses is broken down by age and daily cigarette consumption. Find and analyze an appropriate multivariate model.

TABLE 6.16: Lung Cancer Rates for Smokers and Nonsmokers

		Number Smoked per day					
Age	Nonsmokers	1-9	10-14	15-19	20-24	25-34	> 35
15-20	1 (10366)	0 (3121)	0 (3577)	0 (4319)	0 (5683)	0 (3042)	0 (670)
20-25	0 (8162)	0 (2397)	1 (3286)	0 (4214)	1 (6385)	1 (4050)	0 (1166)
25-30	0 (5969)	0 (2288)	1 (2546)	0 (3185)	1 (5483)	4 (4290)	0 (1482)
30-35	0 (4496)	0 (2015)	2 (2219)	4 (2560)	6 (4687)	9 (4268)	4 (1580)
35-40	0 (3152)	1 (1648)	0 (1826)	0 (1893)	5 (3646)	9 (3529)	6 (1136)
40-45	0 (2201)	2 (1310)	1 (1386)	2 (1334)	12 (2411)	11 (2424)	10 (924)
45-50	0 (1421)	0 (927)	2 (988)	2 (849)	9 (1567)	10 (1409)	7 (556)
50-55	0 (1121)	3 (710)	4 (684)	2 (470)	7 (857)	5 (663)	4 (255)
>55	2 (826)	0 (606)	3 (449)	5 (280)	7 (416	3 (284)	1 (104)

8. Model absences from class where:

School: school 1 or school 2
Gender: female is 1, male is 2
Ethnicity: categories 1 through 6
Math Test: score
Language Test: score
Bilingual: categories 1 through 4

School	Gender	Ethnicity	Math Score	Lang. Score	Bilingual Status	Days Absent
1	2	4	56.98	42.45	2	4
1	2	4	37.09	46.82	2	4
2	1	4	32.37	43.57	2	2
1	1	4	29.06	43.57	2	3
2	1	4	6.75	27.25	3	3
1	1	4	61.65	48.41	0	13
1	1	4	56.99	40.74	2	11
2	2	4	10.39	15.36	2	7
1	2	4	50.52	51.12	2	10
1	2	6	49.47	42.45	0	9

Projects

Project 1. Fit, analyze, and interpret your results for the nonlinear model $y = a\,t^b$ with the data provided below. Produce fit plots and residual graphs with your analysis.

t	7	14	21	28	35	42
y	8	41	133	250	280	297

Project 2. Fit, analyze, and interpret your results for an appropriate model with the data provided below. Produce fit plots and residual graphs with your analysis.

Year	0	1	2	3	4	5	6	7	8	9	10
Quantity	15	150	250	275	270	280	290	650	1200	1550	2750

Project 3. Fit, analyze, and interpret your results for the nonlinear model $y = a\,t^b$ with the data provided by executing the Maple code below. Produce fit plots and residual graphs with your analysis. Use your phone number (no dashes or parentheses) for PN.

```
> randomize(PN) :
> f := unapply(evalf(rand(1.0..9.0)() · x^rand(-2.0..2.0)()), x) :
> t := [(6 · k)$(k = 1..20)] :
  y := f~(t) + [seq(0.001 · rand(50)(), i = 1..20)] :
> data := Matrix([t, y]);
                                   ...
```

6.5 Conclusions and Summary

Along with investigating regression, we've studied some of the common misconceptions decision makers have concerning correlation and regression. Our purpose with this presentation is to help prepare more competent and confident problem solvers. Data can be found using part of a sine curve where the correlation is quite poor, close to zero, but the decision maker can describe and utilize the pattern seeing the relationship in the data as periodic or oscillating. Examples such as these should dispel the idea that correlation very close to zero implies no relationship, and that high linear correlation requires a linear model. Decision makers need to see and understand concepts concerning correlation, linear relationships, and nonlinear, or even no relationship.

Recommended steps for regression analysis:

Step 1. Insure you understand the problem and what answers or predictions are required.

Step 2. Get the data that is available. Identify the dependent and independent variables.

Step 3. Plot the dependent versus each independent variable, and note any apparent trends.

Step 4. If the dependent variable is binary $\{0, 1\}$, then use binary logistic regression. If the dependent variables are counts that follow a Poisson distribution, then use Poisson regression. Otherwise, try linear, multiple, or nonlinear regression as indicated by the situation being studied—science trumps curve fitting.

Step 5. Insure your model produces results that are acceptable. Always use the common-sense test.

References and Further Reading

[AA1979] A. Affi and S. Azen, *Statistical Analysis*, 2nd Ed., Academic Press, London, UK, 1979, pg. 143–144.

[B1997] William C. Bauldry, "Fitting Logistics to the U.S. Population," *MapleTech*, **4**(3), 1997, pg. 73–77.

[D2012] Jay L. Devore, *Probability and Statistics for Engineering and the Sciences*, 8th ed., Cengage learning, 2012.

[FF1996] William P. Fox and Christopher Fowler, "Understanding Covariance and Correlation," *PRIMUS*, **VI** (3), 1996, pp. 235-244.

[Fox2011M] William P. Fox, *Mathematical Modeling with Maple*, Nelson Education, 2011.

[Fox2011E] William P. Fox, "Using the Excel Solver for Nonlinear Regression," *Computers in Education Journal.* October-December, 2011, **2**(4), pg. 77–86.

[Fox2012] William P. Fox, "Issues and Importance of "Good" Starting Points for Nonlinear regression for Mathematical Modeling with Maple: Basic Model Fitting to Make Predictions with Oscillating Data." *J. Computers in Mathematics and Science Teaching*, **31**(1), 2012, pg. 1–16.

[GFH2014] Frank Giordano, William P. Fox, and Steven Horton, *A First Course in Mathematical Modeling*, 5th ed., Nelson Education, 2014.

[J2012] I. Johnson, *An Introductory Handbook on Probability, Statistics, and Excel*, 2012. Available at records.viu.ca/~johnstoi/maybe/maybe4.htm (accessed April 25, 2019).

[M1990] Raymond H. Myers, *Classical and Modern Regression with Applications*, 2nd ed., Duxbury Press, 1990.

[NKNW1996] John Neter, Michael H. Kutner, Christopher J. Nachtsheim, and William Wasserman, *Applied Linear Statistical Models*, Vol. 4, Irwin Chicago, 1996.

7

Problem Solving with Game Theory

Objectives:

(1) Know the concept of formulating a game payoff matrix.

(2) Understand total and partial conflict games.

(3) Understand the use of LP and NLP in game theory.

(4) Understand and interpret the solutions.

In 1943, General Imamura had been ordered to transport Japanese troops across the Bismarck Sea to New Guinea; General Kenney, the United States commander, wanted to bomb the Japanese troop transports prior to their arrival. Imamura had two possible routes to New Guinea: a shorter northern route or a longer southern route. Kenney had to decide where to send his limited number of search planes to find the Japanese fleet. If Kenney sent his planes to the wrong route, he could recall them, but the number of bombing days would be reduced.

Assume both Imamura and Kenney act rationally, each trying to obtain the best outcome. The problem to solve is: What are the strategies each commander should employ?

7.1 Introduction

We begin by studying conflict—an important theme in human history. We assume that conflict arises when two or more individuals with different views, goals, or objectives compete to control the course of future events. *Game theory* studies competition and is used to analyze conflict among two or more opponents. Mathematical tools are used to study situations in which rational players are involved in conflict both with and without cooperation. According to Wiens [Wiens2003], game theory studies situations in which parties compete, and also possibly cooperate, to influence the outcome of interactions to each party's advantage. The situation involves conflict between the participants, called *players*, because some outcomes favor one player at the possible expense of the other players. What each player obtains from a particular outcome is called the player's *payoff*. Each player can choose among a number of strategies to influence payoffs. However, each player's payoff depends on the other players' choices. According to Straffin [Straffin1993], rational players desire to maximize their own payoffs. Game theory is a branch of applied mathematics that is used most notably in economics, and also in business, biology, decision sciences, engineering, political science, international relations, operations research, applied mathematics, computer science, and philosophy. Game theory mathematically captures behavior in strategic situations in which an individual's success in making choices depends on the choices of their opponents. Although initially developed to analyze competitions in which one individual does better at another's expense (*See "zero-sum games"*), game theory has grown to treat a wide class of interactions among players in competition.

Games have many features; a few of the most common are:

Number of Players: Each participant who can make a choice in a game or who receives a payoff from the outcome of those choices is a player. A two-person game has two players. A three or more person is referred to as an N-person game.

Strategies per Player: Each player chooses from a set of possible actions, known as *strategies*. In a two-person game, we can form a grid of strategies. We allow the "row player" to have up to m strategies and the "column player" to have up to n strategies. The choice of a particular strategy by each player determines the payoff to each player.

Pure Strategy Solution: If a player should always choose one strategy over all other strategies to obtain their best outcome in a game, then that strategy represents a *pure strategy solution*. Otherwise if strategies should be played randomly, then the solution is a *mixed strategy solution*.

Nash Equilibrium: A *Nash*[1] *equilibrium* is a set of strategies which represent mutual best responses to the other player's strategies. In other words, if every player is playing their part of Nash equilibrium, no player has an incentive to unilaterally change their strategy. Considering only situations where players play a single strategy without randomizing (a pure strategy), a game can have any number of Nash equilibria.

Sequential Game: A game is *sequential* if one player performs her/his actions after another; otherwise, the game is *simultaneous*.

Simultaneous Game: A game is *simultaneous* if the players each choose their strategy for the game and implement them at the same time.

[1]John Forbes Nash, Jr., the subject of the 2001 movie "A Beautiful Mind," received the John von Neumann Theory Prize in 1978 for his discovery of the *Nash Equilibrium*. He also received both the Nobel Prize in Economics (1994) and the Abel Prize (2015) for his work in game theory.

Perfect Information: A game has *perfect information* if either in a sequential game, every player knows the strategies chosen by the players who preceded them, or in a simultaneous game each player knows the other players' strategies and outcomes in advance.

Constant Sum or **Zero-Sum:** A game is a *constant sum game* if the sums of the payoffs are the same for every set of strategies, and a *zero-sum game* if the payoff sum is always zero. In these games, one player gains if and only if another player loses; otherwise, we have a *variable sum game.*

Extensive Form: The game is presented in a tree diagram.

Normative Form: The game is presented as a payoff matrix. In this chapter we only present the normative form and its associated solution methodologies.

Outcomes: An *outcome* is a set of payoffs resulting from the actions or strategies taken by all players.

Total Conflict Game: A game between players where the sums of the outcomes for all strategy pairs are either the same constant or zero.

Partial Conflict Game: A game whose outcome sums are variable.

The study of game theory has provided many classical and standard games that provide insights into play, tactics, and strategy. Table 7.1 provides a short summary of several classical games; a full list is at:

http://en.wikipedia.org/wiki/List_of_games_in_game_theory

We will primarily be concerned with two-person games. The irreconcilable, conflicting interests between two players in a game surprisingly resemble both parlor games and military encounters between enemy states. Giordano et al. [GFH2014] explain two-person games in the context of mathematical modeling. Players make moves and counter-moves, until the *rules of engagement* declare the game is ended. The rules of engagement determine what each player can or must do at each stage—the available and/or required moves given the circumstances of the game at the current stage—as the game unfolds. For example, in the game Rock, Paper, Scissors, both players simultaneously make one move, with rock beating scissors beating paper beating rock. While this game consists of only one selection of a move (decision choice), games like Chess or Go can require hundreds of moves to end.

Outcomes or payoffs in a game are determined by the strategies players choose and play. These outcomes may come from calculated values or expected values, ordinal rankings, cardinal values developed from a lottery system (See [vNM1944], [Straffin1993]), or cardinal values derived from pairwise comparisons (See [Fox2014]). Here we will assume we have cardinal outcomes (interval or ratio data) or payoffs for our games, since this will allow us to do mathematical calculations.

TABLE 7.1: Classical Two-Player Games in Game Theory

Game	Strategies per Player	Number of Pure Strategy Nash Equilibria	Sequential	Perfect Information	Zero Sum
Battle of the Sexes	2	2	No	No	No
Blotto Games	variable	variable	No	No	Yes
Chicken	2	2	No	No	No
Matching Pennies	2	0	No	No	Yes
Nash Bargaining Game	infinite	infinite	No	No	No
Prisoner's Dilemma	2	1	No	No	No
Rock, Paper, Scissors	3	0	No	No	Yes
Stag Hunt	2	2	No	No	No
Trust Game	infinite	1	Yes	Yes	No

We will present only the *movement diagram* for finding pure strategy solutions, and the linear programming formulation for all solutions of a zero-sum game. There are other methods, also short-cut methods, available to solve many of total-conflict games. For more information on short-cut methods, see [Straffin1993] and the other suggested readings (pg. 336).

For partial conflict games, we will present

- the movement diagram for determining pure strategy solutions, if they exist,
- linear programming formulations for two-person two-strategy games for equalizing strategies,
- nonlinear programming methods for more than two strategies for each player, and
- linear programming methods to find security levels, as all players seek to maximize their preferred outcome.

We use the concept that every partial conflict game has a *Nash's equalizing mixed-strategy solution* even if the game has a pure strategy solution

([GH2009]). We conclude by briefly discussing the *Nash arbitration scheme* and its nonlinear formulation.

Concepts and solution methodologies of N-person games, such as three-person total- and partial-conflict games, will be left to future studies.

7.2 Background of Game Theory

Game theory is the study of strategic decision making; that is, "the study of mathematical models of conflict and cooperation between intelligent rational decision-makers" ([Myerson1991]). Game theory has applications in many areas of business, military, government, networks, and industry. For more information on applications of game theory in these areas, see [CS2001], [Cantwell2003], [EK2010], and [Aiginger1999]. Additionally, [MF2009] discusses game theory in warlord politics which blends military and diplomatic decisions.

The study of game theory began with total conflict games, also known as zero-sum games, such that one person's gain exactly equals the net losses of the other player(s). Game theory continues to grow with application to a wide range of applied problems. A "Google Scholar" search returns over 3 million items.

The Nash equilibrium for a two-player, zero-sum game can be found by solving a linear programming problem and its dual solution ([Dantzig1951] and [Dantzig2002], [Dorfman1951]). In their work, Dantzig and Dortman, respectively, assume that every element of the payoff matrix containing outcomes or payoffs to the row player M_{ij} is positive. More current approachs (e.g., [Fox2008] and [Fox2010]) show the payoff matrix entries can be positive or negative.

7.2.1 Two-Person Total Conflict Games

We begin with characteristics of the two-person total conflict game following [Straffin1993]:
There are two participants: the first, Rose, is the *row player*, and the other, Colin, is the *column player.*

Rose must choose from among her 1 to m strategies, and Colin must choose from among his 1 to n strategies.

If Rose chooses the ith strategy and Colin the jth strategy, then Rose receives a payoff of a_{ij} and Colin loses the amount a_{ij}. In Table 7.2, this is shown as a payoff pair where Rose receives a payoff of M_{ij} and Colin receives a payoff of N_{ij}.

Games are simultaneous and repetitive.

There are two types of possible solutions. A *pure strategy* solution is where each player achieves their best outcomes by always choosing the same strategy in repeated games. A *mixed strategy* solutions is where players play a random selection of their strategies in order to obtain the best outcomes in simultaneous repeated games.

Although we do not address them in this chapter, in sequential games, the players look ahead and reason back.

Table 7.2 shows a generic payoff matrix for simultaneous games.

TABLE 7.2: General Payoff Matrix of a Two-Person Total Conflict Game

		Colin's Strategies			
		Column 1	*Column* 2	...	*Column* n
Rose's Strategies	*Row* 1	$(M_{1,1}, N_{1,1})$	$(M_{1,2}, N_{1,2})$	...	$(M_{1,n}, N_{1,n})$
	Row 2	$(M_{2,1}, N_{2,1})$	$(M_{2,2}, N_{2,2})$	...	$(M_{2,n}, N_{2,n})$
	$\vdots$	$\vdots$	$\vdots$	$\ddots$	$\vdots$
	Row m	$(M_{m,1}, N_{m,1})$	$(M_{m,2}, N_{m,2})$	...	$(M_{m,n}, N_{m,n})$

A game is a *total conflict game* if and only if the sum of the pairs $M_{i,j}+N_{i,j}$ always equals either 0 or the same constant c for all strategies i and j. If the sum equals zero, then we list only the row payoff M_{ij}.

For example, if a player wins x when the other player loses x, then the sum $M_{i,j} + N_{i,j} = x - x = 0$. In a business marketing strategy, if one player gets $x\%$ of the market, then the other player gets $y\% = 100 - x\%$ based upon 100% of the market. We list only $x\%$ as the outcome because when the row player receives $x\%$, the column player loses $x\%$.

Movement Diagrams

A *movement diagram* has arrows in each row (vertical arrow) and column (horizontal arrow) from the smaller payoff to the larger payoff. If there exists one or more payoffs where all arrows point towards it, then those payoffs constitute *pure strategy Nash equilibriums.*

Example 7.1. Baseball Franchises.
Several minor league baseball teams want to enter the market in a new area. The teams can choose to locate in a more densely populated area, or less densely populated town surrounded by other towns. Assume that both the National and American Leagues are interested in the new franchise. Suppose the National League will locate a franchise in either a densely populated area or a less densely populated area. The American League is making the same decision—they will locate either in a densely populated area or a less dense

area. This situation is similar to that of the Cubs and the White Sox both being in Chicago, or the Yankees and Mets both being in New York City. Analysts have estimated the market shares. We place both sets of payoffs in a single game matrix. Listing the row player's payoff, the American League's payoff, as first in the ordered pair, we have the payoff matrix shown in Table 7.3. The payoff matrix represents a constant-sum total-conflict game. Arrows are added to the table to create the movement diagram.

TABLE 7.3: New Baseball Franchise Payoff Matrix and Movement Diagram

		National League Franchise	
		Densely Populated	Less Densely Populated
American League Franchise	Densely Populated	$(65, 35)$ $\Longleftarrow$	$(70, 30)$
		$\Uparrow$	$\Uparrow$
	Less Densely Populated	$(55, 45)$ $\Longleftarrow$	$(60, 40)$

The payoff $(65, 35)$ only has arrows pointing in for the densely populated areas choice for franchise strategies for both players, no arrow exits that outcome. The movement diagram indicates that neither player can unilaterally improve their solution giving a *Nash equilibrium* ([Straffin1993]).

Linear Programming in Total Conflict Games

Von Neumann's *minimax theorem* ([vNeumann1928]) states that for every two-person, zero-sum game with finitely many strategies, there exists an outcome value V and a set of strategies for each player, such that

(a) Given Player 2's strategy, the best payoff possible for Player 1 is V, and

(b) Given Player 1's strategy, the best payoff possible for Player 2 is $-V$.

Equivalently, Player 1's strategy guarantees him a payoff of V regardless of Player 2's strategy, and similarly Player 2 can guarantee a payoff of $-V$ regardless of Player 1's strategy. The name *minimax* arises from each player minimizing the maximum payoff possible for the other; since the game is zero-sum, the Player also minimizes his own maximum loss; i.e., maximize his minimum payoff.

Every total conflict game may be formulated as a linear programming problem ([Dantzig1951] and [Dorfman1951]). Consider a total-conflict two-person game in which maximizing Player X has m strategies and minimizing Player Y has n strategies. The entry (M_{ij}, N_{ij}) from the ith row and jth column of the payoff matrix represents the payoff for those strategies. The following formulation, using only the elements of M_{ij} for the maximizing player, provides results for the value of the game and the probabilities x_i of outcomes ([GFH2014], [Fox2011Maple], and [Winston2002]).

If there are negative entries in the payoff matrix, a slight modification to the linear programming formulation is necessary since all variables must be non-negative when using the simplex method. To obtain a possible negative value solution for the game, use the method described in [Winston2002]: replace any variable that could take on negative values with the difference of two positive variables. Since only V, the value of game, can be positive or negative, replace V with $V = V_j - V_j'$ with both new variables positive. The other values we are looking for are probabilities which are always non-negative. In these games, players want to maximize the value of the game that they receive. The Linear Program (7.1) is a linear programming formulation for finding the optimal strategies and value of the game.

$$
\begin{aligned}
&\text{Maximize } V\\
&\text{subject to}\\
&\qquad M_{1,1}x_1 + M_{2,1}x_2 + \cdots + M_{m,1}x_m - V \geq 0\\
&\qquad M_{1,2}x_1 + M_{2,2}x_2 + \cdots + M_{m,2}x_m - V \geq 0\\
&\qquad \vdots\\
&\qquad M_{1,m}x_1 + M_{2,m}x_2 + \cdots + M_{m,n}x_m - V \geq 0\\
&\qquad x_1 + x_2 + \cdots + x_m = 1\\
&\text{with } V, x_i \geq 0
\end{aligned}
\tag{7.1}
$$

The weights x_i yield Rose's strategy, V is the value of the game to Rose. When the solution to this total conflict game is obtained, we also have the solution to Colin's game through the solution of the dual linear program ([Winston2002]). As an alternative to the dual, we can formulate Colin's game directly as shown in (7.2) using the original N_{ij}s. We call the value of the game for Colin v to distinguish it from Rose's value V. Colin's linear program is

$$
\begin{aligned}
&\text{Maximize } v\\
&\text{subject to}\\
&\qquad N_{1,1}y_1 + N_{1,2}y_2 + \cdots + N_{1,n}y_n - v \geq 0\\
&\qquad N_{2,1}y_1 + N_{2,2}y_2 + \cdots + N_{2,n}y_n - v \geq 0\\
&\qquad \vdots\\
&\qquad N_{m,1}y_1 + N_{m,2}y_2 + \cdots + N_{m,n}y_n - v \geq 0\\
&\qquad y_1 + y_2 + \cdots + y_n = 1\\
&\text{with } V, y_i \geq 0
\end{aligned}
\tag{7.2}
$$

The weights y_i yield Colin's strategy, v is the value of the game to Colin.

Put Example 7.1 into the two formulations and solve to obtain the solution

$$x_1 = 1, x_2 = 0, \text{ yielding } V = 65,$$
$$y_1 = 1, y_2 = 0, \text{ yielding } v = 35$$

The overall solution is that each league should place its new franchise in a densely populated area giving the solution of a $(65, 35)$ market share split.

The primal-dual simplex method only works in the zero-sum game format ([Fox2010]). We may convert this game to a zero-sum form to obtain the solution via linear programming. Since this is a constant sum game, whatever the American League gains, the national League loses. For example out of 100%, if the American League franchise gains 65%, then the National League franchise loses 65% of the market as in Table 7.4

TABLE 7.4: Zero-Sum Game Payoff Matrix for a New Baseball Franchise

		National League Franchise	
		Densely Populated	Less Densely Populated
American League Franchise	Densely Populated	65	67
	Less Densely Populated	55	60

For a zero-sum game, we only need a single formulation of the linear program. The Row Player maximizes and the Column Player minimizes with rows' values constituting a primal and dual relationship. The linear program used in zero-sum games is equivalent to the formulation in (7.1) with a_{ij} for M_{ij} designating the zero-sum outcomes for the Row Player; the linear program is shown in (7.3).

$$\begin{aligned}
&\text{Maximize } V \\
&\text{subject to} \\
&\qquad a_{1,1}x_1 + a_{2,1}x_2 + \cdots + a_{m,1}x_m - V \geq 0 \\
&\qquad a_{1,2}x_1 + a_{2,2}x_2 + \cdots + a_{m,2}y_m - V \geq 0 \\
&\qquad \vdots \\
&\qquad a_{1,m}x_1 + a_{2,m}x_2 + \cdots + a_{m,n}x_m - V \geq 0 \\
&\qquad x_1 + x_2 + \cdots + x_m = 1 \\
&\text{with } V, x_i \geq 0
\end{aligned} \tag{7.3}$$

where V is the value of the game, $a_{i,j}$ are the payoff-matrix entries, and x_i's are the weights (probabilities to play the strategies).

For the baseball franchise example, place the payoffs into (7.3), letting V be the value of the game to the Row Player, the American League, giving

$$\begin{aligned}
&\text{Maximize } V\\
&\text{subject to}\\
&65x_1 + 55x_2 - V \geq 0\\
&70x_1 + 60x_2 - V \geq 0\\
&V, x_1, x_2 \geq 0
\end{aligned}$$

Maple solves this LP easily.

```
> with(Optimization) :
> Obj := V :
> RowConstraints := {65 · x1 + 55 · x2 − V ≥ 0, 70 · x1 + 60 · x2 − V ≥ 0,
    x1 + x2 ≤ 1, x1 + x2 ≥ 1,
    V ≥ 0, x1 ≥ 0, x2 ≥ 0}
> RowSoln := LPSolve(Obj, RowConstraints, maximize);
  fnormal(RowSoln[2], 4);
         RowSoln := [65.0000000671188, [V = 65.0000000671188,
                    x1 = 1.00000000103260, x2 = 0.]]
                    [V = 65.00, x1 = 1.000, x2 = 0.]
```

We applied *fnormal* to the result to eliminate numerical error artifacts.

Now for the other player.

```
> Obj := v :
> ColConstraints := {35 · y1 + 30 · y2 − v ≥ 0, 45 · y1 + 40 · y2 − v ≥ 0,
    y1 + y2 ≤ 1, y1 + y2 ≥ 1,
    V ≥ 0, y1 ≥ 0, y2 ≥ 0}
> ColSoln := LPSolve(Obj, ColConstraints, maximize);
  fnormal(ColSoln[2], 4);
         ColSoln := [35.0000000361409, [v = 35.0000000361409,
                    y1 = 1.00000000103260, y2 = 0.]]
                    [v = 35.00, y1 = 1.000, y2 = 0.]
```

Note that the solutions are not exact; this is due to numerical methods used internally by Maple and why we applied *fnormal* to the result.

The optimal solution strategies found are identical, as before, with both players choosing a more densely populated area as their best strategy. The use of linear programming is quite efficacious for large games between two players each having many strategies ([Fox2010], [Fox2014], and [GFH2014]). We note that the solution to the National League franchise game is found either as the dual solution, (See [Winston2002], Section 11.3) or by simply re-solving the linear program from Column Player's perspective. In Chapter 4 of *Advanced Problem Solving with Maple™: A First Course*, we showed how Maple can be used to solve both the primal and dual linear programs.

The Partial Conflict Game

Partial conflict games are games in which one player wins, but the other player does not have to lose. Both players could win something or both could lose something. Solution methods for partial conflict games include looking for dominance, analyzing movement diagrams, and finding equalizing strategies. Here we present an extension from the total conflict game to the partial conflict game as an application of linear programming; see [Fox2010] and [Fox2014]. Because of the nature of partial conflict games where both players are trying to maximize their outcomes, we can model all players' strategies as their own maximizing linear programs. We treat each player as a separate linear programming maximization problem.

Again, use the payoff matrix of Table 7.2. Now assume that $M_{ij} + N_{ij}$ is not always equal to zero or the same constant for all i and j. In non-cooperative partial-conflict games, we first look for a pure-strategy solution using a movement diagram.

The Row Player, Rose, maximizes payoffs, so she would prefer the highest payoff in each column. Vertical arrows are in columns with Rose's values. Similarly, The Column Player, Colin, maximizes his payoffs, so he would prefer the highest payoff in each row. Draw an arrow to the highest payoff in that row. Horizontal arrows are in rows with Colin's values. If all arrows point to a cell from every direction, then that cell will be a pure Nash equilibrium.

If all the arrows do not point at a value or values, i.e., there is no pure Nash equilibrium, then we must use equalizing strategies to find the weights (probabilities) for each player. For a game with two players having two strategies each, proceed as follows:

Rose's game: Rose maximizing and Colin "equalizing" is a total-conflict game that yields Colin's equalizing strategy.

Colin's game: Colin maximizing and Rose "equalizing" is a total-conflict game that yields Rose's equalizing strategy.

Note: If either side plays its equalizing strategy, the other side cannot unilaterally improve its own situation—the other player is stymied.

This analysis translates into two maximizing linear programming formulations shown in (7.4) and (7.5) below. The LP formulation in (7.4) provides the Nash equalizing solution for Colin with strategies played by Rose, while the LP formulation in (7.5) provides the Nash equalizing solution for Rose with strategies played by Colin.

$$\begin{aligned} &\text{Maximize } V \\ &\text{subject to} \\ &N_{1,1}x_1 + N_{2,1}x_2 - V \geq 0 \\ &N_{1,2}x_1 + N_{2,2}x_2 - V \geq 0 \\ &\qquad V, x_1, x_2 \geq 0 \end{aligned} \tag{7.4}$$

$$\begin{aligned} &\text{Maximize } v \\ &\text{subject to} \\ &M_{1,1}y_1 + M_{1,2}y_2 - v \geq 0 \\ &M_{2,1}y_1 + M_{2,2}y_2 - v \geq 0 \\ &\qquad v, y_1, y_2 \geq 0 \end{aligned} \tag{7.5}$$

If there is a pure strategy solution, it is found through movement diagrams or dominance. The linear programming formulations will not find pure strategy results; LPs only provide the Nash equilibrium using equalizing strategies ([Straffin1993]).

For games with two players and more than two strategies each, Bazarra et al. (See [BSC2013]) presented a nonlinear optimization approach. Consider a two-person game with a standard payoff matrix. Separate the payoff matrix into two matrices M and N for Players I and II. Then solve the nonlinear optimization formulation given in expanded form in (7.6).

$$\text{Maximize } \sum_{i=1}^{n}\sum_{j=1}^{n} x_i\, a_{ij}\, y_j + \sum_{i=1}^{n}\sum_{j=1}^{n} x_i\, b_{ij}\, y_j - p - q$$

subject to

$$\begin{aligned} &\sum_{j=1}^{m} a_{ij}\, y_j \leq p, \quad i = 1, 2, \ldots, n, \\ &\sum_{i=1}^{m} x_i\, b_{ij} \leq q, \quad i = 1, 2, \ldots, m, \\ &\sum_{i=1}^{n} x_i = \sum_{j=1}^{m} y_j = 1, \\ &x_i \geq 0, y_i \geq 0 \end{aligned} \tag{7.6}$$

Example 7.2. A Partial Conflict Equalizing Strategy Game Solution. Table 7.5 shows a partial-conflict game with revised payoff estimates of market shares. The arrows in the movement diagram indicate that the game has no pure strategy solution.

TABLE 7.5: Partial-Conflict Movement Diagram

		Colin	
		Large City	Small City
Rose	Large City	(20, 40) $\Longleftarrow$	(10, 0)
		$\Downarrow$	$\Uparrow$
	Small City	(30, 10) $\Longrightarrow$	(0, 40)

We use (7.4) and (7.5) to formulate and solve these partial conflict games for their Nash equalizing strategies.

Maximize Vc

subject to

$$\begin{aligned} 40x_1 + 10x_2 - Vc &\geq 0 \\ 0x_1 + 40x_2 - Vc &\geq 0 \\ x_1 + x_2 &= 1 \\ Vc, x_1, x_2 &\geq 0 \end{aligned} \tag{7.7}$$

Maximize Vr

subject to

$$\begin{aligned} 20y_1 + 10y_2 - Vr &\geq 0 \\ 30y_1 + \ 0y_2 - Vr &\geq 0 \\ y_1 + y_2 &= 1 \\ Vr, y_1, y_2 &\geq 0 \end{aligned} \tag{7.8}$$

The solutions to this partial conflict game are

(a) $Vc = 22.857$ when $x_1 = 0.571$ and $x_2 = 0.429$, and

(b) $Vr = 15.000$ when $y_1 = 0.500$ and $y_2 = 0.500$.

This game results in Colin playing Large City and Small City each half the time, ensuring a value of 15.00 for Rose. Rose plays a mixed strategy of 4/7 Large City and 3/7 Small City which yields a value of the game of 22.857 for Colin.

To be a solution, the Nash equilibrium must be *Pareto optimal* (*no other solution is better for both players*, northeast region) as defined by Straffin (See [Straffin1993]). Visually, we can plot the coordinates of the outcomes and connect them into a convex set (the convex hull). The Nash equilibrium (15, 22.85) is an interior point, and so it is not Pareto optimal; see Figure 7.1.

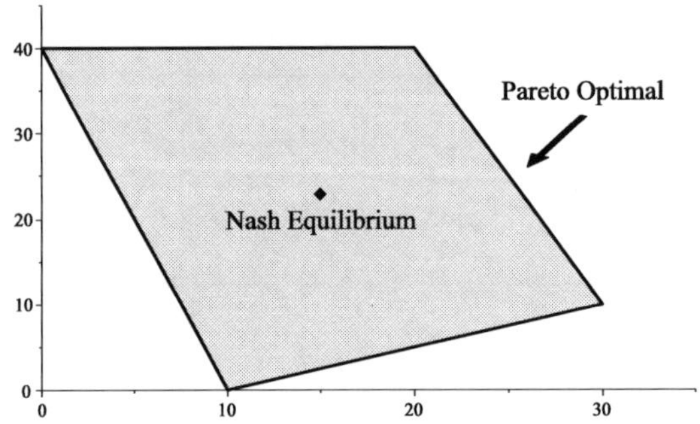

FIGURE 7.1: Payoff Polygon with Nash Equilibrium

When the results are mixed strategies, the implication is that the game is repeated many times in order to achieve that outcome in the long run.

Example 7.3. A 3×3 Nonzero Sum Game.
Rose and Colin each have three strategies with payoffs shown in Table 7.6.

TABLE 7.6: Rose and Colin Strategies

		Colin		
		C_1	C_2	C_3
Rose	R_1	$(-1, 1)$	$(0, 2)$	$(0, 2)$
	R_2	$(2, 1)$	$(1, -1)$	$(0, 0)$
	R_3	$(0, 0)$	$(1, 1)$	$(1, 2)$

First, we use a movement diagram to find two Nash equilibrium points. They are $R_2C_1 = (2, 1)$ and $R_3C_3 = (1, 2)$. These pure strategy solutions are not equivalent and trying to achieve them might lead to other results. We might employ the nonlinear method described earlier to look for other equilibrium solutions, if they exist. We find the nonlinear method does produce another solution when $p = q = 0.667$ when $x_1 = 0$, $x_2 = 0.667$, $x_3 = 0.333$, $y_1 = 0.333$, $y_2 = 0$, and $y_3 = 0.667$. The Maple statements to obtain this solution are:

```
> with(Optimization) :

> M := Matrix([[-1, 0, 0], [2, 1, 0], [0, 1, 1]]) :
  N := Matrix([[1, 2, 2], [1, -1, 0], [0, 1, 2]]) :
```

$> X := Vector[row](3, symbol = x):$
$Y := Vector(3, symbol = y):$

$> Objective := expand(X . M . Y + X . N . Y - p - q):$

$> Constraints := \{seq((M . Y)_i \leq p, i = 1..3),$
$seq((X . N)_i \leq q, i = 1..3),$
$add(x_i, i = 1..3) = 1, add(y_i, i = 1..3) = 1\}$

Since the objective function is quadratic, use Maple's quadratic program solver, *QPSolve*, from the *Optimization* package.

$> Soln := QPSolve(Objective, Constraints, assume = nonnegative, maximize):$

$> fnormal(Soln, 3);$

$$[0., [p = 0.667, q = 0.667, x_1 = 0., x_2 = 0.667, x_3 = 0.333, y_1 = 0.333, y_2 = 0., y_3 = 0.667]]$$

Communications and Cooperation in Partial Conflict Games

Allowing for communication may change the game's solution. A player may consider combinations of moves, threats, and promises to attempt to obtain better outcomes. The strategy of moves is explored in several sources listed in References and Further Reading (pg. 336).

Nash Arbitration Method

When we have not achieved a solution by other methods that is acceptable to the players, then a game may move to arbitration. The Nash Arbitration Theorem (Nash, 1950) states that

> There is a unique *arbitration solution* which satisfies the axioms
>
> **Rationality:** The solution point is feasible.
>
> **Linear Invariance:** Changing scale does not change the solution.
>
> **Symmetry:** The solution does not discriminate against any player.
>
> **Independence of Irrelevant Alternative:** Eliminating solutions that would not be chosen does not change the solution.

Prudential Strategy (Security Levels)

The *security levels* are the payoffs to the players in a partial conflict game where each player attempts to maximize their own payoff. We can solve for these payoffs using a separate linear program for each security level.

The LP formulations are (7.9) and (7.10).

$$\begin{aligned}
&\text{Maximize } V\\
&\text{subject to}\\
&\quad M_{1,1}x_1 + M_{2,1}x_2 + \cdots + M_{m,1}x_m - V \geq 0\\
&\quad M_{1,2}x_1 + M_{2,2}x_2 + \cdots + M_{m,2}x_m - V \geq 0\\
&\quad \vdots\\
&\quad M_{1,m}x_1 + M_{2,m}x_2 + \cdots + M_{m,n}x_m - V \geq 0\\
&\quad x_1 + x_2 + \cdots + x_m = 1\\
&\quad x_i \leq 1 \quad \text{for } i = 1, 2, \ldots, m\\
&\text{with } V, x_i \geq 0
\end{aligned} \tag{7.9}$$

The weights x_i yield Rose's prudential strategy for the security level V.

$$\begin{aligned}
&\text{Maximize } v\\
&\text{subject to}\\
&\quad N_{1,1}y_1 + N_{1,2}y_2 + \cdots + N_{1,n}y_n - v \geq 0\\
&\quad N_{2,1}y_1 + N_{2,2}y_2 + \cdots + N_{2,n}y_n - v \geq 0\\
&\quad \vdots\\
&\quad N_{m,1}y_1 + N_{m,2}y_2 + \cdots + N_{m,n}y_n - v \geq 0\\
&\quad y_1 + y_2 + \cdots + y_n = 1\\
&\quad y_i \leq 1 \text{ for } i = 1, 2, \ldots, m\\
&\text{with } V, y_i \geq 0
\end{aligned} \tag{7.10}$$

The weights y_i yield Colin's prudential strategy for the security level v

Revisit Example 7.2 to illustrate finding security levels. Let SLR and SLC represent the security levels for Rose and Colin, respectively. We use linear programming to find these values using (7.9) and (7.10) yielding

$$\begin{aligned}
&\text{Maximize } SLR\\
&\text{subject to}\\
&\quad 20x_1 + 30x_2 - SLR \geq 0\\
&\quad 10x_1 + \ \ 0x_2 - SLR \geq 0\\
&\quad x_1 + x_2 = 1\\
&\quad x_1, x_2 \leq 1\\
&\quad SLR, x_1, x_2 \geq 0
\end{aligned}
\qquad
\begin{aligned}
&\text{Maximize } SLC\\
&\text{subject to}\\
&\quad 40y_1 + \ \ 0y_2 - SLC \geq 0\\
&\quad 10x_1 + 40y_2 - SLC \geq 0\\
&\quad y_1 + y_2 = 1\\
&\quad y_1, y_2 \leq 1\\
&\quad SLR, y_1, y_2 \geq 0
\end{aligned}$$

The solution yields both how the game is played and the security levels. Rose always plays R_1, and Colin plays 4/7 C_1 and 3/7 C_2. The security level is $(10, 22.86)$.

Using this security level, $(10, 22.86)$, as our status quo point, we can formulate the Nash arbitration scheme. We restate more formally the four axioms stated above that are met using the Nash arbitration scheme.

Axiom 1: *Rationality.* The solution must be in the negotiation set.

Axiom 2: *Linear Invariance.* If either Player 1's or Player 2's utility functions are transformed by a positive linear function, the solution point should be transformed by the same function.

Axiom 3: *Symmetry.* If the polygon happens to be symmetric about the line of slope $+1$ through the status quo point, then the solution should be on this line. No player is favored, no player is discriminated against.

Axiom 4: *Independence of Irrelevant Alternatives.* Suppose N is the solution point for a polygon P with status quo point SQP. Suppose Q is another polygon which contains both SQP and N, and is totally contained in P. Then N should also be the solution point to Q with status quo point SQP, i.e., the solution point is not changed by non-solution points being eliminated from consideration.

Theorem. Nash's Arbitration Theorem (Nash,[2] 1950).
There is one and only arbitration scheme which satisfies the four axioms rationality, linear invariance, symmetry, and independence of irrelevant alternatives. The arbitration scheme is: If the status quo (SQP) point is (x_0, y_0), then the arbitrated solution point N is the point (x, y) in the polygon with $x \geq x_0$, $y \geq y_0$ which maximizes the product $Z = (x - x_0)(y - y_0)$.

We apply Nash's theorem in a nonlinear optimization framework (*Kuhn-Tucker conditions*). The formulation for our example is

$$\text{Maximize } Z = (x - 10)(y - 22.86)$$
$$\text{subject to}$$
$$\begin{aligned} 3x + y &= 100 \\ x &\geq 10 \\ y &\geq 22.86 \\ x, y &\geq 0 \end{aligned}$$

Maple finds the solution easily using *QPSolve.* In this example, for the status quo point $(10, 22.86)$, *QPSolve* gives our Nash arbitration point as $(17.86, 46.43)$.

[2] John F. Nash, Jr., "The Bargaining Problem," *Econometrica*, 18(2), 1950, pg. 155–162.

7.3 Examples of Zero-Sum Games

In this section, we present several illustrative examples of the theory of total-conflict games. We present the scenario, discuss the outcomes used in the payoff matrix, and present a possible solution for the game. In most game theory problems, the solution suggests insights in how to play the game rather than a definite methodology to "winning" the game.

Example 7.4. The Battle of the Bismarck Sea.
The Battle of the Bismarck Sea is set in the South Pacific in 1943. See Figure 7.2. The historical facts are that General Imamura had been ordered to transport Japanese troops to New Guinea. General Kenney, the United States commander in the region, wanted to bomb the troop transports prior to their arrival at their destination. Imamura had two options to choose from as routes to New Guinea: a shorter northern route or a longer southern route. Kenney had to decide where to send his search planes and bombers to find the Japanese fleet. If Kenney sent his planes to the wrong route, he could recall them, but the number of bombing days would be reduced.

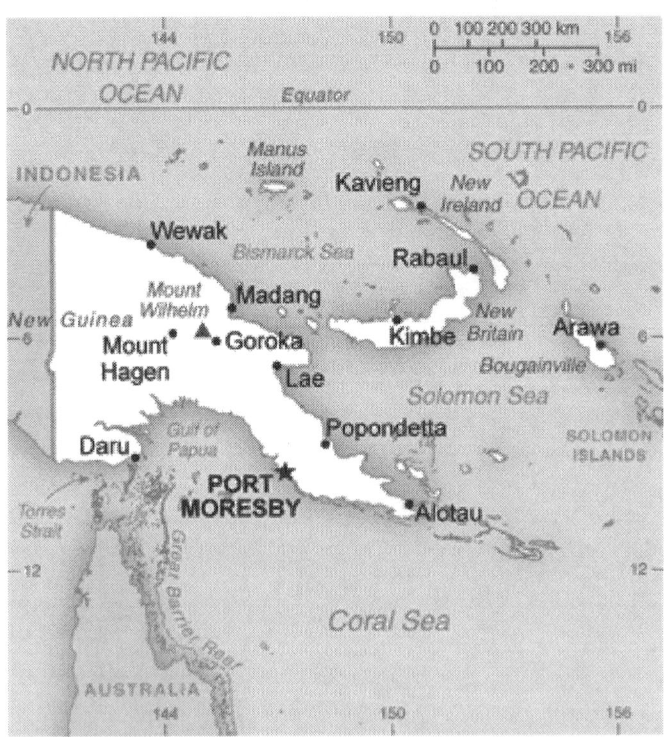

FIGURE 7.2: The Battle of the Bismarck Sea. Japanese troops were being taken from Rabaul to Lae. (Map Source: *The World Fact Book*, CIA)

We assume that both commanders, Imamura and Kenney, are rational players, each trying to obtain his best outcome. Further, we assume that there are no communications or cooperation which may be inferred since the two are enemies engaging in war. Further, each is aware of the intelligence assets that are available to each side and are aware of what the intelligence assets are producing. We assume that the estimates of number of days that US planes can bomb as well as the number of days to sail to New Guinea are accurate.

The players, Kenney and Imamura, both have the same set of strategies for routes: {*North, South*}, and their payoffs, given as the numbers of exposed days for bombing, are shown in Table 7.7. Imamura loses exactly what Kenney gains.

TABLE 7.7: The Battle of the Bismarck Sea with Payoffs (*Kenney*, *Imamura*)

		Imamura	
		North	South
Kenney	North	$(2, -2)$	$(2, -2)$
	South	$(1, -1)$	$(3, -3)$

Graphing the payoffs, Figure 7.3, shows this is a total conflict game.

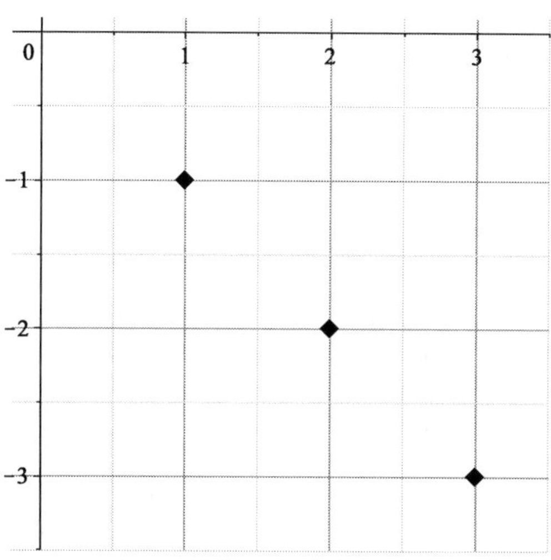

FIGURE 7.3: Graph of Payoffs for the Battle of the Bismarck Sea

As a total conflict game, Table 7.8 only needs to list the outcomes to Kenney in order to find a solution.

TABLE 7.8: The Battle of the Bismarck Sea as a Zero Sum Game

		Imamura	
		North	South
Kenney	North	2	2
	South	1	3

There is a dominant column strategy for Imamura: to sail *North* since the values in the column are correspondingly less than or equal to the values for sailing *South*. The dominant *North* column would eliminate the *South* column. Seeing that as an option, Kenney would search *North*—that option provides a greater outcome than searching *South*, $(2 > 1)$. He could also apply the *minimax* theorem (*saddle point method*) to find a plausible Nash equilibrium as Kenney searches North and Imamura takes the Northern route. See Table 7.9.

TABLE 7.9: Minimax Method (Saddle Point Method)

		Imamura			
		North	South	Row Min	Max of Min
Kenney	North	2	2	2	[2]
	South	1	3	1	
	Column Max	2	3		
	Min of Max	[2]			

Applied to the Battle of the Bismarck Sea, the Nash equilibrium (*North, North*) implies that no player can do unilaterally better by changing their strategy. The solution is for the Japanese to sail *North* and for Kenney to search *North* yielding 2 bombing days. This result, (*North, North*), was indeed the real outcome in 1943.

Next, let's assume that communication is allowed. We will consider first moves by each player. If Kenney moved first, (*North, North*) would remain the outcome. However, *(North, South)* also becomes a valid response with the same value of 2.

If Imamura moved first, (*North, North*) would be the outcome. What is important about moving first in a zero sum game is that, although it gives more information, neither player can do better than the Nash equilibrium from the original zero sum game. We conclude from our brief analysis that moving first does not alter the equilibrium of this game. Moving first in a zero sum games does not alter the equilibrium strategies.

A Maple solution follows. Remember to begin by entering *with*(*Optimization*). First, the solution for Kenney:

$> KenneyObj := V;$

$> KenneyCons := \{2 \cdot x_1 + x_2 - V \geq 0, 2 \cdot x_1 + 3 \cdot x_2 - V \geq 0,$
$x_1 + x_2 \geq 1, x_1 + x_2 \leq 1,$
$V \geq 0, x_1 \geq 0, x_2 \geq 0\}$

$> fnormal(LPSolve(KenneyObj, KenneyCons, maximize), 3);$

$$[2.000, [V = 2.000, x_1 = 1.000, x_2 = 0.]]$$

We used *fnormal* to eliminate artifacts from floating point computations.

Now, the solution for the Japanese commander, Imamura.

$> ImamuraObj := v;$

$> ImamuraCons := \{2 \cdot y_1 + 2 \cdot y_2 - v \geq 0, 1 \cdot y_1 + 3 \cdot y_2 - v \geq 0,$
$y_1 + y_2 \geq 1, y_1 + y_2 \leq 1,$
$v \geq 0, y_1 \geq 0, y_2 \geq 0\}$

$> fnormal(LPSolve(ImamuraObj, ImamuraCons, maximize), 3);$

$$[2.000, [v = 2.000, y_1 = -0., y_2 = 1.000]]$$

Example 7.5. Penalty Kicks in Soccer[3].

A penalty kick in soccer is a game between a kicker and the opposing goalie. The kicker has two alternative strategies: he might kick left or kick right. The goalie will also have two strategies: the goalie can dive left or right to block the kick. We will start with a very simple payoff matrix with a 1 for the player that is successful and a -1 for the player that is unsuccessful, assuming a correct dive blocks the kick. The payoff matrix is in Table 7.10.

TABLE 7.10: Penalty Kick Payoffs

		Goalie	
		Dive Left	Dive Right
Kicker	Kick Left	$(-1, 1)$	$(1, -1)$
	Kick Right	$(1, -1)$	$(-1, 1)$

[3]This example is adapted from Chiappori, Levitt, and Groseclose [CLG2002].

Or, just the kicker's prospective, see Table 7.11.

TABLE 7.11: Kicker's Penalty Kick Payoffs

		Goalie	
		Dive Left	Dive Right
Kicker	Kick Left	−1	1
	Kick Right	1	−1

There is no pure strategy. We find a mixed strategy solution to the zero-sum game using either linear programming or the *method of oddments*[4]. The mixed strategy results are that the kicker randomly kicks 50% left and 50% right, while the goalie randomly dives 50% left and 50% right. The value of the game to each player is 0.

Let's refine the game using real data. A study was done in the Italian Football League in 2002 by Ignacio Palacios-Huerta.[5] As he observed, the kicker can aim the ball to the left or to the right of the goalie, and the goalie can dive either left or right as well. The ball is kicked with enough speed that the decisions of the kicker and goalie are effectively made simultaneously. Based on these decisions, the kicker is likely to score or not score. The structure of the game is remarkably similar to our simplified game. If the goalie dives in the direction that the ball is kicked, then he has a good chance of stopping the goal; if he dives in the wrong direction, then the kicker is likely to score a goal.

After analyzing approximately 1400 penalty kicks, Palacios-Huerta determined the empirical probabilities of scoring for each of four outcomes: the kicker kicks left or right, and the goalie dives left or right. His results led to the payoff matrix in Table 7.12.

TABLE 7.12: Penalty Kick Probabilities of Scoring

		Goalie	
		Dive Left	Dive Right
Kicker	Kick Left	(0.58, −0.58)	(0.95, −0.95)
	Kick Right	(0.93, −0.93)	(0.70, −0.70)

[4]See, e.g., [Straffin1993]

[5]Palacios-Huerta, "Professionals Play Minimax," Review of Economic Studies (2003) 70, 395–415.

Applying our solution method to the linear programming formulation finds the optimal solution as either pure strategy or mixed strategy.

$$> KickerObj := K;$$

$$> KickerCons := \{0.58 \cdot x_1 + 0.93 \cdot x_2 - K \geq 0, 0.95 \cdot x_1 + 0.70 \cdot x_2 - K \geq 0, x_1 + x_2 \geq 1, x_1 + x_2 \leq 1, K \geq 0, x_1 \geq 0, x_2 \geq 0\}$$

$$> fnormal(LPSolve(KickerObj, KickerCons, maximize), 4);$$

$$[0.7958, [K = 0.7958, x_1 = 0.3833, x_2 = 0.6167]]$$

$$> GoalieObj := G;$$

$$> GoalieCons := \{0.42 \cdot y_1 + 0.05 \cdot y_2 - G \geq 0, 0.07 \cdot y_1 + 0.30 \cdot y_2 - G \geq 0, y_1 + y_2 \geq 1, y_1 + y_2 \leq 1, G \geq 0, y_1 \geq 0, y_2 \geq 0\}$$

$$> fnormal(LPSolve(GoalieObj, GoalieCons, maximize), 4);$$

$$[0.2042, [G = 0.2042, y_1 = 0.4167, y_2 = 0.5833]]$$

A short-cut method, the *Method of Oddments*, is shown in Table 7.13.

TABLE 7.13: Method of Oddments

		Goalie			
		Dive Left	Dive Right	Oddments	Probabilities
Kicker	Kick Left	0.58	0.95	0.37	0.23/0.60 = 0.383
	Kick Right	0.93	0.70	0.23	0.37/0.60 = 0.6166
	Oddments	0.35	0.25		
	Probabilities	0.25/0.60 = 0.416	0.35/0.60 = 0.5833		

We find the mixed strategy for the kicker is 38.3% kicking left and 61.7% kicking right, while the goalie dives right 58.3% and dives left 41.7%. If we merely count percentages from the data that was collected by Palacios-Huerta in his study of 459 penalty kicks over 5 years of data, we find the kicker did 40% kicking left, and 60% kicking right, while the goalie dove left 42% and right 58%. Since our model closely approximates the data, our game theory approach adequately models the penalty kick.

The next example, a batter-pitcher duel, continues the theme of technology in sports today.

Example 7.6. Batter-Pitcher Duel.
We extend to four strategies for each player. Consider a batter-pitcher duel between Aaron Judge of the New York Yankees, and various pitchers in the American League where the pitcher throws a fastball, a split-finger fastball, a curve ball, and a change-up. The batter, aware of these pitches, must prepare appropriately for the pitch. We'll consider right- and left-handed pitchers separately in this analysis. Data is available from many websites, such as www.STATS.com.

The data in Table 7.14 has been compiled for an American League right-handed pitcher (RHP) versus Aaron Judge. Let FB = fastball, CB = curve ball, CH = change-up, SF = split-finger fastball.

TABLE 7.14: Aaron Judge vs. a Right-Handed Pitcher

Judge / RHP	FB	CB	CH	SF
FB	0.337	0.246	0.220	0.200
CB	0.283	0.571	0.339	0.303
CH	0.188	0.347	0.714	0.227
SF	0.200	0.227	0.154	0.500

Both the batter and pitcher want the best possible result. We set this up as a linear programming problem. Our decision variables are x_1, x_2, x_3, and x_4 as the percentages to guess FB, CB, CH, SF, respectively, and V represents Judge's batting average.

$$\begin{aligned}
&\text{Maximize } V\\
&\text{subject to}\\
&\quad 0.337 \cdot x_1 + 0.283 \cdot x_2 + 0.188 \cdot x_3 + 0.200 \cdot x_4 - V \geq 0\\
&\quad 0.246 \cdot x_1 + 0.571 \cdot x_2 + 0.347 \cdot x_3 + 0.227 \cdot x_4 - V \geq 0\\
&\quad 0.220 \cdot x_1 + 0.339 \cdot x_2 + 0.714 \cdot x_3 + 0.154 \cdot x_4 - V \geq 0\\
&\quad 0.200 \cdot x_1 + 0.303 \cdot x_2 + 0.227 \cdot x_3 + 0.500 \cdot x_4 - V \geq 0\\
&\qquad\qquad x_1 + x_2 + x_3 + x_4 = 1\\
&\qquad\qquad x_1, x_2, x_3, x_4, V \geq 0
\end{aligned}$$

We solve this linear programming problem with Maple, and find the optimal solution (strategy) is to guess the fastball (FB) 27.49%, guess the curve ball (CB) 64.23%, never guess change-up (CH), and guess split-finger fastball (SF) 8.27% of the time to obtain a 0.291 batting average.

The pitcher also wants to keep the batting average as low as possible. Set up the linear program for the pitcher as follows. The decision variables are y_1, y_2, y_3, and y_4 as the percentages to guess FB, CB, CH, SF, respectively, and V again represents Judge's batting average.

$$\begin{aligned}
&\text{Minimize } V\\
&\text{subject to}\\
&\quad 0.337 \cdot y_1 + 0.246 \cdot y_2 + 0.220 \cdot y_3 + 0.200 \cdot y_4 - V \leq 0\\
&\quad 0.283 \cdot y_1 + 0.571 \cdot y_2 + 0.339 \cdot y_3 + 0.303 \cdot y_4 - V \leq 0\\
&\quad 0.188 \cdot y_1 + 0.347 \cdot y_2 + 0.714 \cdot y_3 + 0.227 \cdot y_4 - V \leq 0\\
&\quad 0.200 \cdot y_1 + 0.227 \cdot y_2 + 0.154 \cdot y_3 + 0.500 \cdot y_4 - V \leq 0\\
&\qquad\qquad y_1 + y_2 + y_3 + y_4 = 1\\
&\qquad\qquad y_1, y_2, y_3, y_4, V \geq 0
\end{aligned}$$

We find the right-handed pitcher (RHP) should randomly throw 65.94% fastballs, no curve balls, 3.24% change-ups, and 30.82% split-finger fastballs for Judge to keep, and not increase, his 0.291 batting average.

Statistics for Judge versus a left-handed pitcher (LHP) are in Table 7.15.

TABLE 7.15: Aaron Judge vs. a Left-Handed Pitcher

Judge / LHP	*FB*	*CB*	*CH*	*SF*
FB	0.353	0.185	0.220	0.244
CB	0.143	0.333	0.333	0.253
CH	0.071	0.333	0.353	0.247
SF	0.300	0.240	0.254	0.450

Set up as before, and solve the linear programming problem.

$$\begin{aligned}
&\text{Maximize } V\\
&\text{subject to}\\
&\quad 0.353 \cdot x_1 + 0.143 \cdot x_2 + 0.071 \cdot x_3 + 0.300 \cdot x_4 - V \geq 0\\
&\quad 0.185 \cdot x_1 + 0.333 \cdot x_2 + 0.333 \cdot x_3 + 0.240 \cdot x_4 - V \geq 0\\
&\quad 0.220 \cdot x_1 + 0.333 \cdot x_2 + 0.353 \cdot x_3 + 0.254 \cdot x_4 - V \geq 0\\
&\quad 0.244 \cdot x_1 + 0.253 \cdot x_2 + 0.247 \cdot x_3 + 0.450 \cdot x_4 - V \geq 0\\
&\qquad\qquad x_1 + x_2 + x_3 + x_4 = 1\\
&\qquad\qquad x_1, x_2, x_3, x_4, V \geq 0
\end{aligned}$$

We find the optimal solution for Judge versus a LHP. Judge should guess as follows: never guess fastball, guess curve ball 24.0%, never guess change-up, and guess split-finger fastball 76.0% for a batting average of .262.

For the left-handed pitchers facing Judge, solve the following LP:

$$
\begin{aligned}
&\text{Minimize } V\\
&\text{subject to}\\
&\quad 0.353 \cdot y_1 + 0.185 \cdot y_2 + 0.220 \cdot y_3 + 0.244 \cdot y_4 - V \leq 0\\
&\quad 0.143 \cdot y_1 + 0.333 \cdot y_2 + 0.333 \cdot y_3 + 0.253 \cdot y_4 - V \leq 0\\
&\quad 0.071 \cdot y_1 + 0.333 \cdot y_2 + 0.353 \cdot y_3 + 0.247 \cdot y_4 - V \leq 0\\
&\quad 0.300 \cdot y_1 + 0.240 \cdot y_2 + 0.254 \cdot y_3 + 0.450 \cdot y_4 - V \leq 0\\
&\qquad\qquad y_1 + y_2 + y_3 + y_4 = 1\\
&\qquad\qquad y_1, y_2, y_3, y_4, V \geq 0
\end{aligned}
$$

The pitcher should randomly throw 26.2% fastballs, 62.8% curve balls, no change-ups, and no split-finger fast balls. Then Judge's batting average will remain at .262, and won't increase.

The manager of the opposing team is in the middle of a close game. There are two outs, runners in scoring position, and Judge is coming to bat. Does the manager keep the LHP in the game or switch to a RHP? The percentages say keep the LHP since $0.262 < 0.291$. Tell the catcher and pitcher to randomly select the pitches to be thrown to Judge.

Judge's manager wants to improve his batting ability against both a curve ball and a LHP. Only by improving against these strategies can he effect change.

Example 7.7. Operation Overlord.
Operation Overlord, the codename for World War II's Battle of Normandy, can be viewed in the context of game theory. In 1944, the Allies were planning an operation for the liberation of Europe; the Germans were planning their defense against it. There were two known possibilities for an initial amphibious landing: the beaches at Normandy, and those at Calais. Any landing would succeed against a weak defense, so the Germans did not want a weak defense at the potential landing site. Calais was more difficult for a landing, but closer to the Allies targets for success.

Suppose the probabilities of an Allied success are as in Table 7.16.

TABLE 7.16: Probabilities of a Successful Allied Landing

		German Defense	
		Normandy	Calais
Allied Landing	Normandy	75%	100%
	Calais	100%	20%

The Allies successfully landing at Calais would earn 100 points, successfully landing at Normandy would earn 80 points, and failure at either landing would earn 0 points. What decisions should be made?

We compute the expected values, placing them in the payoff matrix of Table 7.17.

TABLE 7.17: Payoff Matrix for Allied Landing

		German Defense	
		Normandy	Calais
Allied Landing	Normandy	$0.75 \cdot 80 = 60$	$1. \cdot 80 = 80$
	Calais	$1. \cdot 100 = 100$	$0.2 \cdot 100 = 20$

There are no pure strategy solutions in this example. We use mixed strategies, and determine the game's outcome.

The Allies would employ a mixed strategy of 80% Normandy and 20% Calais to achieve an outcome of 68 points. At the same time, the Germans should employ a strategy of 60% Normandy and 40% Calais for their defenses to keep the Allies at 68 points.

Implementation of the landing was certainly not two-pronged. So what do the mixed strategies imply in strategic thinking? Most likely a strong feint at Calais and lots of information leaks about Calais, while the real landing at Normandy was a secret. The Germans had a choice as to believe the information about Calais, or somewhat equally divide their defenses. Although the true results were in doubt for a while, the Allies prevailed.

Example 7.8. Choosing the Right Course of Action.
The US Army Command and General Staff College presented this approach for choosing the best course of action (COA) for a mission. For a possible battle between two forces, we compute the optimal courses of actions for the two opponents using game theory.[6]

Steps 1 and 2. List the friendly COAs, and rank order them Best to Worst.

COA 1: Decisive Victory

COA 4: Attrition Based Victory

COA 2: Failure by Culmination

COA 3: Defeat in Place

Step 3. The enemy is thought to have six distinct possible courses of action.

Rank best to worst each COA of the enemy where the row represents the

[6]This example is adapted from [Cantwell2003].

friendly COA. For example, the friendly COA 1 is best against the enemy COA 1 and friendly COA 2 is worst against the enemy COA 6.

Step 4. Decide in each case if we think we will Win, Lose, Draw using Table 7.18.

TABLE 7.18: Enemy COA vs. Friendly COA

		Enemy Course of Action					
		COA 1	COA 2	COA 3	COA 4	COA 5	COA 6
Friendly Course of Action	COA 1	Best	Win	Win	Draw	Loss	Loss
	COA 4	Win	Win	Win	Win	Loss	Loss
	COA 2	Win	Win	Loss	Loss	Loss	Worst
	COA 3	Draw	Draw	Draw	Loss	Loss	Loss

Steps 5 and 6. Provide scores. Since there are 4 friendly COAs and 6 enemy COAs, we use scores from 24 (Best) to 1 (Worst). See Table 7.19.

TABLE 7.19: Enemy COA vs. Friendly COA Scores

		Enemy Course of Action					
		COA 1	COA 2	COA 3	COA 4	COA 5	COA 6
Friendly Course of Action	COA 1	Best 24	Win 23	Win 22	Draw –	Loss 3	Loss 2
	COA 4	Win 21	Win 20	Win 19	Win 18	Loss 10	Loss 9
	COA 2	Win 17	Win 16	Loss 11	Loss 7	Loss 8	Worst 1
	COA 3	Draw –	Draw –	Draw –	Loss 6	Loss 5	Loss 4

Steps 7 and 8. Put into numerical order for Loss.

Step 9. Fill in the scores for the draw. See Table 7.20.

TABLE 7.20: Enemy COA vs. Friendly COA Scores with Draws

		Enemy Course of Action					
		COA 1	COA 2	COA 3	COA 4	COA 5	COA 6
Friendly Course of Action	COA 1	Best 24	Win 23	Win 22	Draw 15	Loss 3	Loss 2
	COA 4	Win 21	Win 20	Win 19	Win 18	Loss 10	Loss 9
	COA 2	Win 17	Win 16	Loss 11	Loss 7	Loss 8	Worst 1
	COA 3	Draw 14	Draw 13	Draw 12	Loss 6	Loss 5	Loss 4

Step 10. Put the courses of action back in their original order. Add *minimax* data. See Table 7.21.

TABLE 7.21: Enemy COA vs. Friendly COA Minimax

		Enemy Course of Action (COA)						
		1	2	3	4	5	6	*Min*
Friendly Course of Action (COA)	1	24	23	22	15	3	2	2
	2	17	16	11	7	8	1	1
	3	14	13	12	6	5	4	4
	4	21	20	19	18	10	9	[9]
	Max	24	23	22	15	[10]	18	*No saddle*

There is no pure strategy solution. Because of the size of the payoff matrix, we did not use the movement diagram, but instead used the Minimax theorem. Basically, we find the minimum in each row, and then the maximum of these minimums. Then we find the maximums in each column, then the minimum of those maximums. If the maximum of the row minimum's equals the minimum of the column maximums, then we have a pure strategy solution. If not, we have to find the mixed strategy solution. In Step 10 above, the maximum of the minimums is 9, while the minimum of the maximums is 10. They are not equal.

Linear programming may be used in zero-sum games to find the solutions whether they are pure strategy or mixed strategy solutions. So, we solve this game using a linear program. Let V be the value of the game, x_1 to x_4 be the probabilities in which to play strategies (COAs) 1 through 4 for the friendly side. The values y_1 to y_6 represent the probabilities the enemy should employ COA 1 to COA 6, respectively, to obtain their best results.

$$
\begin{aligned}
&\text{Maximize } V\\
&\text{subject to}\\
&\quad 24x_1 + 16x_2 + 13x_3 + 21x_4 - V \geq 0\\
&\quad 23x_1 + 17x_2 + 12x_3 + 20x_4 - V \geq 0\\
&\quad 22x_1 + 11x_2 + 6x_3 + 19x_4 - V \geq 0\\
&\quad 3x_1 + 7x_2 + 5x_3 + 10x_4 - V \geq 0\\
&\quad 15x_1 + 8x_2 + 4x_3 + 9x_4 - V \geq 0\\
&\quad 2x_1 + 1x_2 + 14x_3 + 18x_4 - V \geq 0\\
&\quad x_1 + x_2 + x_3 + x_4 = 1\\
&\quad x_1, x_2, x_3, x_4, V \geq 0
\end{aligned}
$$

The linear program for the enemy is

$$
\begin{aligned}
&\text{Minimize } v\\
&\text{subject to}\\
&\quad 24y_1 + 23y_2 + 22y_3 + 3y_4 + 15y_5 + 2y_6 - v \leq 0\\
&\quad 16y_1 + 17y_2 + 11y_3 + 7y_4 + 8y_5 + y_6 - v \leq 0\\
&\quad 13y_1 + 12y_2 + 6y_3 + 5y_4 + 4y_5 + 14y_6 - v \leq 0\\
&\quad 21y_1 + 20y_2 + 19y_3 + 10y_4 + 9y_5 + 18y_6 - v \leq 0\\
&\quad y_1 + y_2 + y_3 + y_4 + y_5 + y_6 = 1\\
&\quad y_1, y_2, y_3, y_4, y_5, y_6, v \geq 0
\end{aligned}
$$

Maple gives the solution as $V = 9.462$ when "friendly" chooses $x_1 = 7.7\%$, $x_2 = 0$, $x_3 = 0$, $x_4 = 92.3\%$, while the "enemy" best results come when $y_1 = 0$, $y_2 = 0$, $y_3 = 0$, $y_4 = 46.2\%$ and $y_5 = 53.9\%$ holding the "friendly" to 9.462.

Interpretation: At 92.3%, we see we should defend along the Vistula River (COA 4) almost all the time. The value of the game, $V = 9.462$, is greater than the pure strategy solution of 9 for always picking to defend the Vistula River (COA 4). This implies that we benefit from secrecy and employing deception. We can benefit by "selling the enemy" on our "attack North and fix in the South" (COA 1). A negative 9.462 for the enemy does not mean the enemy loses. We need to further consider the significance of the values and mission analysis.

Exercises

Solve the problems using any method.

1. What should each player do according to the payoff matrix in Table 7.22?

TABLE 7.22: Attack and Defense

Attack and Defense Tableau		Colonel Blotto	
		Defend City I D_1	Defend City II D_2
Colonel Sotto	Attack City I A_1	30	30
	Attack City II A_2	20	0

Use Table 7.23 below for Exercises 2. to 10.

TABLE 7.23: Payoff Table

Payoff Tableau		Colin	
		C_1	C_2
Rose	R_1	a	b
	R_2	c	d

2. What assumptions have to be true for a at R_1C_1 to be the pure strategy solution?

3. What assumptions have to be true for b at R_1C_1 to be the pure strategy solution?

4. What assumptions have to be true for c at R_1C_1 to be the pure strategy solution?

5. What assumptions have to be true for d at R_1C_1 to be the pure strategy solution?

6. What assumptions have to be true for there not to be a saddle point solution in the game?

7. Show the value of the game is

$$x = \frac{ad - bc}{(a - c) + (d - b)}.$$

8. Set $a = 1/4$, $b = 1/4$, $c = 2$, and $d = 0$ in Table 7.23. What is the pure strategy solution for this game?

9. Set $a = -3$, $b = 5$, $c = 4$, and $d = -3$ in Table 7.23. What is the solution?

10. Let $a > d > b > c$. Show that Colin should play C_1 and C_2 with probabilities x and $(1 - x)$ where

$$x = \frac{d - b}{(a - c) + (d - b)}.$$

11. Consider Table 7.24 of a batter-pitcher duel. All the entries in the payoff matrix reflect the percent of hits off the pitcher, the batting average. What strategies are optimal for each player?

TABLE 7.24: Batter-Pitcher Duel Payoffs

Payoff Tableau		*Pitcher*	
		Throw Fast Ball (C_1)	Throw Knuckle Ball (C_2)
Batter	Guess Fast Ball (R_1)	.360	.160
	Guess Knuckle Ball (R_2)	.310	.260

12. Find the solution in the game shown in Table 7.25.

TABLE 7.25: Two by Three Game

Payoff Tableau		*Colin*		
		C_1	C_2	C_3
Rose	R_1	85	45	75
	R_2	75	35	35

13. Find the solution in the game given in Table 7.26.

TABLE 7.26: Four by Four Game

Payoff Tableau		*Colin*			
		C_1	C_2	C_3	C_4
Rose	R_1	40	80	35	60
	R_2	55	90	55	70
	R_3	55	40	45	75
	R_4	45	25	50	50

14. Table 7.27 represents a game between a professional athlete (Rose) and management (Colin) for contract decisions. The athlete has two strategies and management has three strategies. The values are in 1,000s. What decision should each make?

TABLE 7.27: Two by Three Game

Payoff Tableau		*Colin*		
		C_1	C_2	C_3
Rose	R_1	490	220	195
	R_2	425	350	150

15. Solve the game given in Table 7.28.

TABLE 7.28: Game Payoff Table

		Colin	
		C_1	C_2
Rose	R_1	$(2, -2)$	$(1, -1)$
	R_2	$(3, -3)$	$(4, -4)$

16. The predator has two strategies for catching the prey: ambush or pursuit. The prey has two strategies for escaping: hide or run. The game matrix appears in Table 7.29. Solve the game.

TABLE 7.29: Predator-Prey Payoffs

Payoff Tableau		*Predator*	
		Ambush C_1	Pursue C_2
Prey	Hide R_1	0.26	0.33
	Run R_2	0.72	0.56

17. A professional football team has collected data for certain plays against certain defenses. The payoff matrix in Table 7.30 shows the yards gained or lost for a particular play against a particular defense. Find the best strategies.

TABLE 7.30: Yards Gained or Lost

Payoff Tableau		*Team B*		
		C_1	C_2	C_3
Team A	R_1	0	−1	5
	R_2	7	5	10
	R_3	15	−4	−5
	R_4	5	0	10
	R_5	−5	−10	10

18. Solve the Jeter-Romero batter-pitcher duel using the payoff matrix given in Table 7.31.

TABLE 7.31: Jeter-Romero Batter-Pitcher Duel Payoffs

Payoff Tableau		*Ricky Romero*	
		Throw Fast Ball (C_1)	Throw Split-Finger (C_2)
Derek Jeter	Guess Fast Ball (R_1)	.343	.267
	Guess Split-Finger (R_2)	.195	.406

19. Solve the game with the payoff matrix of Table 7.32.

TABLE 7.32: Payoff Tableau

Payoff Tableau		*Colin*		
		C_1	C_2	C_3
Rose	R_1	0.5	0.9	0.9
	R_2	0.1	0	0.1
	R_3	0.9	0.9	0.5

20. Solve the game with the payoff matrix of Table 7.33.

TABLE 7.33: Payoff Tableau

Payoff Tableau		*Colin*	
		C_1	C_2
Rose	R_1	6	5
	R_2	1	4
	R_3	8	5

21. Solve the game with the payoff matrix of Table 7.34.

TABLE 7.34: Payoff Tableau

Payoff Tableau		*Colin*	
		C_1	C_2
Rose	R_1	3	-1
	R_2	2	4
	R_3	6	2

22. Find the solution for the game using the payoff matrix in Table 7.35.

TABLE 7.35: Payoff Tableau

Payoff Tableau		*Colin*			
		C_1	C_2	C_3	C_4
Rose	R_1	1	-1	2	3
	R_2	2	4	0	5

23. In Table 7.36 of a batter-pitcher duel, the entries reflect the percent of hits off the pitcher, the batting average. Find the optimal strategies for each player?

TABLE 7.36: Batter-Pitcher Duel Payoffs

Payoff Tableau		*Pitcher*	
		Throw Fast Ball (C_1)	Throw Curve Ball (C_2)
Batter	Guess Fast Ball (R_1)	.300	.200
	Guess Curve Ball (R_2)	.100	.500

24. Solve the game shown in Table 7.37.

TABLE 7.37: Payoff Tableau

Payoff Tableau		Colin	
		C_1	C_2
Rose	R_1	3	−4
	R_2	1	3
	R_3	−5	10

25. Solve the game shown in Table 7.38.

TABLE 7.38: Payoff Tableau

Payoff Tableau		Colin		
		C_1	C_2	C_3
Rose	R_1	1	1	10
	R_2	2	3	−4

26. Solve the game shown in Table 7.39.

TABLE 7.39: Payoff Tableau

Payoff Tableau		Colin		
		C_1	C_2	C_3
Rose	R_1	1	2	2
	R_2	2	1	2
	R_3	2	2	0

27. Determine the solution to the following zero-sum games shown in Tables 7.40 to 7.44 by any method or methods, but show/state work. State the value of the game for both Rose and Colin, and what strategies each player should choose.

    a)

TABLE 7.40: Payoff Tableau

Payoff Tableau		Colin		
		C_1	C_2	C_3
Rose	R_1	10	20	14
	R_2	5	21	8
	R_3	8	22	0

    b)

TABLE 7.41: Payoff Tableau

Payoff Tableau		Colin	
		C_1	C_2
Rose	R_1	−8	12
	R_2	2	6
	R_3	0	−2

    c)

TABLE 7.42: Payoff Tableau

Payoff Tableau		Colin	
		C_1	C_2
Rose	R_1	8	2
	R_2	5	16

d)

TABLE 7.43: Payoff Tableau

Payoff Tableau		*Colin*		
		C_1	C_2	C_3
Rose	R_1	15	12	11
	R_2	14	16	17

e)

TABLE 7.44: Payoff Tableau

Payoff Tableau		*Colin*			
		C_1	C_2	C_3	C_4
	R_1	3	2	4	1
Rose	R_2	−9	1	−1	0
	R_3	6	4	7	3

28. For the game of Table 7.45 between Rose and Colin, write the linear programming formulation for Rose. Using Maple's *simplex* or *Optimization* packages, find and state the complete solution to the game in context of the game.

TABLE 7.45: Payoff Tableau

Payoff Tableau		*Colin*			
		C_1	C_2	C_3	C_4
	R_1	0	1	2	6
	R_2	2	4	1	2
Rose	R_3	1	−1	4	−1
	R_4	−1	1	−1	3
	R_5	−2	−2	2	2

Projects

Project 7.1. Research the solution methodologies for the three-person games. Analyze the three-person, zero-sum game between Rose, Colin, and Larry shown in Tables 7.46 and 7.47.

TABLE 7.46: Payoff Tableau for Larry's D_1

Larry D_1		*Colin*	
		C_1	C_2
Rose	R_1	$(4, 4, -8)$	$(-2, 4, -2)$
	R_2	$(4, -5, 1)$	$(3, -3, 0)$

TABLE 7.47: Payoff Tableau for Larry's D_2

Larry D_2		*Colin*	
		C_1	C_2
Rose	R_1	$(-2, 0, 2)$	$(-2, -1, 3)$
	R_2	$(-4, 5, -1)$	$(1, 2, -3)$

a) Draw the movement diagram for this zero-sum game. Find any and all equilibria. If this game were played without any possible coalitions, what would you expect to happen?

b) If Colin and Rose were to form a coalition against Larry, set up and solve the resulting 2×4 game. What are the strategies to be played and the payoffs for each of the three players? Solve by hand (*Show your work!*), then check using the 3-Person Template.

c) Given the results of the coalitions of Colin vs. Rose-Larry and Rose vs. Colin-Larry as follows:

   Colin vs. Rose-Larry: $(3, -3, 0)$ and Rose vs. Colin-Larry: $(-2, 0, 2)$

   If no *side payments* were allowed, would any player be worse off joining a coalition than playing alone? Briefly explain (include the values to justify decisions).

d) What is (are) the preferred coalition(s), if any?

e) Briefly explain how *side payments* could work in this game.

7.4 Examples of Partial Conflict Games

We present examples of partial conflict games with their solutions.

Example 7.9. Cuban Missile Crisis—A Classic Game of Chicken[7]. "We're eyeball to eyeball, and I think the other fellow just blinked," were the eerie words of Secretary of State Dean Rusk at the height of the Cuban Missile Crisis in October, 1962. Secretary Rusk was referring to signals by the Soviet Union that it desired to defuse the most dangerous nuclear confrontation ever to occur between the superpowers, which many analysts have interpreted as a classic instance of a nuclear "Game of Chicken."

We will highlight the scenario from 1962. The Cuban Missile Crisis was precipitated in October 1962, by the Soviet's attempt to install medium- and intermediate-range nuclear-armed ballistic missiles in Cuba that were capable of hitting a large portion of the United States. The range of the missiles from Cuba allowed for major political, population, and economic centers to become targets. The goal of the United States was immediate removal of the Soviet missiles. U.S. policy makers seriously considered two strategies to achieve this end: naval blockade or airstrikes.

President Kennedy, in his speech to the nation, explained the situation as well as the goals for the United States. He set several initial steps. First, to halt the offensive build-up, a strict quarantine on all offensive military equipment under shipment to Cuba was being initiated. He went on to say that any launch of missiles from Cuba at anyone would be considered an act of war by the Soviet Union against the United States resulting in a full retaliatory nuclear strike against the Soviet Union. He called upon Soviet Premier Krushchev to end this threat to the world, and restore world peace.

We will use the Cuban Missile Crisis to illustrate parts of the theory—not just an abstract mathematical model, but one that mirrors the real-life choices, and underlying thinking of flesh-and-blood decision makers. Indeed, Theodore Sorensen, special counsel to President Kennedy, used the language of "moves" to describe the deliberations of EXCOM, the Executive Committee of key advisors to President Kennedy during the Cuban Missile Crisis. Sorensen said,

> We discussed what the Soviet reaction would be to any possible move by the United States, what our reaction with them would have to be to that Soviet action, and so on, trying to follow each of those roads to their ultimate conclusion.

Problem: Build a mathematical model that allows for consideration of alternative decisions by the two opponents.

---

[7]Adapted from Brams, "Game theory and the Cuban missile crisis," available at http://plus.maths.org/issue13/features/brams/index.html.

Assumption: We assume the two opponents are rational players.

Model Choice: Game Theory.

The goal of the United States was the immediate removal of the Soviet missiles; U.S. policy makers seriously considered two strategies to achieve this end:

1. *A naval blockade* (B), or "quarantine" as it was euphemistically called, to prevent shipment of more missiles, possibly followed by stronger action to induce the Soviet Union to withdraw the missiles already installed.
2. *A "surgical" air strike* (A) to destroy the missiles already installed, insofar as possible, perhaps followed by an invasion of the island.

The alternatives open to Soviet policy makers were:

1. *Withdrawal* (W) of their missiles.
2. *Maintenance* (M) of their missiles.

We set (x, y) as (*payoffs to the United States, payoffs to the Soviet Union*) where 4 is the best result, 3 is next best, 2 is next worst, and 1 is the worst. Table 7.48 shows the payoffs.

TABLE 7.48: Cuban Missile Crisis Payoffs

		Soviet Union	
		Withdraw Missiles (W)	Maintain Missiles (M)
United States	Blockade (B)	$(3, 3)$	$(2, 4)$
	Air Strike (A)	$(4, 2)$	$(1, 1)$

We show the movement diagram in Table 7.49 where we have equilibria at $(4, 2)$ and $(2, 4)$. The Nash equilibria are boxed. Note both equilibria, $(4, 2)$ and $(2, 4)$ are found by our arrow diagram.

As in Chicken, as both players attempt to get to their equilibrium, the outcome of the games end up at $(1, 1)$. This is disastrous for both countries and their leaders. The best solution is the $(3, 3)$ compromise position. However, $(3, 3)$ not stable. This choice will eventually put us back at $(1, 1)$. In this situation, one way to avoid the “chicken dilemma” is to try strategic moves.

Both sides did not choose their strategies simultaneously or independently. Soviets responded to our blockade after it was imposed. The U.S. held out the chance of an air strike as a viable choice even after the blockade. If the U.S.S.R. would agree to remove the weapons from Cuba, the U.S. would agree

TABLE 7.49: Cuban Missile Crisis Movement Diagram

		Soviet Union		
		Withdraw Missiles (W)		Maintain Missiles (M)
Rose	Blockade (B)	(3, 3)	$\Longrightarrow$	**(2, 4)**
		$\Downarrow$		$\Uparrow$
	Air Strike (A)	**(4, 2)**	$\Longleftarrow$	(1, 1)

to (a) remove the quarantine and (b) agree not to invade Cuba. If the Soviets maintained their missiles, the U.S. preferred the airstrike to the blockade. Attorney General Robert Kennedy said, "if they do not remove the missiles, then we will." The U.S. used a combination of promises and threats. The Soviets knew our credibility in both areas was high (strong resolve). Therefore, they withdrew the missiles, and the crisis ended. Khrushchev and Kennedy were wise.

Needless to say, the strategy choices, probable outcomes, and associated payoffs shown in Table 7.48 provide only a skeletal picture of the crisis as it developed over a period of thirteen days. Both sides considered more than the two alternatives listed, as well as several variations on each. The Soviets, for example, demanded withdrawal of American missiles from Turkey as a *quid pro quo* for withdrawal of their own missiles from Cuba, a demand publicly ignored by the United States.

Nevertheless, most observers of this crisis believe that the two superpowers were on a collision course, which is actually the title of one book[8] describing this nuclear confrontation. Analysts also agree that neither side was eager to take any irreversible step, such as one of the drivers in Chicken might do by defiantly ripping off the steering wheel in full view of the other driver, thereby foreclosing the option of swerving.

Although in one sense the United States "won" by getting the Soviets to withdraw their missiles, Premier Nikita Khrushchev of the Soviet Union at the same time extracted from President Kennedy a promise not to invade Cuba, which seems to indicate that the eventual outcome was a compromise of sorts. But this is not game theory's prediction for Chicken because the strategies associated with compromise do not constitute a Nash equilibrium.

To see this, assume play is at the compromise position (3, 3), that is, the U.S. blockades Cuba, and the U.S.S.R. withdraws its missiles. This strategy

[8]Henry M. Pachter, *Collision Course: The Cuban Missile Crisis and Coexistence*, Praeger, NY, 1963.

is not stable because both players would have an incentive to defect to their more belligerent strategy. If the U.S. were to defect by changing its strategy to airstrike, play would move to $(4, 2)$, improving the payoff the U.S. received; if the U.S.S.R. were to defect by changing its strategy to Maintenance, play would move to $(2, 4)$, giving the U.S.S.R. a payoff of 4. (This classic game theory setup gives us no information about which outcome would be chosen, because the table of payoffs is symmetric for the two players. This is a frequent problem in interpreting the results of a game theoretic analysis, where more than one equilibrium position can arise.) Finally, should the players be at the mutually worst outcome of $(1, 1)$, that is, nuclear war, both would obviously desire to move away from it, making the strategies associated with $(1, 1)$, like those with $(3, 3)$, unstable.

Example 7.10. Writer's Guild Strike of 2007–2008.
Game Theory Approach[9]
Let us begin by stating strategies for each side. Our two rational players will be the *Writer's Guild* and the *Management*. We develop strategies for each player.

Strategies:

- Writer's Guild: Their strategies are to strike (S) or not to strike (NS).
- Management: Salary Increase and revenue sharing (IN) or status quo (SQ).

First, we rank order the outcomes for each side in order of preference. (The rank orderings are *ordinal utilities.*)

Alternatives and Rankings

- Strike vs. Status Quo = (S, SQ):
  Writer's worst case (1); Management's next to best case (3)
- No Strike vs. Status Quo = (NS, SQ):
  Writer's next to worst case (2); Management's best case (4)
- Strike vs. Salary Increase and Revenue Sharing = (S, IN):
  Writers next to best case (3); Management's next to worst case (2)
- No Strike vs. Salary Increase and Revenue Sharing = (NS, IN):
  Writer's best case (4); Management's worst case (1)

This list provides us with a payoff matrix consisting of ordinal values; see Table 7.50. We will refer to the Writer's Guild as *Rose* and the Management as *Colin.*

The movement diagram in Table 7.51 finds $(2, 4)$ as the likely outcome.

[9]From [Fox2008].

TABLE 7.50: Payoff Matrix for Writers and Management

		Management (*Colin*)	
		Status Quo (SQ)	Increase Salary (IN)
Writer's Guild (*Rose*)	Strike (S)	$(1,3)$	$(3,2)$
	No Strike (NS)	$(2,4)$	$(4,1)$

TABLE 7.51: Writer's Guild and Management's Movement Diagram

		Management (*Colin*)		
		Status Quo (SQ)		Increase Salary (IN)
Writer's Guild (Rose)	Strike (S)	$(1,3)$	$\Longleftarrow$	$(3,2)$
		$\Downarrow$		$\Downarrow$
	No Strike (NS)	$\boxed{(2,4)}$	$\Longleftarrow$	$(4,1)$

The movement arrows point towards $(2,4)$ as the pure Nash equilibrium. We also note that this result is not satisfying to the Writer's Guild, and that they would like to have a better outcome. Both $(3,2)$ and $(4,1)$ within the payoff matrix provide better outcomes to the Writers.

The Writers can employ several options to try to secure a better outcome. They can first try Strategic Moves, and if that fails to produce a better outcome, then they can move to Nash Arbitration. Both of these methods employ communications in the game. In strategic moves, we examine the game to see if the outcome is changed by "moving first", threatening our opponent, or making promises to our opponent, or whether a combination of threats and promises changes the outcome.

Examine the strategic moves. If the writers move first; their best result is again $(2,4)$. If management moves first, the best result is $(2,4)$. First moves keep us at the Nash equilibrium. The writers consider a threat: they tell management that if they choose SQ, they will strike putting us at $(1,3)$. This result is indeed a threat, as it is worse for both the writers and management. However, the options for management under IN are both worse than $(1,3)$, so they do not accept the threat. The writers do not have a promise to offer. At this point we might involve an arbiter using the Nash arbitration method as suggested earlier.

The Nash arbitration formulation then is

$$\begin{aligned}&\text{Maximize } Z = (x-2)\cdot(y-3)\\&\text{subject to}\\&\qquad \tfrac{3}{2}x + y = 7\\&\qquad\qquad x \geq 2\\&\qquad\qquad y \geq 3\end{aligned}$$

Writers and management security levels are found from prudential strategies using Equations (7.9) and (7.10). The security levels are calculated to be $(2, 3)$. We show this in Figure 7.4.

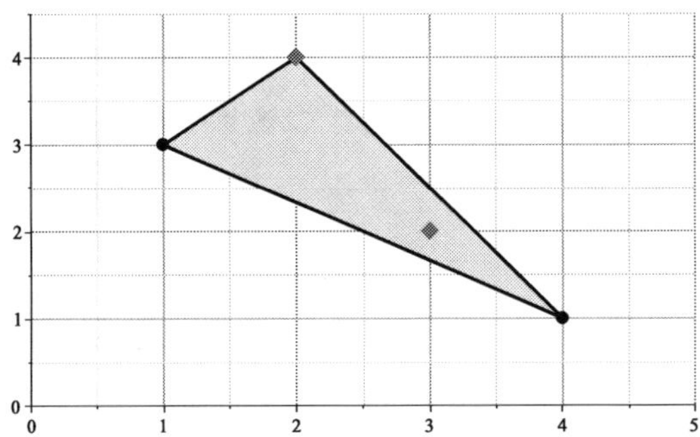

FIGURE 7.4: The Payoff Polygon for Writer's Guild Strike

The Nash equilibrium value $(2, 4)$ lies along the Pareto Optimal line segment (from $(4, 1)$ to $(2, 4)$). But the Writers can do better by going on strike and forcing arbitration, which is what they did.

In this example, we consider "binding arbitration" where the players have a third party work out the outcomes that best meet their desires and is acceptable to all players. Nash found that this outcome can be obtained by:

> The status quo point is formed from the security levels of each side. We find the value $(2, 3)$ using prudential strategies. The function for the Nash Arbitration scheme is *Maximize* $(x-2)(y-3)$. Using Maple's *QPSolve* from the *Optimization* package, we find the desired solution to our quadratic program is $x = 2.33$ and $y = 3.5$.

We have the $(x, y) = (2.33, 3.5)$ as our arbitrated solution. We can now determine how the arbiters should proceed. We solve the following simultaneous equations

$$2p_1 + 4p_2 = 2.33$$
$$4p_1 + p_2 = 3.5.$$

We find that the probabilities to be played are 5/6 and 1/6. Further, we see that Player I, the Writers, should always play R_2, so the management arbiter plays $5/6 \cdot C_1$ and $1/6 \cdot C_2$ during the arbitration.

Example 7.11. Game Theory Applied to a Dark Money Network. "Dark money" originally referred to funds given to nonprofit organizations who can spend the money to influence elections and policy while not having to disclose their donors.[10] The term has extended to encompass nefarious groups seeking to undermine a government. Discussing the strategies for defeating a Dark Money Network (DMN) leads naturally into the Game Theory analysis of the strategies for the DMN and for the State trying to defeat the DMN. When conducting this game theory analysis, we originally limited the analysis by using ordinal scaling, and ranking each of the four strategic options one through four. The game was set up in Table 7.52. Strategy A is for the state to pursue a *non-kinetic strategy*, B is a *kinetic strategy*. Strategy C is for the DMN to maintain its organization and D is for it to decentralize.

TABLE 7.52: Dark Money Network

Payoff Tableau		*DMN*	
		C	D
State	A	(2,1)	(4,2)
	B	(1,4)	(3,3)

This ordinal scaling worked when allowing communications and strategic moves; however, without a way of determining interval scaling it was impossible to conduct analysis of prudential strategies, Nash Arbitration, or Nash's equalizing strategies. Here we show an application of *Analytic Hierarchy Process* (AHP) (See Chapter 8) in order to determine the interval scaled payoffs of each strategy for both the DMN and the State. We will use Saaty's standard nine point preference in the pairwise comparison of combined strategies. For the State's evaluation criteria, we chose four possible outcomes: how well the strategy degraded the DMN, how well it maintained the state's own ability to raise funds, how well the strategy would rally their base, and finally how well it removed nodes from the DMN. The evaluation criteria we chose for

[10]Adapted from Couch et al. [CFE2016].

the DMN's four possible outcomes were: how anonymity was maintained, how much money the outcome would raise, and finally how well the DMN's leaders could maintain control of the network.

After conducting an AHP analysis, we obtained a new payoff matrix in Table 7.53 with cardinal utility values.

TABLE 7.53: Dark Money Network Revised

Payoff Tableau		*DMN*	
		C	D
State	A	$(0.168, 0.089)$	$(0.366, 0.099)$
	B	$(0.140, 0.462)$	$(0.323, 0.348)$

With the cardinal scaling, it is now possible to conduct a proper analysis that might include mixed strategies or arbitrated results such as finding prudential strategies, Nash's Equalizing Strategies, and Nash Arbitration. Using a series of game theory solvers developed by Feix[11] we obtain the following results.

- Nash Equilibrium: A pure strategy Nash equilibrium was found at (A,D) of $(0.366, 0.099)$ using strategies of Non-Kinetic and Decentralize.
- Mixed Nash Equalizing Strategies: The State plays Non-Kinetic 91.9% of the time and Kinetic 8.1% of the time; DMN always plays Maintain Organization.
- Prudential Strategies: The Security Levels are $(0.168, 0.099)$ when the State plays C and the DMN plays A.

Since there is no equalizing strategy for the DMN, should the State attempt to equalize the DMN. The result is as follows. This is a significant departure from our original analysis prior to including the AHP pairwise comparisons. The recommendations for the state were to use a kinetic strategy 50% of the time and a non-kinetic strategy 50% of the time. However, it is obvious that with proper scaling the recommendation should have been to execute a non-kinetic strategy the vast majority, 92%, of the time, and only occasionally, 8%, conduct kinetic targeting of network nodes. This greatly reinforces the recommendation to execute a non-kinetic strategy to defeat the DMN.

Finally, if the State and the DMN could enter into arbitration, the result would be at BD, which was the same prediction as before the proper scaling.

[11] Feix, *Game Theory: Toolkit and Workbook for Defense Analysis Students*, Defense Tech. Info. Center, 2007. Available at http://www.dtic.mil/dtic/tr/fulltext/u2/a470073.pdf.

Example 7.12. Course of Actions Decision Process for Partial Conflict Games.
Return to the zero-sum game of Example 7.8 of Section 7.3. In many real world analyses, a player winning is not necessarily another player losing. Both might win or both might lose based upon the mission and courses of action played. Again, assume Player I has four strategies and Player II has six strategies that they can play. We use an AHP approach (see Fox [Fox2014]) to obtain cardinal values of the payoff to each player. Although this example is based on combat courses of action, this methodology can be used when competing players both have courses of action that they could employ. As in Example 7.11, we use AHP to convert the ordinal rankings of the combined COAs in order to obtain cardinal values.

The game's payoff matrices become

```
> with(Optimization) :
> A := Matrix([[6, 5.75, 5.5, 0.75, 3.75, 0.5],
     [4, 4.25, 2.75, 1.75, 2, 0.25],
     [3.25, 3, 1.5, 1.25, 1, 3.5],
     [5.25, 5, 4.75, 2.5, 2.25, 4.5]]) :
> B := Matrix([[0.167, 0.333, 0.5, 0.6336, 3.667, 3.833],
     [1.5, 1.333, 2.333, 0.8554, 2.833, 4],
     [2.1667, 2, 3.1667, 0.3574, 3.5, 1.833],
     [0.667, 0.833, 1, 5.95, 2.667, 1.1667]] :
> X := Vector[row](4, symbol = x) :
  Y := Vector(6, symbol = y) :
> Constraints := {seq((A . Y)_i ≤ p, i = 1..4),
     seq((X . B)_i ≤ q, i = 1..4),
     add(x_i, i = 1..4) = 1, add(y_i, i = 1..6) = 1}
> Objective := expand(X . A . Y + X . B . Y − p − q) :
```

The solutions, depending on the starting conditions for P and Q, are found using Maple's *QPSolve* since *Objective* is quadratic or *NLPSolve* since it is nonlinear. As before, use *fnormal* to eliminate numerical artifacts.

```
> theProgram := (Objective, Constraints, assume = nonnegative,
     maximize) :
> fnormal(QPSolve(theProgram), 3);
  [3.08, [p = 2.86, q = .634, x1 = 1., x2 = 0., x3 = 0., x4 = 0., y1 = 0., y2 = 0.,
     y3 = 0., y4 = 0., y5 = 0.727, y6 = 0.273]]
> fnormal(NLPSolve(theProgram, initialpoint = {p = 3, q = 6}), 3);
  [0., [p = 2.50, q = 5.95, x1 = 0., x2 = 0., x3 = 0., x4 = 1., y1 = 0., y2 = 0.,
     y3 = 0., y4 = 1., y5 = 0., y6 = 0.]]
```

```
> fnormal(NLPSolve(theProgram, initialpoint = {p = 2, q = 1}), 3);
```

$[3.08, [p = 2.86, q = 0.634, x_1 = 1., x_2 = 0., x_3 = 0., x_4 = 0., y_1 = 0., y_2 = 0., y_3 = 0., y_4 = 1., y_5 = 0.727, y_6 = 0.273]]$

We find Player I should play a pure strategy of COA 4, while Player 2 should play either a mixed strategy of $y_5 = 8/11$ and $y_6 = 3/11$, or a pure strategy of always y_4. We also employed sensitivity analysis by varying the criteria weights which represent the cardinal values. We found not much change in the results from our solution analysis presented here.

Exercises

Solve the problems using any method.

1. Find all the solutions for Table 7.54.

TABLE 7.54: U.S. vs. State-Sponsored Terrorism

	Payoff Tableau	*State Sponsor*	
		Sponsor Terrorism	Stop Sponsoring Terrorism
U.S.	Strike Militarily	$(2, 4)$	$(1.5, 0)$
	Do Not Strike Militarily	$(1, 1.5)$	$(0, 4)$

2. Find all the solutions for Table 7.55.

TABLE 7.55: U.S. vs. State-Sponsored Terrorism

	Payoff Tableau	*State Sponsor*	
		Sponsor Terrorism	Stop Sponsoring Terrorism
U.S.	Strike Militarily	$(3, 5)$	$(2, 1)$

3. Find all the solutions for Table 7.56.

TABLE 7.56: U.S. vs. State-Sponsored Terrorism

Payoff Tableau		*Colin*	
		Arm	Disarm
Rose	Arm	$(2, 2)$	$(1, 1)$
	Disarm	$(4, 1)$	$(3, 3)$

4. Consider the following classical games of Chicken in Tables 7.57 to 7.59. Find the solutions.

   (a)

TABLE 7.57: A Classical Game of Chicken I

Payoff Tableau		*Colin*	
		C	D
Rose	A	$(3, 3)$	$(2, 4)$
	B	$(4, 2)$	$(1, 1)$

   (b)

TABLE 7.58: A Classical Game of Chicken II

Payoff Tableau		*Colin*	
		C	D
Rose	A	$(2, 3)$	$(4, 1)$
	B	$(1, 2)$	$(3, 4)$

(c)

TABLE 7.59: A Classical Game of Chicken III

Payoff Tableau		*Colin* C	D
Rose	A	$(4,3)$	$(3,4)$
	B	$(2,1)$	$(1,2)$

5. Consider the classical games of Prisoner's Dilemma in Tables 7.60 and 7.61. Find the solutions.

(a)

TABLE 7.60: A Classical Game of Prisoner's Dilemma I

Payoff Tableau		*Colin* C	D
Rose	A	$(3,3)$	$(-1,5)$
	B	$(5,-1)$	$(0,0)$

(b)

TABLE 7.61: A Classical Game of Prisoner's Dilemma II

Payoff Tableau		*Colin* C	D
Rose	A	$(3,3)$	$(1,5)$
	B	$(4,0)$	$(0,2)$

Projects

Project 7.1. Corporation XYZ consists of Companies Rose and Colin. Company Rose can make Products R_1 and R_2. Company Colin can make Products C_1 and C_2. These products are not in strict competition with one another, but there is an interactive effect depending on which products are on the market at the same time as reflected in Table 7.62 below. The table reflects profits in millions of dollars per year. For example, if products R_2 and C_1 are produced and marketed simultaneously, Rose's profits are 4 million and Colin's 5 million annually. Rose can make any mix of R_1 and R_2, and Colin can make any mix of C_1 and C_2. Assume the information below is known to each company.

NOTE: The CEO is *not satisfied* with just summing the total profits. He might want the *Nash Arbitration Point* to award each company proportionately based on their strategic positions, if other options fail to produce the results he desires. Further, he does not believe a dollar to Rose has the same importance to the corporation as a dollar to Colin.

TABLE 7.62: Corporate Payoff Matrix

Payoff Tableau		*Company Colin*	
		C_1	C_2
Company Rose	R_1	$(3,7)$	$(8,5)$
	R_2	$(4,5)$	$(5,6)$

a. Suppose the companies have perfect knowledge and implement market strategies independently without communicating with one another. What are the likely outcomes? Justify your choice.

b. Suppose each company has the opportunity to exercise a strategic move. Try *first moves* for each player; determine if a first move improves the results of the game.

c. In the event things turn "hostile" between Rose and Colin, find, state, and then interpret

   i. Rose's Security Level and Prudential Strategy.

   ii. Colin's Security Level and Prudential Strategy.

   Now suppose that the CEO is disappointed with the lack of spontaneous cooperation between Rose and Colin, and decides to intervene and dictate the "best" solution for the corporation. The CEO employs an arbiter to determine an "optimal production and marketing schedule" for the corporation. What is this strategy?

d. Explain the concept of "Pareto Optimal" from the CEO's point of view. Is the "likely outcome" you found in question (1) at or above Pareto Optimal? Briefly explain and provide a payoff polygon plot.

e. Find and state the Nash Arbitration Point using the security levels found above.

f. Briefly discuss how you would implement the Nash Point. In particular, what mix of the products R_1 and R_2 should Rose produce and market, and what mix of the products C_1 and C_2 should Colin produce? Must their efforts be coordinated, or do they simply need to produce the "optimal mix"? Explain briefly.

g. How much annual profit will Rose and Colin each make when the CEO's dictated solution is implemented?

7.5 Conclusion

We have presented some basic material concerning applied game theory and its uses in business, government, and industry. We presented some solution methodologies to solve these simultaneous total- and partial-conflict games. For analysis of sequential games, games with cooperation, and N-person games, please see the additional readings.

References and Further Reading

[Aiginger1999] Karl Aiginger, "The use of game theoretical models for empirical industrial organization," in *Competition, Efficiency, and Welfare*, Springer, 1999, pp. 253–277.

[Aumann1987] R.J. Aumann, *Game theory.* in *The New Palgrave: A Dictionary of Economics*, Ed John Eatwell, Murray Milgate, and Peter Newman, Macmillan Press, 1987.

[BSC2013] Mokhtar S. Bazaraa, Hanif D. Sherali, and Chitharanjan M. Shetty, *Nonlinear Programming: Theory and Algorithms*, John Wiley & Sons, 2013.

[Brams1994] S. Brams, "Theory of Moves," American Scientist (1994), no. 81, 562-570.

[Cantwell2003] Gregory L. Cantwell, *Can two person zero sum game theory improve military decision-making course of action selection?*,

Tech. Report, Army Command and General Staff College, Fort Leavenworth, KS, School Of Advanced Military Studies, 2003.

[CLG2002] P-A. Chiappori, Steven Levitt, and Timothy Groseclose, "Testing mixed-strategy equilibria when players are heterogeneous: The case of penalty kicks in soccer," American Economic Review (92), 2002, no. 4, 1138–1151.

[CS2001] Kalyan Chatterjee and William Samuelson, *Game Theory and Business Applications*, Springer, 2001.

[CFE2016] Christopher Couch, William P. Fox, and Sean F. Everton, "Mathematical modeling and analysis of a dark money network," Journal of Defense Modeling and Simulation **13** (2016), no. 3, 343–354.

[Dantzig1951] George B. Dantzig, *Maximization of a linear function of variables subject to linear inequalities*, New York (1951).

[Dantzig2002] George B. Dantzig, "Linear Programming," Operations Research **50** (2002), no. 1, 42–47.

[Dixit1993] Avinash K. Dixit and Barry J. Nalebuff, *Thinking strategically: The Competitive Edge in Business, Politics, and Everyday Life*, Norton & Co., 1993.

[Dorfman1951] Robert Dorfman, "Application of the simplex method to a game theory problem," in *Activity Analysis of Production and Allocation, Proceedings of a Conference*, 1951, pp. 348–358.

[EK2010] David Easley and Jon Kleinberg, *Networks, Crowds, and Markets: Reasoning About a Highly Connected World*, Cambridge University Press, 2010.

[Fox2008] William P. Fox, "Mathematical modeling of conflict and decision making: The Writers' Guild strike 2007-2008," Computers in Education Journal **18** (2008), no. 3, 2–11.

[Fox2011Maple] William P. Fox, *Mathematical Modeling with Maple*, Nelson Education, 2011.

[Fox2010] William P. Fox, "Teaching the applications of optimization in game theory's zero sum and non-zero sum games," International Journal of Data Analysis Techniques and Strategies **2** (2010), no. 3, 258–284.

[Fox2014] William P. Fox, "Using multi-attribute decision methods in mathematical modeling to produce an order of merit list of high valued terrorists," American Journal of Operations Research (2014), no. 4, 365–374.

[GH2009] Rick Gillman and David Housman, *Models of Conflict and Cooperation*, American Mathematical Soc., 2009.

[GFH2014] Frank Giordano, William P. Fox, and Steven Horton, *A First Course in Mathematical Modeling*, 5th ed., Nelson Education, 2014.

[Koopmans1951] Tjallinged Koopmans, *Activity analysis of production and allocation: proceedings of a conference*, Tech. report, 1951.

[KT1951] H. W. Kuhn and A. W. Tucker, "Nonlinear programming," in *Proceedings of the 2nd Berkeley Symposium on Mathematical Statistics and Probability*, 1951, pp. 481–492.

[MF2009] Gordon H. McCormick and Lindsay Fritz, *The logic of warlord politics*, Third World Quarterly **30** (2009), no. 1, 81–112.

[Myerson1991] Roger B. Myerson, *Game theory: Analysis of conflict*, Harvard Univ. Press, Cambridge, MA, 1991.

[Nash2009NPS] John Nash, *Lecture at NPS*, Feb 19, 2009.

[Nash1951] John Nash, "Non-cooperative games," Annals of mathematics (1951), 286–295.

[Nash1950] John Nash *et al.*, "Equilibrium points in n-person games," Proceedings of the National Academy of Sciences **36** (1950), no. 1, 48–49.

[Nash1950BP] John Nash, "The bargaining problem," Econometrica: Journal of the Econometric Society (1950), 155–162.

[vNeumann1928] John von Neumann, *Zur theorie der gesellschaftsspiele*, Mathematische Annalen **100** (1928), no. 1, 295–320.

[vNM1944] John von Neumann and Oskar Morgenstern, *Theory of Games and Economic Behavior*, Princeton University Press, 1944.

[SP1973] J. Maynard Smith and George R. Price, "The logic of animal conflict," Nature **246** (1973), no. 5427, 15.

[Straffin1993] Philip D. Straffin, *Game theory and strategy*, vol. 36, MAA, 1993.

[Williams1986] J. D. Williams, *The Compleat Strategyst: Being a Primer on the Theory of Games of Strategy*, Courier Corporation, 1986.

[Wiens2003] E. Wiens, "Game Theory", http://www.egwald.ca/operationsresearch/gameintroduction.php. (Accessed May 18, 2019.)

[WikiGameTheory] *Wikipedia*, "Game Theory," http://en.wikipedia.org/wiki/List_of_games_in_game_theory. (Accessed May 18, 2019.)

[Winston2002] W. L. Winston, *Introduction to Mathematical Programming Applications and Algorithms*, 4th ed., Duxbury Press/Brooks-Cole Pub., 2002.

8

Introduction to Problem Solving with Multi-Attribute Decision Making

Objectives:

(1) Know the types of multi-attribute decision techniques.

(2) Understand the different weighting schemes and how to implement them.

(3) Know which technique or techniques to use.

(4) Know the importance of sensitivity analysis.

(5) Recognize the importance of technology in the solution process.

The Department of Homeland Security (DHS) has a limited number of assets and a finite amount of time to conduct investigations, thus action priorities must be established. The Risk Assessment Office has collected the data shown in Table 8.1 for the morning Daily Briefing. Your Operations Research Team must analyze the information and provide a priority list to the Risk Assessment Team in time for the briefing.

TABLE 8.1: DHS Risk Assessment Data

Threat Nature	Reliability Assess.	Approx. deaths ($\times 10^3$)	Damage Estimate ($\times 10^6$)	Pop. Density ($\times 10^6$)	Psych-Factor	Number Intel. Tips
Dirty Bomb	0.40	10	150	4.5	9	3
Bio-Terror	0.45	0.8	10	3.2	7.5	12
DC Roads or Bridges	0.35	0.005	300	0.85	6	8
NY Subway	0.73	12	200	6.3	7	5
DC Metro	0.69	11	200	2.5	7	5
Major Bank Robbery	0.81	0.0002	10	0.57	2	16
Air Traffic Control	0.70	0.001	5	0.15	4.5	15

(Note: *Psych-factor = Destructive psychological influence of the act.*)

TASK. Build a model that ranks the incidents in a priority order.

ASSUMPTIONS. The main suppositions are:

- Past decisions will give insights into the decision maker's process.
- Table 8.1 holds the only data available: reliability, approximate number of deaths anticipated, approximate remediation costs, location of the action, destructive psychological influence on the citizenry, and the number of intelligence tips gathered.
- The listed factors will form the criteria for the analysis.
- The data is accurate and precise.

The problem will be solved with the SAW (exercise) and TOPSIS methods.

8.1 Introduction

Multiple-attribute decision making (MADM) concerns making decisions when there are multiple, but finite, alternatives and criteria. This topic is sometimes called multi-criteria decision analysis or MCDA. These problems differ from analysis where we have only one criteria such as cost with several alternatives. We address problems such as in the DHS scenario where there are six criteria with seven alternatives that impact the decision.

Consider a problem where management needs to prioritize or rank order alternative choices such as: identifing key nodes in a supply chain, choosing a contractor or sub-contractor, selecting airports, ranking recruiting efforts, ranking banking facilities, ranking schools or colleges, etc. How can setting relative priorities or choosing rank orders be accomplished analytically?

We will present four methodologies for prioritizing or rank ordering alternatives based upon multiple criteria. The methodologies are

- Data Envelopment Analysis (DEA)
- Simple Average Weighting (SAW)
- Analytical Hierarchy Process (AHP) with Objective Data[1]
- Technique of Order Preference by Similarity to Ideal Solution (TOPSIS)

[1]Our discussion of AHP will be restricted to data that is real, and not subjective. For further study, see Saaty's *Fundamentals of Decision Making and Priority Theory With the Analytic Hierarchy Process.*

For each technique, we describe its methodology, discuss strengths and limitations, offer tips for conducting sensitivity analysis, and present illustrative examples using Maple.

8.2 Data Envelopment Analysis

Charnes, Cooper, and Rhodes [CharnesCooperRhodes1978] described data envelopment analysis (DEA) as a mathematical programming model applied to observational data, providing a new way of obtaining empirical estimates of relationships among *decision making units* (DMUs) that take multiple inputs and produce multiple outputs. They were inspired by "relative efficiency" in combustion engineering. The definition of a DMU is generic and very flexible. Any object to be ranked can be a DMU, from individuals to government ministries. DEA has been formally defined as a methodology directed to frontier analysis, rather than to central tendencies. The technique is a "data input-output driven" approach for evaluating the relative performance of DMUs. DEA has been used to evaluate the performance or *efficiencies* of hospitals, schools, departments, university faculty, US Air Force Wings, armed forces recruiting agencies, universities, cities, courts, businesses, banking facilities, countries, regions, SOF airbases, key nodes in networks, ...; the list goes on. A Google Scholar search on "data envelopment analysis" returns over 330,000 results in under 0.1 seconds. According to Cooper ([CooperLSTTZ2001], the first item returned by the search), DEA has been used to gain insights into activities that were not able to be obtained by other quantitative or qualitative methods.

Data Envelopment Analysis as a Linear Program

A DEA model, in simplest terms, may be formulated and solved as a linear programming problem (Winston [Winston2002], Callen [Callen1991]). Although there are several representations for DEA, we'll use the most straightforward formulation for maximizing the efficiency of the kth DMU as constrained by inputs and outputs (shown in (8.1)). As an option, we may wish to normalize the metric inputs and outputs for the alternatives if the values are poorly scaled within the data. We will call this data matrix $\mathbf{X}$ with entries x_{ij}. Define each DMU or efficiency unit as E_i for $i = 1, 2, \ldots, n$ for n DMUs. Let w_i be the weights or coefficients for the linear combinations. Further, restrict any efficiency value from being larger than one (100%). Thus, the largest efficient DMU will have efficiency value 1. These requirements give the following linear programming formulation for DMUs with multiple inputs yielding a single output.

$$\begin{aligned} &\text{Maximize } E_k \\ &\quad \text{subject to} \\ &\sum_{i=1}^{n} w_i x_{ij} - E_i = 0, \quad j = 1, 2, \ldots \\ &\qquad\qquad E_i \leq 1 \quad \text{for all } i \end{aligned} \tag{8.1}$$

For multiple inputs and outputs, we recommend using (8.2), the formulations provided by Winston ([Winston2002]) and Trick ([Trick1996]). Let X_i be the inputs array and Y_i be the outputs array for DMU$_i$. Let X_0 and Y_0 be for DMU$_0$, the DMU being modeled, then

$$\begin{aligned} &\text{Minimize } \theta \\ &\quad \text{subject to} \\ &\qquad \sum_{i=1}^{n} \lambda_i X_i \leq \theta X_0 \\ &\qquad \sum_{i=1}^{n} \lambda_i Y_i \leq Y_0 \\ &\qquad\qquad \lambda_i \geq 0 \quad \text{for all } i \end{aligned} \tag{8.2}$$

Strengths and Limitations of DEA

DEA is a very useful tool when used wisely [Trick1996]. Strengths that make DEA very useful include:

1. DEA can handle multiple input and multiple output models;
2. DEA doesn't require an assumption of a functional form relating inputs to outputs;
3. DMUs are directly compared against a peer or combination of peers; and
4. inputs and outputs can have very different units.

For example, X_1 could be in units of lives saved, while X_2 could be in units of dollars spent without requiring any a priori tradeoff between the two.

The same characteristics that make DEA a powerful tool can also create limitations. An analyst should keep these limitations in mind when choosing whether or not to use DEA. Limitations include:

1. DEA is an extreme point technique, thus noise in the data, such as measurement error, can cause significant problems.

2. DEA is good at estimating relative efficiency of a DMU, but it does not directly measure absolute efficiency. In other words, DEA can show how well a DMU is doing compared to its peers, but not compared to a theoretical maximum.

3. Since DEA is a nonparametric technique, statistical hypothesis tests are difficult—they are the focus of ongoing research.

4. Since a standard formulation of DEA with multiple inputs and outputs creates a separate linear program for each DMU, large problems can be computationally extremely intensive.

5. Linear programming does not ensure all weights are considered. We find that the values for weights are only for those that optimally determine an efficiency rating. If having all criteria (all inputs and outputs) weighted is essential to the decision maker, then DEA is not appropriate.

Sensitivity Analysis

Sensitivity analysis is always an important element in every modeling project. According to Nerali ([Neralic1998]), an increase in any output cannot worsen an efficiency rating, nor can a decrease in inputs alone worsen an already achieved efficiency rating. As a result, in our examples we only decrease outputs and increase inputs. We will briefly illustrate sensitivity analysis, as applicable, in the examples.

Example 8.1. Manufacturing Units.
A manufacturing process involves three DMUs each having two inputs and three outputs. Management wishes to assess the efficiency of each DMU in order to target resources to improve performance. The data appears in Table 8.2.

TABLE 8.2: Manufacturing DMU Data

DMU	Input 1	Input 2	Output 1	Output 2	Output 3
I	5	14	9	4	16
II	8	15	5	7	10
III	7	12	4	9	13

Since no units are given and the values have similar scales, the data doesn't have to be normalized.

Define the variables

$$\begin{aligned}
t_i &= \text{value of a single unit of output of DMU}_i \\
w_i &= \text{cost or weights for one unit of inputs to DMU}_i \\
\mathbf{X} &= \text{matrix of input data} \\
\mathbf{Y} &= \text{matrix of output data} \\
DMU_i &= \text{objective function for DMU}_i\text{'s linear program} \\
Eff_i &= \text{relative efficiency of DMU}_i\text{, with a vector of weights}
\end{aligned}$$

all for $i = 1$, 2, and 3.

Assume that

- No DMU can have an efficiency of more than 100%.
- If any efficiency is less than 100%, then that DMU is *inefficient.*
- The costs are scaled so that the costs of the inputs equals 1 for each linear program. For example, use $5w_1 + 14w_2 = 1$ in the LP for DMU_1.
- All values and weights must be strictly positive. (We may have to use a constant such as 0.0001 in lieu of 0 in inequalities to help numeric routines converge.)

To calculate the efficiency of DMU_1, use the linear program

$$\begin{aligned}
&\text{Maximize } DMU_1 = 9t_1 + 4t_2 + 16t_3 \\
&\text{subject to} \\
&\quad -9t_1 - 4t_2 - 16t_3 + 5w_1 + 14w_2 \geq 0, \\
&\quad -5t_1 - 7t_2 - 10t_3 + 8w_1 + 15w_2 \geq 0, \\
&\quad -4t_1 - 9t_2 - 13t_3 + 7w_1 + 12w_2 \geq 0, \\
&\quad 5w_1 + 14w_2 = 1, \\
&\quad t_i, w_j \geq 0 \quad \text{for all } i,\ j.
\end{aligned}$$

To calculate the efficiency of DMU_2, use the linear program

$$\begin{aligned}
&\text{Maximize } DMU_2 = 5t_1 + 7t_2 + 10t_3 \\
&\text{subject to} \\
&\quad -9t_1 - 4t_2 - 16t_3 + 5w_1 + 14w_2 \geq 0, \\
&\quad -5t_1 - 7t_2 - 10t_3 + 8w_1 + 15w_2 \geq 0, \\
&\quad -4t_1 - 9t_2 - 13t_3 + 7w_1 + 12w_2 \geq 0, \\
&\quad 8w_1 + 15w_2 = 1, \\
&\quad t_i, w_j \geq 0 \quad \text{for all } i,\ j.
\end{aligned}$$

To calculate the efficiency of DMU_3, use the linear program

$$\begin{aligned}
&\text{Maximize } DMU_3 = 4t_1 + 9t_2 + 13t_3\\
&\text{subject to}\\
&\quad -9t_1 - 4t_2 - 16t_3 + 5w_1 + 14w_2 \geq 0,\\
&\quad -5t_1 - 7t_2 - 10t_3 + 8w_1 + 15w_2 \geq 0,\\
&\quad -4t_1 - 9t_2 - 13t_3 + 7w_1 + 12w_2 \geq 0,\\
&\quad 7w_1 + 12w_2 = 1,\\
&\quad t_i, w_j \geq 0 \quad \text{for all } i,\ j.
\end{aligned}$$

Use Maple to solve the three linear programs.

> $with(Optimization):$

Define the input and output data matrices. Using the *Matrix Palette* makes entry easier and less error prone. Set the size, then click *Insert Matrix.*

> $Inputs := \begin{bmatrix} 5 & 14 \\ 8 & 15 \\ 7 & 12 \end{bmatrix}:$
>
> $Outputs := \begin{bmatrix} 9 & 4 & 16 \\ 5 & 7 & 10 \\ 4 & 9 & 13 \end{bmatrix}:$

Define the decision variables and weights.

> $T := \langle t_i \$ i = 1..3\rangle:$
> $W := \langle w_i \$ i = 1..2\rangle:$

Compute the LP's objective functions.

> $DMUObj := Outputs\,.\,T;$

$$\begin{bmatrix} 9\,t_1 + 4\,t_2 + 16\,t_3 \\ 5\,t_1 + 7\,t_2 + 10\,t_3 \\ 4\,t_1 + 9\,t_2 + 13\,t_3 \end{bmatrix}$$

Set up the constraints.

> $Inputs\,.\,W - Outputs\,.\,T \geq\sim 0;$ # '$\geq\sim 0$' *applies* '≥ 0' *element-wise.*
> $MainConstraints := convert(\%, set):$

$$\begin{bmatrix} 0 \leq 5\,w_1 + 14\,w_2 - 9\,t_1 - 4\,t_2 - 16\,t_3 \\ 0 \leq 8\,w_1 + 15\,w_2 - 5\,t_1 - 7\,t_2 - 10\,t_3 \\ 0 \leq 7\,w_1 + 12\,w_2 - 4\,t_1 - 9\,t_2 - 13\,t_3 \end{bmatrix}$$

```
> Inputs.W =~ 1 :    # '=~ 1' applies '= 1' to each element.
  DMUcons := convert(%, list) :
```

$$\begin{bmatrix} 5\,w_1 + 14\,w_2 = 1 \\ 8\,w_1 + 15\,w_2 = 1 \\ 7\,w_1 + 12\,w_2 = 1 \end{bmatrix}$$

Now solve the three linear programs.

```
> NumDMU := 3 :
  for i to NumDMU do
     Constraints_i := MainConstraints union {DMUcons_i} :
     LPSolve(DMUObj_i, Constraints_i, assume = nonnegative,
        maximize) :
     Eff_i := fnormal(%, 4) :
     end do :
  Matrix(convert(Eff, list));
```

$$\begin{bmatrix} 1.0 & [t_1 = 0.0, t_2 = 0.0, t_3 = 0.06250, w_1 = 0.0, w_2 = 0.07143] \\ 0.7733 & [t_1 = 0.08000, t_2 = 0.05333, t_3 = 0.0, w_1 = 0.0, w_2 = 0.06667] \\ 1.0 & [t_1 = 0.0, t_2 = 0.0, t_3 = 0.07065, w_1 = 0.0, w_2 = 0.08333] \end{bmatrix}$$

The linear program solutions show the relative efficiencies of DMU_1 and DMU_3 are 100%, while DMU_2's is 77.3%.

INTERPRETATION. DMU_2 is operating at 77.3% of the efficiency of DMU_1 and DMU_3. Management could concentrate on improvements for DMU_2 by taking best practices from DMU_1 or DMU_3.

To compute the shadow prices for the linear programs, paying special attention to those of DMU_2, solve the dual LPs. Maple's *dual* command in the *simplex* package does not handle equality constraints, so replace all equalities with two inequalities $\leq$ and $\geq$.

```
> for i to NumDMU do
     DualConstraints_i := MainConstraints union
        {1 ≤ lhs(DMUcons_i), 1 ≥ lhs(DMUcons_i)} :
     DualLP_i := simplex:-dual(DMUObj_i, DualConstraints_i, λ) :
     LPSolve(DualLP_i, assume = nonnegative) :
     DualEff_i := fnormal(%, 4) :
     end do :
  Matrix(convert(DualEff, list));
```

$$\begin{bmatrix} 1.0 & [\lambda 1 = 0.0, \lambda 2 = 1.0, \lambda 3 = 0.0, \lambda 4 = 0.0, \lambda 5 = 1.0] \\ 0.7733 & [\lambda 1 = 0.0, \lambda 2 = 0.7733, \lambda 3 = 0.0, \lambda 4 = 0.6615, \lambda 5 = 0.2615] \\ 1.0 & [\lambda 1 = 0.0, \lambda 2 = 1.0, \lambda 3 = 0.0, \lambda 4 = 1.0, \lambda 5 = 0.0] \end{bmatrix}$$

Examining the shadow prices from the dual linear program for DMU_2 shows $\lambda 5 = 0.26$, $\lambda 4 = 0.66$, and $\lambda 3 = 0$. The average output vector for DMU_2 can be written as

$$0.26 \begin{bmatrix} 9 \\ 4 \\ 16 \end{bmatrix} + 0.66 \begin{bmatrix} 4 \\ 9 \\ 13 \end{bmatrix} = \begin{bmatrix} 5 \\ 7 \\ 12.785 \end{bmatrix},$$

and its average input vector is

$$0.26 \begin{bmatrix} 5 \\ 14 \end{bmatrix} + 0.66 \begin{bmatrix} 7 \\ 12 \end{bmatrix} = \begin{bmatrix} 5.94 \\ 11.6 \end{bmatrix}.$$

Output 3 in Table 8.2 is 10 units. Thus, the inefficiency is in Output 3 where 12.785 units are required. We find that they are short 2.785 units $(= 12.785 - 10)$. This calculation helps focus on treating the inefficiency found for Output 3.

Sensitivity Analysis. In linear programming, sensitivity analysis is sometimes referred to as "what if" analysis. Assume that without management providing some additional training, DMU_2's Output 3 value dips from 10 to 9 units, while Input 2 increases from 15 to 16. We find that these changes in the *technology coefficients* are easily handled when re-solving the LPs. Since DMU_2 is affected, we might only modify and solve the LP for DMU_2. With these changes, DMU_2's efficiency is now only 74% of DMU_1 or DMU_3.

Example 8.2. Ranking Five Departments in a College.
Five science departments in the College of Arts & Sciences are scheduled for review. The dean has provided the data in Table 8.3 and asked for relative efficiency ratings.[2]

TABLE 8.3: Arts & Sciences Department's Data

Department	Inputs	Outputs		
	No. Faculty	Student Cr. Hr.	No. Students	Total Degrees
Biology	25	18,341	9,086	63
Chemistry	15	8,190	4,049	23
Comp. Sci.	10	2,857	1,255	31
Math.	33	22,277	6,102	31
Physics	12	6,830	2,910	19

[2] Adapted from [Bauldry2009a].

Since the data values differ by orders of magnitude, divide both student credit hours and number of students by 1,000.

Follow the same sequence of Maple commands as in the previous example.

$> Inputs := Matrix(5, 1, [25, 15, 10, 33, 13]) :$

$$Outputs := \begin{bmatrix} 18.341 & 9.086 & 63 \\ 8.190 & 4.049 & 23 \\ 2.857 & 1.255 & 31 \\ 22.277 & 6.102 & 31 \\ 6.830 & 2.910 & 19 \end{bmatrix} :$$

$> T := \langle t_i \$ i = 1..3)\rangle :$

$W := \langle w_i \$ i = 1..1)\rangle :$

$> DMUObj := Outputs\,.\,T;$

$$\begin{bmatrix} 18.341\, t_1 + 9.086\, t_2 + 63\, t_3 \\ 8.190\, t_1 + 4.049\, t_2 + 23\, t_3 \\ 2.857\, t_1 + 1.255\, t_2 + 31\, t_3 \\ 22.277\, t_1 + 6.102\, t_2 + 31\, t_3 \\ 6.830\, t_1 + 2.910\, t_2 + 19\, t_3 \end{bmatrix}$$

$> Inputs\,.\,W - Outputs\,.\,T \geq\sim 0;$

$MainConstraints := convert(\%, set) :$

$$\begin{bmatrix} 0 \leq 25\, w_1 - 18.341\, t_1 - 9.086\, t_2 - 63\, t_3 \\ 0 \leq 15\, w_1 - 8.190\, t_1 - 4.049\, t_2 - 23\, t_3 \\ 0 \leq 10\, w_1 - 2.857\, t_1 - 1.255\, t_2 - 31\, t_3 \\ 0 \leq 33\, w_1 - 22.277\, t_1 - 6.102\, t_2 - 31\, t_3 \\ 0 \leq 13\, w_1 - 6.830\, t_1 - 2.910\, t_2 - 19\, t_3 \end{bmatrix}$$

$> Inputs\,.\,W =\sim 1;$

$DMUcons := convert(\%, list) :$

$$\begin{bmatrix} 25\, w_1 = 1 \\ 15\, w_1 = 1 \\ 10\, w_1 = 1 \\ 33\, w_1 = 1 \\ 13\, w_1 = 1 \end{bmatrix}$$

```
> NumDMU := 5 :
  for i to NumDMU do
    Constraints := MainConstraints union {DMUcons[i]} :
    LPSolve(DMUObj[i], Constraints, assume = nonnegative,
      maximize) :
    Eff[i] := fnormal(%, 7) :
  end do :
  Matrix(convert(Eff, list));
```

$$\begin{bmatrix} 1.0 & [t_1 = 0.01492614, t_2 = 0.0, t_3 = 0.01152761, w_1 = 0.04000000] \\ 0.7442342 & [t_1 = 0.09087109, t_2 = 0.0, t_3 = 0.0, w_1 = 0.06666667] \\ 1.0 & [t_1 = 0.0, t_2 = 0.0, t_3 = 0.03225806, w_1 = 0.1000000] \\ 0.9201524 & [t_1 = 0.04130504, t_2 = 0.0, t_3 = 0.0, w_1 = 0.03030303] \\ 0.7161341 & [t_1 = 0.1048513, t_2 = 0.0, t_3 = 0.0, w_1 = 0.07692308] \end{bmatrix}$$

We see the DMUs are ranked as: Biology and Computer Science: 100%; Mathematics: 92%; Chemistry: 74%; and Physics: 71%.

Examine the results from the dual LPs.

```
> for i to NumDMU do
    DualConstraints := MainConstraints union
      {lhs(DMUcons[i]) ≤ 1, lhs(DMUcons[i]) ≥ 1} :
    DualLP[i] := simplex:-dual(DMUObj[i], DualConstraints, λ) :
    LPSolve(DualLP, assume = nonnegative,
    DualEff[i] := fnormal(%, 7) :
  end do :
  Matrix(convert(DualEff, list));
```

$$\begin{bmatrix} 1.0 & [\lambda1 = 0.0, \lambda2 = 1.0, \lambda3 = 0.0, \lambda4 = 1.0, \lambda5 = 0.0, \lambda6 = 0.0, \lambda7 = 0.0] \\ 0.744 & [\lambda1 = 0.0, \lambda2 = 0.744, \lambda3 = 0.0, \lambda4 = 0.447, \lambda5 = 0.0, \lambda6 = .0, \lambda7 = 0.0] \\ 1.0 & [\lambda1 = 0.0, \lambda2 = 1.0, \lambda3 = 0.0, \lambda4 = 0.0, \lambda5 = 0.0, \lambda6 = 0.0, \lambda7 = 1.0] \\ 0.920 & [\lambda1 = 0.0, \lambda2 = 0.920, \lambda3 = 0.0, \lambda4 = 1.215, \lambda5 = 0.0, \lambda6 = .0, \lambda7 = 0.0] \\ 0.716 & [\lambda1 = 0.0, \lambda2 = 0.716, \lambda3 = 0.0, \lambda4 = 0.372, \lambda5 = 0.0, \lambda6 = 0.0, \lambda7 = 0.0] \end{bmatrix}$$

Comparing the values of the λs, improving Output 2, Numbers of Students, will provide the largest gains in efficiency for both Chemistry and Physics.

Exercises

1. Table 8.4 lists data for three hospitals where inputs are number of beds and labor hours in thousands per month, and outputs, all measured in

hundreds, are patient-days for patients under 14, between 14 and 65, and over 65. Determine the relative efficiency of the three hospitals.

TABLE 8.4: Three Hospitals' Data

Hospital	Inputs		Outputs		
	No. Beds	Labor Hr.	≤ 14	14–65	≥ 65
I	5	14	9	4	16
II	8	15	5	7	10
III	7	12	4	9	13

2. The three hospitals of Exercise 1 have revised procedures. Reanalyze their relative efficiencies using the new data of Table 8.5.

TABLE 8.5: Three Hospitals' Revised Data

Hospital	Inputs		Outputs		
	No. Beds	Labor Hr.	≤ 14	14–65	≥ 65
I	4	16	6	5	15
II	9	13	10	6	9
II	5	11	5	10	12

3. The First National Bank of Spruce Pine, NC, has four branches in the greater Spruce Pine metropolitan area. The CEO directed an efficiency study be undertaken. The data to be collected is:

   INPUT 1: labor hours (hundred per month)

   INPUT 2: space used for tellers (hundreds of square feet)

   INPUT 3: supplies used (dollars per month)

   OUTPUT 1: loan applications per month

   OUTPUT 2: deposits (thousands of dollars per month)

   OUTPUT 3: checks processed (thousands of dollars per month)

   The data for the bank branches appears in Table 8.6.

TABLE 8.6: Bank Branch Data

Branch	Inputs			Outputs		
	Labor Hr.	Space.	Supplies	Loans	Deposits	Checks
I	15	20	50	200	15	35
II	14	23	51	220	18	45
III	16	19	51	210	17	20
IV	13	18	49	199	21	35

(a) Determine the branches' relative efficiencies.

(b) What "best practices" might you suggest to the branches that are less efficient?

8.3 Simple Additive Weighting

Simple Additive Weighting (SAW) was developed in the 1950s by Churchman and Ackoff [ChurchmanAckoff1954]; it is the simplest of the MADM methods, yet still one of the most widely used since it is easy to implement. SAW, also called the *weighted sum method* [Fishburn1967], is a straightforward and easily executed process.

Methodology

Given a set of n alternatives and a set of m criteria for choosing among the alternatives, SAW creates a function for each alternative rating its overall utility. Each alternative is assessed with regard to every criterion (attribute) giving the matrix $M = [m_{ij}]$ where m_{ij} is the assessment of alternative i with respect to criterion j. Each criterion is given a weight w_i; the sum of all weights must equal 1; i.e., $\sum_i w_i = 1$. If the criteria are equally weighted, then we merely need to sum the alternative values. The overall or composite performance score P_i of the ith alternative with respect to the m criteria is given by

$$P_i = \frac{1}{m} \sum_{j=1}^{m} w_j m_{ij}. \tag{8.3}$$

for $i = 1, \ldots, n$. Write $\vec{P} = [P_i]$ using matrix/vector notation as

$$\vec{P} = \frac{1}{m} \mathbf{M} \vec{w}.$$

where $\vec{w} = [w_i]$. The alternative with the highest value of P_i is the best *relative to the chosen criteria weighting.*

Originally, all the units of the criteria had to be identical, such as dollars, pounds, seconds, etc. A normalization process making the values unitless relaxes the requirement. We recommend always normalizing the data.

Strengths and Limitations

The main strengths are (1) ease of use, and (2) normalized data allows for comparisons across many differing criteria. Limitations include "larger is always better" or "smaller is always better." The method lacks flexibility in stating which criterion should be larger or smaller to achieve better performance, thus making gathering useful data with the same relational schema (larger or smaller) essential.

Sensitivity Analysis

Sensitivity analysis should be used to determine how sensitive the model is to the chosen weights. A decision maker can choose arbitrary weights, or, choose weights using a method that performs pairwise comparisons (as done with the *analytic hierarchy process* discussed later in this chapter). Whenever weights are chosen subjectively, sensitivity analysis should be carefully undertaken. Later sections investigate techniques for sensitivity analysis that can be used for individual criteria weights.

Examples of SAW

A Maple procedure, *SAW*, to compute rankings using simple additive weighting is included in the book's *PSMv2* package. The parameters are the matrix of criteria data for each alternative and the weights, either as a vector or a *comparison matrix.*

```
> with(PSMv2) :
  Describe(SAW);
  # Usage: SAW(M:data matrix [rows:alternatives, columns:criteria],
  # weights: vector or comparison matrix)
  SAW( AltM::Matrix, w )
```

To examine the program's code, use *print*(*SAW*).

Example 8.3. Selecting a Car.
It's time to purchase a new car. Six cars have made the final list: Ford Fusion, Toyota Prius, Toyota Camry, Nissan Leaf, Chevy Volt, and Hyundai Sonata. There are seven criteria for our decision: cost, mileage (city and highway), performance, style, safety, and reliability. The information in Table 8.7 has been collected online from the *Consumer's Report* and *US News and World Report* websites.

TABLE 8.7: Car Selection Data

Cars	Cost ($1000)	MPG City	MPG Highway	Perfor-mance	Interior & Style	Safety	Reliability
Prius	27.8	44	40	7.5	8.7	9.4	3
Fusion	28.5	47	47	8.4	8.1	9.6	4
Volt	38.7	35	40	8.2	6.3	9.6	3
Camry	25.5	43	39	7.8	7.5	9.4	5
Sonata	27.5	36	40	7.6	8.3	9.6	5
Leaf	36.2	40	40	8.1	8.0	9.4	3

Initially, we assume all criteria are weighted equally to obtain a baseline ranking. Even though the different criteria values are relatively close, let's normalize the data for illustration. There are three typical methods to use:

$$m_{ij} \to \frac{m_{ij}}{M_j}, \quad m_{ij} \to \frac{m_{ij} - m_j}{M_j - m_j}, \quad m_{ij} \to \mathrm{rank}_j(m_{ij})$$

where M_j and m_j are the maximum and minimum values in the jth column, and rank_j is the rank order of the jth column. Our *SAW* program from the *PSMv2* package uses the first method.

We'll exclude the cost data from our first baseline ranking since larger cost is worse, whereas all the other data has larger is better. SAW requires consistent criteria ranking.

$> CarModel := Vector[row]([Prius, Fusion, Volt, Camry, Sonata, Leaf]) :$

$$> CarData := \begin{bmatrix} 44 & 40 & 7.5 & 8.7 & 9.4 & 3 \\ 47 & 47 & 8.4 & 8.1 & 9.6 & 4 \\ 35 & 40 & 8.2 & 6.3 & 9.6 & 3 \\ 43 & 39 & 7.8 & 7.5 & 9.4 & 5 \\ 36 & 40 & 7.6 & 8.3 & 9.6 & 5 \\ 36.2 & 40 & 8.1 & 8.0 & 9.4 & 3 \end{bmatrix} :$$

$Price := \langle 27.8, 28.5, 38.7, 25.5, 27.5, 36.2 \rangle :$

$> equalWeights := \langle 1/6.\$6 \rangle :$

The *SAW* function will return a matrix of rankings. The first row is a raw ranking, the second row is normalized so the largest ranking is 1. We'll apply *fnormal* to the rankings and add a legend to make it easier to read the results.

```
> SAW(CarData, equalWeights) :
  ⟨CarModel, fnormal(%, 3)⟩;
```

$$\begin{bmatrix} \textit{Prius} & \textit{Fusion} & \textit{Volt} & \textit{Camry} & \textit{Sonata} & \textit{Leaf} \\ 0.877 & 0.955 & 0.816 & 0.919 & 0.913 & 0.847 \\ 0.918 & 1.0 & 0.854 & 0.962 & 0.955 & 0.887 \end{bmatrix}$$

Our rank ordering with equal weighting is Fusion, Camry, Sonata, Prius, Volt, and Leaf.

Let's add cost to the ranking. In order to match "larger is better," invert cost by $c_i \to 1/c_i$, then append this criterion to the data.

```
> PriceR := Price~(-1) :    # the '~' applies the power to each element
> ⟨CarModel, fnormal(SAW(⟨PriceR | Car⟩, ⟨1./7$7⟩), 3)⟩;
```

$$\begin{bmatrix} \textit{Prius} & \textit{Fusion} & \textit{Volt} & \textit{Camry} & \textit{Sonata} & \textit{Leaf} \\ 0.882 & 0.947 & 0.794 & 0.931 & 0.915 & 0.827 \\ 0.932 & 1.0 & 0.838 & 0.983 & 0.966 & 0.874 \end{bmatrix}$$

Adding cost to the criteria considered changed the ranking to Fusion, Camry, Sonata, Prius, Leaf, and Volt. Only Leaf and Volt changed places. Should cost weigh more than other criteria?

A weighting vector can be created from pairwise preference assessments. This technique was introduced by Saaty in 1980 when he developed the *analytic hierarchy process* that we'll study in Section 8.4. Decide which item of the pair is more important and by how much using the scale of Table 8.8. If

TABLE 8.8: Saaty's Nine-Point Scale

Importance level	Definition
1	Equal importance
2	*Intermediate*
3	Moderate importance
4	*Intermediate*
5	Strong importance
6	*Intermediate*
7	Very Strong importance
8	*Intermediate*
9	Extreme importance

$Criterion_i$ is k times as important as $Criterion_j$, then $Criterion_j$ is $1/k$, the reciprocal, times as important as $Criterion_i$.

Begin with comparing cost to the other criteria.

to	Cost	City	Hwy	Perform.	Style	Safety	Reliable
Cost	1	4	5	3	7	2	3

Make all the pairwise assessments creating an upper triangular *comparison matrix.*

to	Cost	City	Hwy	Perform.	Style	Safety	Reliable
Cost	1	4	5	3	7	2	3
City		1	3	1/2	6	1/3	1/3
Hwy			1	1/4	3	1/4	1/4
Perform.				1	6	1/2	1/2
Style					1	1/6	1/5
Safety						1	2
Reliable							1

Note that if cost is 3 (*weakly more important*) to performance, then performance is 1/3, the reciprocal value, to cost. Use reciprocals to fill in the lower triangle of the matrix, so that if $m_{ij} = k$, then $mji = 1/k$.

to	Cost	City	Hwy	Perform.	Style	Safety	Reliable
Cost	1	4	5	3	7	2	3
City	1/4	1	3	1/2	6	1/3	1/3
Hwy	1/5	1/3	1	1/4	3	1/4	1/4
Perform.	1/3	2	4	1	6	1/2	1/2
Style	1/7	1/6	1/3	1/6	1	1/6	1/5
Safety	1/2	3	4	2	6	1	2
Reliable	1/3	3	4	2	5	1/2	1

A comparison matrix is a *positive reciprocal matrix.* This type of matrix has a dominant positive eigenvalue and eigenvector. That eigenvector will be our weighting vector. The *SAW* program will compute (approximate) the weighting vector from a comparison matrix using an abbreviated power method.

We must ensure that the comparison assessments are consistent; i.e., if a is preferred to b and b is preferred to c, then a is preferred to c; according to Saaty's scheme, compute the *consistency ratio* CR as a test. The value of CR must be less than or equal to 0.1 to be considered consistent. If $CR > 0.1$, the preference choices must be revisited and adjusted. First compute the largest eigenvalue λ of the comparison matrix. Then calculate the *consistency index* CI

$$CI = \frac{\lambda - n}{n-1}.$$

Now determine the consistency ratio $CR = CI/RI$ where RI, the random index (see [Saaty1980]), is taken from

n	1	2	3	4	5	6	7	8	9	10
RI	0.0	0.0	0.58	0.90	1.12	1.24	1.32	1.41	1.45	1.49

Enter the comparison matrix in Maple, calling it *CM*. Use the *LinearAlgebra* package to find the dominant eigenvector of *CM* is $\lambda \approx 7.392$. Thus, the consistency ratio is

$$CR = \frac{7.392 - 7}{7 - 1} \cdot \frac{1}{1.35} \approx 0.05.$$

Our *CR* is well below 0.1; we have a consistent prioritization of our criteria.

Use *SAW* once more to find our new rankings.

$> \langle CarModel, fnormal(SAW(\langle PriceR \mid Car\rangle, CM), 3)\rangle;$

$$\begin{bmatrix} Prius & Fusion & Volt & Camry & Sonata & Leaf \\ 0.880 & 0.931 & 0.779 & 0.966 & 0.934 & 0.798 \\ 0.911 & 0.963 & 0.806 & 1.0 & 0.967 & 0.826 \end{bmatrix}$$

Preference ranking the criteria changed the result to Camry, Sonata, Fusion, Prius, Leaf, and Volt. The leaders changed places.

Since the importance values chosen are subjective judgments, sensitivity analysis is a must. The sensitivity analysis for this example is left as an exercise.

A Krackhardt "Kite network," shown in Figure 8.1, is a simple graph with 10 vertices that has three different answers to the question, "Which vertex is central?" depending on the definition of "central." Krackhardt introduced the graph in 1990 as a fictional social network.[3]

Example 8.4. Krackhardt's Kite Network.
In the Kite network, Susan is "central" as she has the most connections, Steve and Sarah are "central" as they are closest to all the others, and Claire is "central" as a critical connection between the largest disjoint subnetworks. *ORA-PRO*,[4] "a tool for network analysis, network visualization, and network forecasting," returns the data in Table 8.9 for the kite. Use *SAW* to rank the nodes.

After consulting with several network experts and combining their comparison matrices,[5] we have the weighting vector $w = [TC, BTW, EC, INC]$

$$w = \begin{bmatrix} 0.47986 & 0.26215 & 0.155397 & 0.102592 \end{bmatrix}.$$

[3] D. Krackhardt, "Assessing the Political Landscape: Structure, Cognition, and Power in Organizations." Admin. Sci. Quart. 35, 1990, pg. 342 369.

[4] Available from Netanomics, http://netanomics.com.

[5] A standard method uses the harmonic means of the experts' comparison matrices.

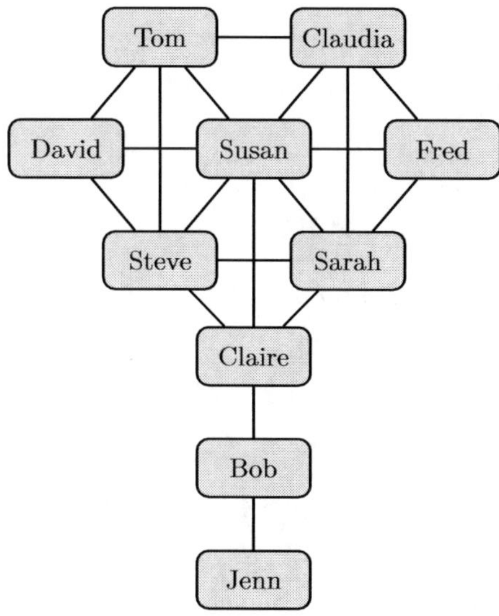

FIGURE 8.1: Krackhardt's "Kite Network"

TABLE 8.9: ORA Metric Measures for the Kite Network

	TC	BTW	EC	INC
Susan	0.1806	0.2022	0.1751	0.1088
Steve	0.1389	0.1553	0.1375	0.1131
Sarah	0.1250	0.1042	0.1375	0.1131
Tom	0.1111	0.0194	0.1144	0.1009
Claire	0.1111	0.0194	0.1144	0.1009
Fred	0.0833	0.0000	0.0938	0.0975
David	0.0833	0.0000	0.0938	0.0975
Claudia	0.0833	0.3177	0.1042	0.1088
Bob	0.0556	0.1818	0.0241	0.0885
Jenn	0.0278	0.0000	0.0052	0.0707

Table Legend: TC - Total Centrality; BTW - Betweenness; EC - Eigenvector Centrality; INC - Information Centrality

The consistency ratio for the combined comparison matrix is 0.003; that is well less than 0.1.

It's time for Maple.

```
> Labels := Vector[row]([Susan, Steve, Sarah, Tom, Claire, Fred, David,
     Claudia, Bob, Jenn]) :
  Kite := Matrix([[0.1806, 0.2022, 0.1751, 0.1088],
     [0.1389, 0.1553, 0.1375, 0.1131],
     [0.1250, 0.1042, 0.1375, 0.1131],
     [0.1111, 0.0194, 0.1144, 0.1009],
     [0.1111, 0.0194, 0.1144, 0.1009],
     [0.0833, 0.0000, 0.0938, 0.0975],
     [0.0833, 0.0000, 0.0938, 0.0975],
     [0.0833, 0.3177, 0.1042, 0.1088],
     [0.0556, 0.1818, 0.0241, 0.0885],
     [0.0278, 0.0000, 0.0052, 0.0707]]) :
  w := ⟨0.47986, 0.26215, 0.155397, 0.102592⟩ :
> Results := ⟨Labels, fnormal(SAW(Kite, w), 3)⟩;
```

Susan	*Steve*	*Sarah*	*Tom*	*Claire*	*Fred*	*David*	*Claudia*	*Bob*	*Jenn*
0.901	0.722	0.643	0.504	0.504	0.393	0.393	0.675	0.399	0.143
1.0	0.801	0.714	0.560	0.560	0.436	0.436	0.749	0.443	0.158

The results are easier to parse when we sort the array. The program *MatrixSort* is in the *PSMv2* package with syntax *MatrixSort*(*Matrix*, ⟨*row/col*⟩, ⟨*options*⟩). The options are *sortby='row'/'column'* and *order='ascending'/'descending'* with defaults *'row'* and *'ascending'*.

```
> MatrixSort(Results, 3, order = 'descending');
```

Susan	*Steve*	*Claudia*	*Sarah*	*Claire*	*Tom*	*Bob*	*David*	*Fred*	*Jenn*
0.901	0.722	0.675	0.643	0.504	0.504	0.399	0.393	0.393	0.143
1.0	0.801	0.749	0.714	0.560	0.560	0.443	0.436	0.436	0.158

We see the resulting rank order for "overall centrality" is

$$\text{Susan} > \text{Steve} > \text{Claudia} > \text{Sarah} > \begin{matrix}\text{Claire}\\ \text{Tom}\end{matrix} > \text{Bob} > \begin{matrix}\text{David}\\ \text{Fred}\end{matrix} > \text{Jenn}$$

Sensitivity Analysis. We can apply sensitivity analysis to the weights to determine how changes impact the final rankings. We recommend using an algorithmic method to modify the weights. For example, if we reverse the weights for TC (total centrality) and BTW (betweenness), the rankings change to

Susan	*Claudia*	*Steve*	*Sarah*	*Bob*	*Claire*	*Tom*	*David*	*Fred*	*Jenn*
0.822	0.792	0.661	0.563	0.457	0.384	0.384	0.293	0.293	0.109
1.0	0.964	0.804	0.686	0.556	0.467	0.467	0.356	0.356	0.133

Susan is still the "top node," but Claudia and Steve have swapped for second; Bob is now above Claire, rather than tied.

Exercises

Use SAW in each problem to find the ranking under the weight:

(a) Equal weighting.

(b) Choose and state weights.

1. Rank order Hospital A's procedures using the data below.

Hospital A	Procedure			
	I	II	III	IV
X-Ray Time	6	5	4	3
Laboratory Time	5	4	3	2
Profit	\$200	\$150	\$100	\$80

2. Rank order Hospital B's procedures using the data below.

Hospital B	Procedure			
	I	II	III	IV
X-Ray Time	6	5	5	3
Laboratory Time	5	4	3	3
Profit	\$190	\$150	\$110	\$80

3. A college student is planning to move to a new city after graduation. Rank the cities in order of best-to-move-to given the following data.

City	Housing Affordability	Cultural Opportunity	Crime Rate	Quality of Schools
I	250	5	10	0.75
II	325	4	12	0.60
III	676	6	9	0.81
IV	1,020	10	6	0.80
V	275	3	11	0.35
VI	290	4	13	0.41
VII	425	6	12	0.62
VIII	500	7	10	0.73
IX	300	8	9	0.79

LEGEND: Housing Affordability: avg. home cost in \$100,000s; Cultural Opportunity: events per month; Crime Rate: # crimes reported per month in 100's; Quality of Schools: index in [0, 1].

4. Rank order the threat information collected by the Risk Assessment Office that is shown in Table 8.1 (pg. 339) for the Department of Homeland Security.

8.4 Analytical Hierarchy Process

The *analytical hierarchy process* (AHP) is a multi-objective decision analysis tool first proposed by Thomas Saaty [Saaty1980]. The technique has become very popular—a Google Scholar search for "analytical hierarchy process" returns over 1.5 *million* items in under a tenth of a second.

Description and Uses

The process is designed for using either objective and subjective measures to evaluate a set of alternatives based upon multiple criteria, organized in a hierarchical structure as shown in Figure 8.2. The goal or objective is at the

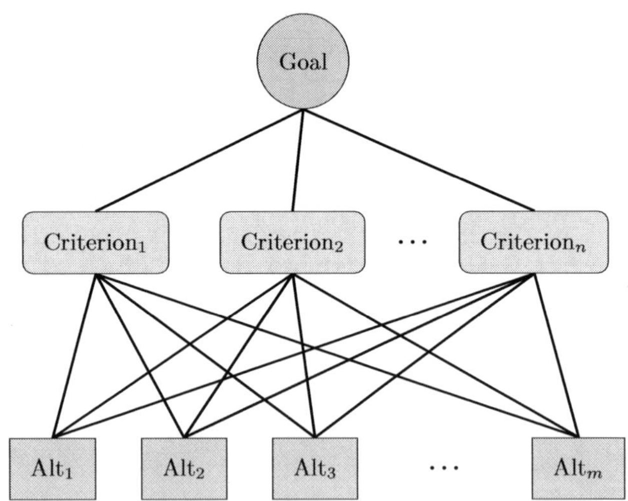

FIGURE 8.2: A Generic Three-Layer AHP Hierarchy

top level. The next layer holds the criteria that are evaluated or weighted. The bottom level has the alternatives which are measured against each criterion. The decision maker makes pairwise comparisons of the criteria in which every pair is subjectively or objectively compared. Subjective assessments use Saaty's nine-point scale that we introduced with SAW. (See Table 8.8.)

The AHP process can be described as a method to decompose a problem into sub-problems. In most cases, the decision maker has a choice among many alternatives. Each alternative has a set of attributes or characteristics, AHP calls criteria, that can be measured, either subjectively or objectively. The attribute elements of the hierarchical process can relate to any aspect of the decision problem, either tangible or intangible, carefully measured or roughly estimated, well- or poorly understood—essentially anything that applies to the decision at hand.

To perform an AHP, we need a goal or an objective and a set of alternatives, each with criteria (attributes) to compare. Once the hierarchy is built, the decision makers systematically pairwise-evaluate the various elements (comparing them to one another two at a time), with respect to their impact on an element above them in the hierarchy. The decision makers can use concrete data about the elements or subjective judgments concerning the elements' relative meaning and importance for making the comparisons. Since subjective judgments are imperfect, sensitivity analysis will be very important.

The process converts subjective evaluations to numerical values that can be processed and compared over the entire range of the problem. A *priority* or numerical weight is derived for each element of the hierarchy, allowing diverse and often incommensurable elements to be rationally and consistently compared to one another. The final step of the process calculates a numerical priority score for each decision alternative. These scores represent the alternatives' relative ability to achieve the decision's goal; they allow a straightforward consideration of the various courses of action.

AHP can be used by individuals for simple decisions or by teams working on large, complex problems. The method has unique advantages when important elements of the decision are difficult to quantify or compare, or where communication among team members is impeded by their different specializations, lexicons, or perspectives.

Methodology of the Analytic Hierarchy Process

The procedure for AHP can be summarized as:

STEP 1. Build the hierarchy for the decision following Figure 8.2.

Goal: *Select the best alternative*

Criteria: *List* $c_1, c_2, c_3, \ldots, c_m$

Alternatives: *List* $a_1, a_2, a_3, \ldots, a_n$

STEP 2. Judgments and Comparison.
Build comparison matrices using a 9-point scale of pairwise comparisons shown in Table 8.10 for the criteria (attributes) and the alternatives relative to each criterion. A problem with m criteria and n alternatives will require $n + 1$

matrices. Find the dominant eigenvector of each matrix. The power method is often used to calculate eigenvectors; see, e.g., [BurdenFaires2005]. The goal, in AHP, is to obtain a set of eigenvectors of the system that measure the importance of alternatives with respect to the criteria.

TABLE 8.10: Saaty's Nine-Point Scale

Importance level	Definition
1	Equal importance
3	Moderate importance
5	Strong importance
7	Very Strong importance
9	Extreme importance

TABLE NOTES: Even numbers represent intermediate importance levels which should only be used as compromises. If the importance level of A to B is 3, then that of B to A is $1/3$, the reciprocal.

Saaty's *consistency ratio* CR measures the consistency of the pairwise assessments of relative importance. The value of CR must be less than or equal to 0.1 to be considered valid. To compute CR, start by approximating the largest eigenvalue λ of the comparison matrix. Then calculate the *consistency index* CI with

$$CI = \frac{\lambda - n}{n-1}.$$

Finally, $CR = CI/RI$, where RI is the *random index* (from [Saaty1980]) found from

n	1	2	3	4	5	6	7	8	9	10
RI	0.0	0.0	0.58	0.90	1.12	1.24	1.32	1.41	1.45	1.49

If $CR > 0.1$, we must go back to our pairwise comparisons and repair the inconsistencies. In general, consistency ensures that if $A > B$ and $B > C$, then $A > C$ for all A, B, and C.

STEP 3. Combine all the alternative comparison eigenvectors into a matrix, then multiply by the criteria matrix's eigenvector to obtain an overall comparative ranking.

STEP 4. After the m criterion weights are combined into an $n \times m$ normalized matrix (for n alternatives by m criteria), multiply by the criteria ranking vector to obtain the final rankings.

STEP 5. Interpret the order presented by the final ranking.

Strengths and Limitations

AHP is very widely used in business, industry, and government. The technique is quite useful for discriminating between competing options for a range of objectives needing to be met. Even though AHP relies on what might be seen as obscure mathematics—eigenvectors and eigenvalues of positive reciprocal matrices—the calculations are not complex, and can be carried out with a spreadsheet. A decision maker doesn't need to understand linear algebra theory to use the technique, but must be aware of its strengths and limitations.

AHP's main strength is producing a ranking of alternatives ordered by their effectiveness relative to the criteria's weighting to meet the project goal. The calculations of AHP logically lead to the alternatives' ranking as a consequence of preference judgments on the relative importance of the criteria and on how the alternatives satisfy each of the criteria. Making accurate and good-faith relative importance assessments is critical to the method. Manually adjusting the pairwise judgments to obtain a predetermined result is quite hard, but not impossible.

A further strength is that AHP provides a heuristic for detecting inconsistent judgments in pairwise comparisons. When there are a large number of criteria and/or alternatives, inconsistencies can easily be hidden by the problem's size; AHP highlights hidden inconsistencies.

A main limitation of AHP comes from being based on eigenvectors of positive reciprocal matrices. This basis requires that symmetric judgments must be reciprocal: if A is 3 times more important than B, then B is $1/3$ as important as A.[6] This restriction can lead to problems such as "rank reversal," a change in the ordering of alternatives when criteria or alternatives are added to or deleted from the initial set compared. Several modifications to AHP have been proposed to ameliorate this and other related issues. Many of the enhancements involved ways of computing, synthesizing pairwise comparisons, and/or normalizing the priority and weighting vectors. In the next section, we'll see TOPSIS, a method that corrects rank reversal.

Another limitation is implied scaling in the results. The final ranking indicates that one alternative is *relatively better* than another, not by how much. For example, suppose that rankings for alternatives $\langle A, B, C\rangle$ are $\langle 0.392, 0.406, 0.204\rangle$. The values only imply that alternatives A and B are about equally good (≈ 0.4), and C is worse (≈ 0.2). The ranking does not mean that A and B are *twice as good* as C.

Hartwich [Hartwich1999] criticized AHP for not providing sufficient guidance about structuring the problem to be solved; that is, how to form the levels of the hierarchy for criteria and alternatives. When project team members carry out rating items individually or as a group, guidance on aggregating separate criteria assessments is necessary. As the number of levels in the hierarchy increases, the complexity of AHP increases faster; n criteria require $\mathcal{O}(n^2)$ comparisons.

[6] See Saaty's book [Saaty1980] for his rationale.

Nevertheless, AHP is a very powerful and useful decision tool when used intelligently.

Sensitivity Analysis

Using subjective pairwise comparisons in AHP makes sensitivity analysis extremely important. How often do we change our minds about the relative importance of objects, places, or things? Often enough that we should test the pairwise comparison values to determine the robustness of AHP's rankings. Test the decision maker weights to find the "break point" values, if they exist, that change the alternatives' rankings. At a minimum, perform trial-and-error sensitivity analysis using a *numerical incremental analysis* of the weights. Numerical incremental analysis works by incrementally changing one parameter at a time (OAT), finding the new solution, and graphically showing how the rankings change. Several variations of this method are given in [BarkerZabinsky2011] and [Hurley2001].

Chen [ChenKocaoglu2008] divided sensitivity analysis into three main groups: numerical incremental analysis, probabilistic simulations, and mathematical models. Probabilistic simulation employs the use of Monte Carlo simulation ([ButlerJiaDyer1997]) that makes random changes in the weights and simultaneously explores the effect on rankings. Modeling may be used when it is possible to express the relationship between the input data and the solution results as a mathematical model. Leonelli's [Leonelli2012] master's thesis outlines these three procedures.

We prefer numerical incremental analysis weight adjusting with the new weight w_j' given by

$$w_j' = \frac{1 - w_p'}{1 - w_p} \cdot w_j \tag{8.4}$$

where w_p is the original weight of the criterion to be adjusted, and w_p' is the value after the criterion was adjusted [AlinezhadAmini2011].

Whichever method is chosen, sensitivity analysis is always an important part of an AHP solution.

Examples Using AHP

Let's look at a selection of examples using AHP starting with a quite simple 3-criteria for 3 alternatives.

Example 8.5. Selecting a VHF Transceiver Antenna.
An amateur radio operator wants to build and install a new VHF antenna choosing from designs for a vertical 1/4-wave antenna, a 3-element Yagi antenna, and a J-pole antenna. See Figure 8.3. The three main criteria will

be size, antenna gain (how much the antenna focuses the signal), and ease of assembly.[7]

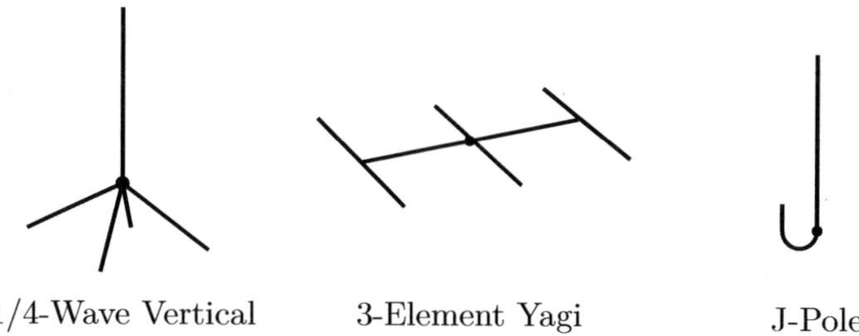

FIGURE 8.3: VHF Antenna Types

We'll use the *AHP* and *AnalyzeComparisonMat* programs from the book's Maple package *PSMv2*. After loading the package via *with*, use the *Describe* command to see a brief description and a list of arguments.

STEP 1. Build the 3-level hierarchy listing the criteria and alternatives.

Goal:	*Select the best antenna*
Criteria:	*Size, Gain, Assembly*
Alternatives:	*Vertical, Yagi, J-Pole*

```
> CritLabel := ["Size", "Gain", "Assembly"] :
  AltLabel := ["Vertical", "Yagi", "J-Pole"] :
```

STEP 2. Perform the pairwise comparisons using Saaty's 9-point scale.
Use the row and column orders specified by the lists in Step 1. The ham operator's choices are as follows.
Criteria comparison matrix:

$$> CritPM := \begin{bmatrix} 1 & 0.2 & 0.333 \\ 5 & 1 & 3 \\ 3 & 0.333 & 1 \end{bmatrix} :$$

AnalyzeComparisonMat(%);

$$[[\ 0.1047 \quad 0.6370 \quad 0.2582\ ], 0.03276]$$

The consistency ratio 0.03 is very good.
Alternatives by Size comparison matrix:

[7] Adapted from W. Bauldry, "Choosing an Antenna Mathematically," 2018.

$$> SizePM := \begin{bmatrix} 1 & 5 & 0.5 \\ 0.2 & 1 & 0.2 \\ 2 & 5 & 1 \end{bmatrix}:$$

AnalyzeComparisonMat(%);

$$[\begin{bmatrix} 0.3522 & 0.08875 & 0.5591 \end{bmatrix}, 0.04655]$$

The consistency ratio 0.05 is very good.

Alternatives by Gain comparison matrix:

$$> GainPM := \begin{bmatrix} 1 & 0.143 & 0.33 \\ 7 & 1 & 5 \\ 3 & 0.2 & 1 \end{bmatrix}:$$

AnalyzeComparisonMat(%);

$$[\begin{bmatrix} 0.08080 & 0.7308 & 0.1884 \end{bmatrix}, 0.05431]$$

The consistency ratio 0.05 is very good.

Alternatives by Assembly comparison matrix:

$$> AssemblyPM := \begin{bmatrix} 1 & 5 & 3 \\ 0.2 & 1 & 0.333 \\ 0.33 & 3 & 1 \end{bmatrix}:$$

AnalyzeComparisonMat(%);

$$[\begin{bmatrix} 0.6374 & 0.1048 & 0.2578 \end{bmatrix}, 0.03017]$$

The consistency ratio 0.03 is very good.

STEP 3. The *AHP* program will combine the eigenvectors from the three alternatives' priority matrices to form the overall alternative priority matrix.

STEP 4. Obtain the AHP rankings.

> *Results* := *AHP*(*CritLabels*, *CritPM*, *AltLabels*, *SizePM*, *GainPM*, *AssemblyPM*);

$$\left[\begin{array}{c} CriteriaWeights = \begin{bmatrix} \text{"Size"} & \text{"Gain"} & \text{"Assembly"} \\ 0.1047 & 0.6370 & 0.2582 \end{bmatrix} \\ ConsistencyRatio = 0.03276 \\ AlternativesRanking = \begin{bmatrix} \text{"Vertical"} & \text{"Yagi"} & \text{"J-Pole"} \\ 0.2529 & 0.5020 & 0.2451 \end{bmatrix} \\ AltPriorityMatrix = \begin{bmatrix} 0.3522 & 0.08080 & 0.6374 \\ 0.08875 & 0.7308 & 0.1048 \\ 0.5591 & 0.1884 & 0.2578 \end{bmatrix} \end{array}\right]$$

Interpretation of Results. The Yagi is the best choice antenna, with the vertical and J-pole at essentially the same rating. Since the Yagi was a clear preference for gain, and gain was the highest rated criteria, the rankings pass an initial "common-sense test."

The necessary sensitivity analysis will be left to the reader. The first step will be to find the break-even points for the criteria ratings.

Example 8.6. Selecting a Car Redux.
Revisit Example 8.3 "Selecting a Car" with the data presented in Table 8.7 (pg. 353), but now use AHP to rank the models.

STEP 1. Build the 3-level hierarchy and list the criteria from the highest to lowest priority (mainly for convenience).

Goal: *Select the best car*

Criteria: *Cost, MPG-City, MPG-Hwy, Safety, Reliab., Perform., Style*

Alternatives: *Prius, Fusion, Volt, Camry, Sonata, Leaf*

$> Criteria := [Cost, MPG\text{-}City, MPG\text{-}Hwy, Safety, Reliab., Perform., Style]:$

$> Model := [Prius, Fusion, Volt, Camry, Sonata, Leaf]:$

STEP 2. Perform the pairwise comparisons using Saaty's 9-point scale.
We chose the priority order as: Cost, Safety, Reliability, Performance, MPG-City, MPG-Hwy, and Interior & Style. Putting the criteria in priority order allows for easier pairwise comparisons. A spreadsheet similar to Figure 8.4 organizes the pairwise comparisons nicely. Enter the comparisons in Maple as the matrix *PCM.*

$$> PCM := \begin{bmatrix} 1.0 & 2.0 & 2.0 & 3.0 & 4.0 & 5.0 & 6.0 \\ 0.500 & 1.0 & 2.0 & 3.0 & 4.0 & 5.0 & 5.0 \\ 0.500 & 0.500 & 1.0 & 2.0 & 2.0 & 3.0 & 3.0 \\ 0.333 & 0.333 & 0.500 & 1.0 & 1.0 & 2.0 & 3.0 \\ 0.250 & 0.250 & 0.500 & 1.0 & 1.0 & 2.0 & 3.0 \\ 0.200 & 0.200 & 0.333 & 0.500 & 0.500 & 1.0 & 2.0 \\ 0.167 & 0.200 & 0.333 & 0.333 & 0.333 & 0.500 & 1.0 \end{bmatrix}$$

$AnalyzeComparisonMat(PCM);$

$$[[\, 0.3178\ \ 0.2545\ \ 0.1515\ \ 0.09450\ \ 0.08783\ \ 0.05443\ \ 0.03939 \,], 0.01932]$$

The consistency ratio 0.02 is very good.

STEP 3. We enter the *AltM* matrix with columns listed in the priority order we chose in Step 1; the order must match the *PCM* matrix. The cost data does not follow the rubric "larger is better"; therefore, the reciprocal of cost

	A	B	C	D	E
1	Element			Comparison	
2	A		B	More Important	Intensity
3	Cost	to	MPG-City	A	2
4			MPG-Hwy	A	2
5			Safety	A	3
6			Reliability	A	4
7			Performance	A	5
8			Interior & Style	A	6
9	MPG-City	to	MPG-Hwy	A	2
10			Safety	A	3
11			Reliability	A	4
12			Performance	A	5
13			Interior & Style	A	5
14	MPG-Hwy	to	Safety	A	2
15			Reliability	A	2
16			Performance	A	3
17			Interior & Style	A	3
18	Safety	to	Reliability	A	1
19			Performance	A	2
20			Interior & Style	A	3
21	Reliability	to	Performance	A	2
22			Interior & Style	A	3
23	Performance	to	Interior & Style	A	2

FIGURE 8.4: Pairwise Comparison of Criteria

is used for the first column.

$$> AltM := \begin{bmatrix} 0.0360 & 9.4 & 3.0 & 7.5 & 44.0 & 40.0 & 8.7 \\ 0.0351 & 9.6 & 4.0 & 8.4 & 47.0 & 47.0 & 8.1 \\ 0.0258 & 9.6 & 3.0 & 8.2 & 35.0 & 40.0 & 6.3 \\ 0.0392 & 9.4 & 5.0 & 7.8 & 43.0 & 39.0 & 7.5 \\ 0.0364 & 9.6 & 5.0 & 7.6 & 36.0 & 40.0 & 8.3 \\ 0.0276 & 9.4 & 3.0 & 8.1 & 36.2 & 40.0 & 8.0 \end{bmatrix} :$$

Standard methods for dealing with a variable like *cost* include: (1) replace *cost* with 1/*cost*, (2) use a pairwise comparison with the nine-point scale, (3) use a pairwise comparison with ratios $cost_i/cost_j$, or (4) remove *cost* as a criteria and a variable, perform the analysis, and then use a *benefit*/*cost* ratio to re-rank the results. Many analysts prefer (4) when cost figures are large and dominate the procedure.

For the alternatives, we either use the raw data, or we can use pairwise comparisons by criteria for how each alternative fares versus its competitors. Here, we take the raw data replacing *cost* by *1/cost*, then normalize the columns.

$$> AltMN := evalf\left(Matrix\left(\left[seq\left(\frac{AltM[..,i]}{add(AltM[..,i])}, i = 1..7\right)\right]\right), 4\right):$$

STEP 4. Execute the procedure and obtain the output to interpret.
Note: Our *AHP* program will accept either the alternatives' comparison matrices or the eigenvectors merged into a single alternative comparison matrix.

$$> Result := AHP(Criteria, PCM, Model, AltMN);$$

CriteriaWeights =

"Cost"	"MPG-City"	"MPG-Hwy"	"Safety"	"Reliab."	"Perform."	"Style"
0.3178	0.2545	0.1515	0.09450	0.08783	0.05443	0.03939

$$ConsistencyRatio = 0.01932$$

AlternativesRanking =

"Prius"	"Fusion"	"Volt"	"Camry"	"Sonata"	"Leaf"
0.1659	0.1759	0.1468	0.1833	0.1776	0.1504

Once again, use *MatrixSort* to make the results easier to parse.

$$> MatrixSort(rhs(Result[3]), 2, order = \text{'descending'});$$

"Camry"	"Sonata"	"Fusion"	"Prius"	"Leaf"	"Volt"
0.1833	0.1776	0.1759	0.1659	0.1504	0.1468

We see the resulting rank order for the best car is

$$\text{Camry} > \begin{matrix}\text{Sonata}\\ \text{Fusion}\end{matrix} > \text{Prius} > \text{Leaf} > \text{Volt}.$$

SENSITIVITY ANALYSIS. We alter our pairwise comparison values to obtain a new set of weights and obtain new results: Camry, Fusion, Sonata, Prius, Leaf, and Volt. First we adjust the weights and place the weights into a new comparison matrix.

We altered *cost*, the largest decision criteria, by lowering its value incrementally. Then we created a matrix of the new decision weights that included the original set of weights as a reference. We then multiplied it by the transpose of the normalized matrix of alternatives still in criteria order. Using *Statistics:-Rank* finds the ordering. With the changes in the decision weights, the cars were ranked the same. That is, the resulting values have changed, but not the relative rankings of the cars. The Maple work follows.

$$> with(LinearAlgebra):$$

$$> rhs(Result[1]);$$
$$w_1 := \%[2]:$$

"Cost"	"MPG-City"	"MPG-Hwy"	"Safety"	"Reliab."	"Perform."	"Style"
0.3178	0.2545	0.1515	0.09450	0.08783	0.05443	0.03939

```
> AdjVec := fnormal(Vector[row]([-1, 1./6$6]), 4);
        [ -1  0.1667  0.1667  0.1667  0.1667  0.1667  0.1667 ]
> w2 := fnormal(w1 + 0.025 · AdjVec, 4);
    [ 0.2928  0.2587  0.1557  0.09867  0.09200  0.05860  0.04356 ]
> for i from 3 to 6 do
    w_i := fnormal(w_{i-1} + 0.025 · AdjVec, 4) :
    end do :
  NewW := Matrix(convert(convert~(w, list), listlist));
  NewW :=
    [ 0.3178  0.2545  0.1515  0.09450  0.08783  0.05443  0.03939 ]
    [ 0.2928  0.2587  0.1557  0.09867  0.09200  0.05860  0.04356 ]
    [ 0.2678  0.2629  0.1599  0.1028   0.09617  0.06277  0.04773 ]
    [ 0.2428  0.2671  0.1641  0.1070   0.1003   0.06694  0.05190 ]
    [ 0.2178  0.2713  0.1683  0.1112   0.1045   0.07111  0.05607 ]
    [ 0.1928  0.2755  0.1725  0.1154   0.1087   0.07528  0.06024 ]
> fnormal(Transpose(AltM . Transpose(NewW))), 4) :
  ⟨Statistics:-Rank~(convert(%, listlist))⟩;
                                [ [4, 6, 1, 5, 3, 2] ]
                                [ [4, 6, 1, 5, 3, 2] ]
                                [ [4, 6, 1, 5, 3, 2] ]
                                [ [4, 6, 1, 5, 3, 2] ]
                                [ [4, 6, 1, 5, 3, 2] ]
                                [ [4, 6, 1, 5, 3, 2] ]
```

Again, we recommend using sensitivity analysis to try to find break points, if any exist.

In the next sensitivity analysis, take the smallest criteria, Interior and Style, and increase its value by 0.1, 0.2, and 0.25, adjusting the other weights proportionally. The new results show a change in rank ordering between the 3rd and 4th increments. Thus the break point is adding between 0.2 and 0.25 to the criteria weight for Interior and Style. *Verify these computations with Maple!*

Example 8.7. The Kite Network Redux.
Revisit Krackhardt's Kite social network; search for the key influencer nodes. According to Newman[8] there are four metrics that contribute to identifying

---

[8]See Chapter 4 of M. Newmann, *Networks: An Introduction*, Oxford Univ Press, Oxford, 2010. Available at http://dx.doi.org/10.1093/acprof:oso/9780199206650.001.0001.

the key nodes of a network. In our priority order, the key criteria are:

$$\textit{Total Centrality}, \textit{Betweenness}, \textit{Eigenvector Centrality}, \textit{Closeness Centrality}$$

Assume we have the outputs from the network analysis program *ORA-Pro* (which is not shown here due to the volume of output). Take the metrics from *ORA-Pro* and normalize each column. The columns for each criterion are placed in the matrix $\mathbf{X} = [x_{ij}]$. Define w_j to be the weight for each criterion. Limit the size of $\mathbf{X}$ matrix to 8 alternatives with the four criteria for this example.

Next, assume we have obtained the criteria pairwise comparison matrix from the decision maker. Using the output from *ORA-Pro* and normalizing the results, we are ready for AHP to rate the alternatives within each criterion. We provide a sample pairwise comparison matrix with Saaty's nine-point scale for weighting the Kite network criteria. The consistency ratio is $CR = 0.01148$, which is much less than 0.1, so our pairwise comparison matrix is consistent. We continue with Maple.

$> \textit{CritLabels} := [TC, BTW, EC, CC] :$
$\textit{AltLabels} := [\textit{Bob}, \textit{Claire}, \textit{Fred}, \textit{Sarah}, \textit{Susan}, \textit{Steve}, \textit{Claudia}, \textit{Tom}] :$

$$> PCM := \begin{bmatrix} 1 & 2 & 3 & 4 \\ 0.5 & 1 & 2 & 3 \\ 0.3333 & 0.5 & 1 & 2 \\ 0.25 & 0.3333 & 0.5 & 1 \end{bmatrix} :$$

$$> \textit{AltM} := \begin{bmatrix} 0.11111111 & 0.019407559 & 0.114399403 & 0.100733634 \\ 0.11111111 & 0.019407559 & 0.114399403 & 0.100733634 \\ 0.08333333 & 0.0 & 0.093757772 & 0.09734763 \\ 0.125 & 0.104187947 & 0.137527978 & 0.100733634 \\ 0.180555556 & 0.202247191 & 0.175080826 & 0.122742664 \\ 0.138888889 & 0.15526047 & 0.137527978 & 0.112866817 \\ 0.083333333 & 0.317671093 & 0.104202935 & 0.108634312 \\ 0.055555556 & 0.181818182 & 0.024123352 & 0.088318284 \end{bmatrix} :$$

> *Result* := *AHP*(*CritLabels*, *PCM*, *AltLabels*, *AltM*);

Result :=

CriteriaWeights =

TC	BTW	EC	CC
0.4673	0.2772	0.1601	0.09543

ConsistencyRatio = 0.01148

AlternativesRanking =

Bob	Claire	Fred	Sarah	Susan	Steve	Claudia	Tom
0.09303	0.09303	0.06903	0.1298	0.1967	0.1536	0.1681	0.09676

As before, sort the results to make them easier to read.

> *MatrixSort*(*rhs*(*Result*[3]), 2, *order* = *'descending'*);

Susan	Claudia	Steve	Sarah	Tom	Claire	Bob	Fred
0.1967	0.1681	0.1536	0.1298	0.09676	0.09303	0.09303	0.06903

AHP gives Susan as the key node. However, the bias of the decision maker is important in the analysis of the weights of the criteria. The *Betweenness* criterion is rated 2 to 3 times more important than the others.

SENSITIVITY ANALYSIS. Changes in the pairwise comparisons for the criteria cause fluctuations in the key nodes. The reader should change the pairwise comparisons given above so that *Total Centrality* is not so dominant, and rerun the AHP as we did with the previous example.

Exercises

1. Redo Section 8.3's Exercises (pg. 359) using AHP. Compare with your previous results using SAW.
2. In each of the problems above, perform sensitivity analysis by changing the weight of your dominant criteria until it is no longer the highest. Did the change affect your rankings?
3. In each of the problems above, find break points, if any exist, for the weights.
4. Suppose a criterion has no break point. What does this indicate about the sensitivity of the solution to that criterion's weight?

8.5 Technique of Order Preference by Similarity to the Ideal Solution

In 1981, Hwang and Yoon [HwangYoon1981] introduced the Technique of Order Preference by Similarity to the Ideal Solution (TOPSIS) as a multi-criteria decision analysis method that is based on comparing the relative "distances" of alternatives from a theoretical best solution and a theoretical worst solution. The optimal alternative will have the shortest geometric distance from the best or positive ideal solution, and the longest geometric distance from the worst, or negative ideal solution. The method is a *compensatory aggregation* that compares a set of alternatives by identifying weights for each criterion, normalizing the scores for each criterion, and calculating a distance between each alternative and the theoretical ideal alternative based on the best score in each criterion. TOPSIS requires that the criteria are monotonically increasing or decreasing. Normalization is usually required as the criteria often have incompatible dimensions. Compensatory methods allow trade-offs between criteria, where a poor result in one criterion can be negated by a good result in another criterion. This compensation provides a more realistic form of modeling than non-compensatory methods which often include or exclude alternative solutions based on strict cut-off values.

A 2012 survey by Behzadian et al.[9] finds the main areas of application of TOPSIS include

- Supply Chain Management and Logistics,
- Design, Engineering and Manufacturing Systems,
- Business and Marketing Management,
- Health, Safety and Environment Management,
- Human Resources Management,
- Energy Management,
- Chemical Engineering and
- Water Resources Management.

We begin with a brief discussion of the framework of TOPSIS as a method of decomposing a problem into sub-problems. Typically, a decision maker must choose from many alternatives each having a set of attributes or characteristics that can be measured subjectively or objectively. The attributes can relate to any tangible or intangible aspect of the decision problem. Attributes can be

[9] Behzadian, Khanmohammadi, Yazdani, and Ignatius, "A state-of the-art survey of TOPSIS applications," *Expert systems with Applications*, 39 (2012): 13051-13069.

carefully measured or roughly estimated, or be well or poorly understood. Basically, anything that applies to the decision at hand can be used in the TOPSIS process.

Methodology

The TOPSIS process is carried out as follows.

STEP 1. Create an evaluation matrix $\mathbf{X} = [x_{ij}]$ consisting of m alternatives and n criteria where x_{ij} is alternative i's value for criterion j.

$$\mathbf{X} = \begin{array}{c} \\ A_1 \\ A_2 \\ \vdots \\ A_m \end{array} \begin{array}{c} \begin{array}{cccc} c_1 & c_2 & \cdots & c_n \end{array} \\ \begin{bmatrix} x_{11} & x_{12} & \cdots & x_{1n} \\ x_{21} & x_{22} & \cdots & x_{2n} \\ \vdots & \vdots & \ddots & \vdots \\ x_{m1} & x_{m2} & \cdots & x_{mn} \end{bmatrix} \end{array}$$

STEP 2. $\mathbf{X}$ is normalized to form $\mathbf{R} = [r_{ij}]_{m \times n}$ using the normalization

$$r_{ij} = \frac{x_{ij}}{\sqrt{\sum_{k=1}^{m} x_{kj}^2}}$$

for $i = 1..m$ and $j = 1..n$.

STEP 3. Calculate the weighted normalized decision matrix $\mathbf{T}$. Weights must total 1 (100%), and can come from either the decision maker directly, or by computation such as from the eigenvector of a comparison matrix using Saaty's nine-point scale. $\mathbf{T}$ is given by

$$\mathbf{T} = [w_j r_{ij}]_{m \times n};$$

i.e., multiply each column by its weight.

STEP 4. Determine each criterion's best alternative A_b and worst alternative A_w. Examine each attribute's column and select the largest and smallest values. If the criterion's values imply larger is better (e.g., profit), then the best alternatives are the largest values; if the values imply smaller is better (such as cost), the best alternative is the smallest value. (Whenever possible, define all criteria in terms of positive impacts.) Separate the index set of the criteria into two classes:

$$J_+ = \{\text{indices of criteria having positive impact}\},$$

and

$$J_- = \{\text{indices of criteria having a negative impact}\}.$$

Now define the best, the ideal positive alternative, as

$$\begin{aligned} A_b &= \{\langle \min(t_{ij}|i = 1..m; j \in J_- \rangle, \langle \max(t_{ij}|i = 1..m; j \in J_+ \rangle\} \\ &\equiv \{t_{bj}|j = 1..n\}, \end{aligned}$$

and the worst, the ideal negative alternative, as

$$\begin{aligned} A_w &= \{\langle \max(t_{ij}|i = 1..m; j \in J_- \rangle, \langle \min(t_{ij}|i = 1..m; j \in J_+ \rangle\} \\ &\equiv \{t_{wj}|j = 1..n\}, \end{aligned}$$

STEP 5. Calculate the Euclidean distances between each alternative and the ideal positive alternative

$$d_{ib} = \sqrt{\sum_{j=1}^{n} (t_{ij} - t_{bj})^2}, \quad \text{for } i = 1..m,$$

then the ideal negative alternative

$$d_{iw} = \sqrt{\sum_{j=1}^{n} (t_{ij} - t_{wj})^2}, \quad \text{for } i = 1..m.$$

STEP 6. Calculate each alternative's *similarity to the worst condition*

$$s_{iw} = \frac{d_{iw}}{d_{iw} + d_{ib}} \quad \text{for } i = 1..m.$$

Note that $0 \leq s_{iw} \leq 1$ for each i, and that

$s_{iw} = 1$ iff alternative i equals the ideal positive alternative, $d_{ib} = 0$; and
$s_{iw} = 0$ iff alternative i equals the ideal negative alternative, $d_{iw} = 0$.

STEP 7. Rank the alternatives by their s_{iw} values.

Normalization

Wojciech Sałabun[10] presents four methods of normalization: 3 methods of linear normalization and the vector method we used in Step 2. Each method has variants for "profit" (larger is better) and "cost" (smaller is better). Vector normalization, which uses nonlinear distances between single dimension scores and ratios, should produce smoother trade-offs [HwangYoon1981].

[10] W. Sałaban, "Normalization of attribute values in TOPSIS method," Chapter 4 in Behzadian et al, *Nowe Trendy w Naukach Inżynieryjnych*, CREATIVETIME, 2012.

Strengths and Limitations

TOPSIS is based on the concept that the chosen alternative should have the shortest geometric distance from the positive ideal solution and the longest geometric distance from the negative ideal solution. See Figure 8.5.

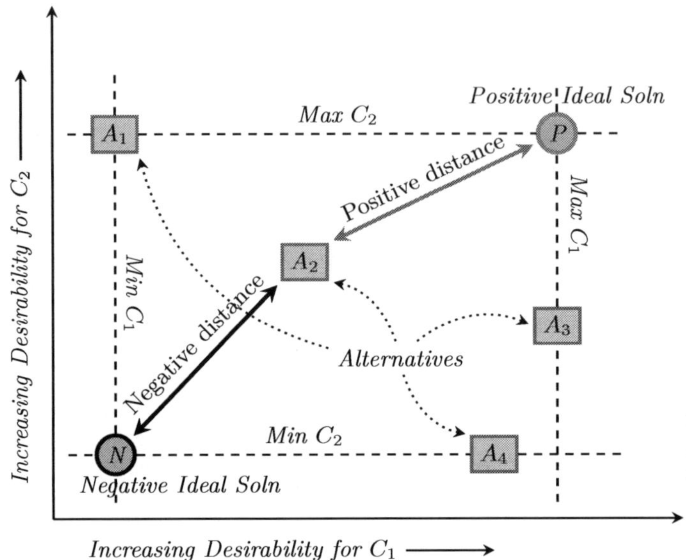

FIGURE 8.5: TOPSIS with Two Criteria

Two main advantages of TOPSIS are its ease of use and ease of implementation. A standard spreadsheet can handle the computations, and deep mathematical expertise is not required.

The main weakness of TOPSIS is that subjectivity in setting criteria and weights can inordinately influence the rankings produced. As always, sensitivity analysis is a must.

Sensitivity Analysis

Sensitivity analysis is essential to good modeling. The criterion weights are the main target of sensitivity analysis to determine how they affect the final ranking. The same procedures discussed for AHP are all useful for TOPSIS. We will again use Equation 8.4 (pg. 364) to adjust weights in our sensitivity analysis [AlinezhadAmini2011].

Examples using TOPSIS

We'll revisit examples from AHP and SAW so as to compare results, method efficacy, and computational complexity.

Example 8.8. Selecting a Car Redux.
The decision maker's weights used for AHP will be used here slightly modified with *cost* not inverted. Use the *PCM* matrix again, but with *cost* not inverted. The input data for the alternatives must be in the same order as the prioritized criteria.

> *Labels* := ["Prius", "Fusion", "Volt", "Camry", "Sonata", "Leaf"] :

$$AltM := \begin{bmatrix} 27.8 & 9.4 & 3 & 7.5 & 44 & 40 & 8.7 \\ 28.5 & 9.6 & 4 & 8.4 & 47 & 47 & 8.1 \\ 38.7 & 9.6 & 3 & 8.2 & 35 & 40 & 6.3 \\ 25.5 & 9.4 & 5 & 7.8 & 43 & 39 & 7.5 \\ 27.5 & 9.6 & 5 & 7.6 & 36 & 40 & 8.3 \\ 36.2 & 9.4 & 3 & 8.1 & 36.2 & 40 & 8.0 \end{bmatrix} :$$

DMWR := [0.3612, 0.2093, 0.1446, 0.1167, 0.0801, 0.0530, 0.0351] :

Use the *TOPSIS* program from the book's *PSMv2* package.

> *with*(*PSMv2*) :

> *Describe*(*TOPSIS*);

```
# TOPSIS(AltM, Weights, <Best=[max/min,...]>) returns the
# ranking of the alternatives given the Weights. Optional
# argument Best=[max/min,...] lists type of 'best' element
# for each criterion.
TOPSIS( Labels::Vector, list, A::Matrix, W::Vector, list,
{Best := [max$15] } )
```

> [*min*, *max*\$6];
Result := *TOPSIS*(*Labels*, *AltM*, *DMWR*, *Best* = %);

[*min*, *max*, *max*, *max*, *max*, *max*, *max*]

$$Result := \begin{bmatrix} \text{“}Prius\text{''} & \text{“}Fusion\text{''} & \text{“}Volt\text{''} & \text{“}Camry\text{''} & \text{“}Sonata\text{''} & \text{“}Leaf\text{''} \\ 0.6156 & 0.7159 & 0.06138 & 0.9087 & 0.8098 & 0.1766 \end{bmatrix}$$

The ranked order of alternatives are: Camry (0.9087), Sonata (0.8098), Fusion (0.7159), Prius (0.6156), Leaf (0.1766) and Volt (0.06138). *How does this compare to the AHP rankings?*

To begin sensitivity analysis, reduce the weight of *cost* by steps of 0.05, modifying the other weights linearly to keep $\sum w_i = 1$, until *cost* is overtaken as the highest weighted criteria. We find no changes in the rank-ordering of our alternatives until 10 steps.

```
> AdjVec := [−1, 1./6$6] :
  DMWR;
  NewDMWR := fnormal(DMWR + 0.05 · AdjVec, 4);
            [0.3612, 0.2093, 0.1446, 0.1167, 0.0801, 0.0530, 0.0351]
  NewDMWR := [0.3112, 0.2176, 0.1529, 0.1250, 0.08843, 0.06133, 0.04343]
> NewResult[1] := TOPSIS(Labels, AltM, NewDMWR,
      Best = [min, max$6])[2, ..];
   NewResult1 := [0.5728  0.6992  0.07139  0.8888  0.7947  0.1744]
> for i from 2 to 10 do
     NewDMWR := fnormal(NewDMWR + 0.05 · AdjVec, 4);
     NewResult[i] := TOPSIS(Labels, AltM, NewDMWR,
        Best = [min, max$6])[2, ..];
     end do :
> NewResult := Matrix(convert(convert~(NewResult, list), list)) :
> Statistics:-LineChart(NewResult, gridlines = true, symbolsize = 14,
     symbol = solidbox, title = "Sensitivity Analysis Output", font =
     [TIMES, 16], labels = ["New Weights", "Alternative Rankings"],
     labeldirections = [horizontal, vertical]);
```

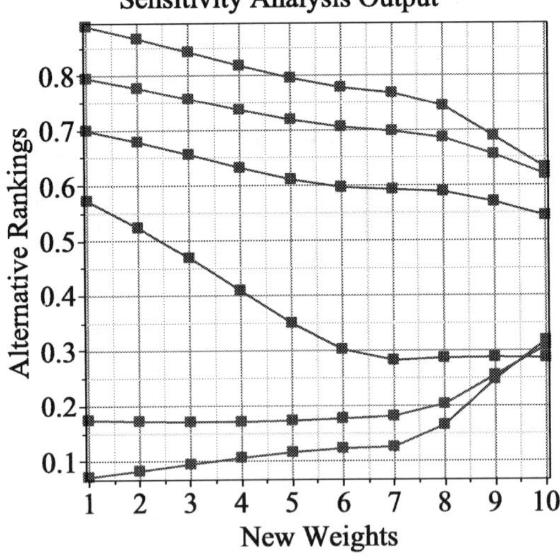

We see the rankings are stable until some switching at the 10th step.

Modifying the other weights and determining the results' sensitivity to those changes is left to the exercises.

Example 8.9. The Kite Network Redux.
Revisit analyzing the Kite Network, this time with TOPSIS, to find the main influencers in the network.

Use the same four criteria as Example 8.7 (pg. 370):

1. *Total Centrality*(*TC*) 2. *Betweenness*(*BTW*)
3. *Eigenvector Centrality*(*EC*) 4. *Closeness Centrality*(*CC*)

$$> CritLabels := [TC, BTW, EC, CC] :$$
$$AltLabels := [Bob, Claire, Fred, Sarah, Susan, Steve, Claudia, Tom] :$$
$$AltM := \begin{bmatrix} 0.11111111 & 0.019407559 & 0.114399403 & 0.100733634 \\ 0.11111111 & 0.019407559 & 0.114399403 & 0.100733634 \\ 0.08333333 & 0.0 & 0.093757772 & 0.09734763 \\ 0.125 & 0.104187947 & 0.137527978 & 0.100733634 \\ 0.180555556 & 0.202247191 & 0.175080826 & 0.122742664 \\ 0.138888889 & 0.15526047 & 0.137527978 & 0.112866817 \\ 0.083333333 & 0.317671093 & 0.104202935 & 0.108634312 \\ 0.055555556 & 0.181818182 & 0.024123352 & 0.088318284 \end{bmatrix} :$$

Use the same weights as in AHP which have a good CR of 0.01.

$$> DMWR := [0.4673, 0.2772, 0.1601, 0.09543] :$$

$$> TOPSIS(AltLabels, AltM, DMWR);$$
$$Result[1] := convert(\%[2, ..], list) :$$
$$MatrixSort(\%\%, 2, order = \text{'descending'});$$
$$\begin{bmatrix} Susan & Claudia & Steve & Sarah & Tom & Claire & Bob & Fred \\ 0.7648 & 0.5850 & 0.5801 & 0.4576 & 0.3457 & 0.3033 & 0.3033 & 0.1766 \end{bmatrix}$$

To begin sensitivity analysis, change the weight for the largest criteria, *Total Centrality.* Adjust the other weights linearly to keep $\sum w_i = 1$.

```
> AdjVec := [−1, 1./3$3] :
  NewDMWR := fnormal(DMWR + 0.05 · AdjVec, 4) :
  Result[2] := convert(TOPSIS(AltLabels, AltM, NewDMWR)[2, ..], list) :
> for i from 3 to 8 do
    NewDMWR := fnormal(NewDMWR + 0.05 · AdjVec, 4);
    Result[i] := convert(TOPSIS(AltLabels, AltM, NewDMWR)[2, ..], list);
  end do :
> NewResult := Matrix(convert(Result, list)) :
```

```
> Statistics:-LineChart(NewResult, gridlines = true, symbolsize = 14,
    symbol = solidbox, title = "Sensitivity Analysis Output", font =
    [TIMES, 16], labels = ["New Weights", "Alternative Rankings"],
    labeldirections = [horizontal, vertical]);
```

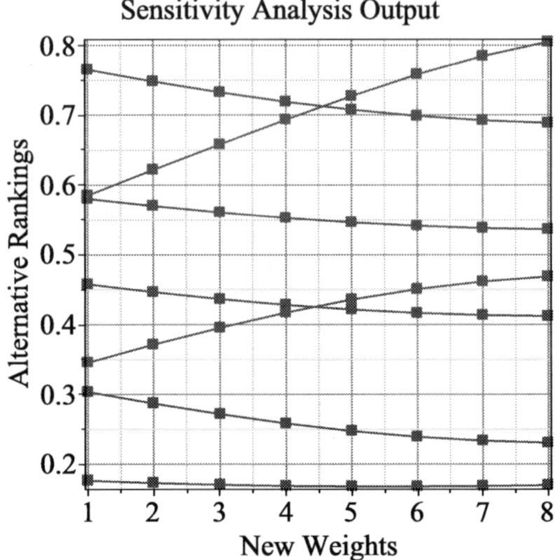

The plot of the results from sensitivity analysis shows that two sets of alternatives change place. This shift is a significant change, and again emphasizes the importance of sensitivity analysis.

Exercises

1. Redo Section 8.3's Exercises (pg. 359) using TOPSIS. Compare with your previous results using SAW and using AHP.

2. In each of the problems above, perform sensitivity analysis by changing the weight of your dominant criteria until it is no longer the highest. Did the change affect your rankings?

3. In each of the problems above, find break points, if any exist, for the weights.

4. Suppose the dominant criterion has no break point. What does this indicate about the sensitivity of the solution to that criterion's weight?

5. Suppose the least weighted criterion has no break point. Can this criterion be eliminated without affecting the rankings?

Projects

Project 1. Write a program using the technology of your choice to implement:

(a) SAW,

(b) AHP, and

(c) TOPSIS.

Project 2. Enhance your program to perform sensitivity analysis.

Project 3. Perform and discuss a comparative analysis using the Kite network and each MADM method.

8.6 Methods of Choosing Weights

Last, we consider several standard methods for choosing the weights for SAW, AHP, and TOPSIS.

Rank Order Centroid Method

The *rank order centroid* (ROC) method is a simple way of giving weights to a number of items ranked according to their importance. In general, decision makers can rank items much more easily and quickly than assigning weights. This method takes those ranks as inputs, and converts them to weights for each of m items based on the following schema.

1. List the m items in rank-order from most important to least important.

2. Assign the ith item in the list the weight w_i where

$$w_i = \frac{1}{m}\sum_{k=i}^{m}\frac{1}{k}$$

.

E.g., for four items:

$> m := 4 :$

$weights := \left[seq\left(w_i = \frac{1}{m} \cdot sum\left(\frac{1}{k}, k = i..m\right), i = 1..m\right)\right];$

$add(weights);$

$$weights := \left[w_1 = \frac{25}{48}, w_2 = \frac{13}{48}, w_3 = \frac{7}{48}, w_4 = \frac{1}{16}\right]$$

$$w_1 + w_2 + w_3 + w_4 = 1$$

ROC is simple and easy to follow, but it gives weights which are highly dispersed. As an example, consider weighting the factors in the example below.

Example 8.10. Agency Project.
Suppose an agency has selected four main criteria and put them in rank order as

1. Shortening Schedule, 2. Project Cost, 3. Agency Control, 4. Competition

Using ROC to rank the four criteria based on their importance and influence as listed gives the ranks calculated above:

$$w_1 = 0.52, w_2 = 0.27, w_3 = 0.15, \text{ and } w_4 = 0.06.$$

These weights almost eliminate the effect of the fourth factor, encouraging competition. This ranking could prove to be an issue.

Ratio Method

The Ratio Method is another simple way of calculating weights for a selection of critical factors. The decision maker first rank-orders all items according to their importance. Next, each item is given an initial weight based on its rank beginning with the lowest ranked item given an initial weight of 10. The decision maker moves up the ranking giving each successive item a weight that is a multiple of 10 and is an amount greater than the previous indicating the item's relative importance. Last, the raw weights are normalized by dividing each by the sum of all. Each increase in weight is based on the differences between the items' importance and is a subjective judgment of the decision maker. Ranking the items in the first step helps to assign more accurate weights.

Rating the factors from Example 8.10 using the ratio method with initial weights assigned by the Project Manager gives

Criterion	1	2	3	4
Raw Weight	50	40	20	10
Normalized Weight	0.417	0.333	0.167	0.083

While just as easy as ROC, the ratio method also suffers from dispersed weightings.

Pairwise Comparison

We've used pairwise comparison in several examples of SAW, AHP, and TOPSIS. The main task of pairwise comparison is to rank each item's relative importance to other items individually, building a pairwise comparison matrix. We have used Saaty's ranking system of 1 (equally important) to 9 (extremely more important) to assess relative importance. (See Table 8.10, pg. 362.) The reverse relation uses reciprocals. For instance, if A is 3 times more important than B, then B must be $1/3$ as important as A. Using reciprocals makes the pairwise comparison matrix a *positive reciprocal matrix* which must have a dominant eigenvalue with an associated eigenvector. That eigenvector, when normalized, gives a normalized set of weights. The eigenvalue helps determine the consistency of the decision maker's ratings. Further analysis of methods and applications of consistency measurement appears in [Temesi2006].

Suppose we have the following pairwise comparison matrix from Example 8.10 with the criteria listed in rank order.

$$> PCM := \begin{bmatrix} 1 & 5 & 5/2 & 8 \\ 1/5 & 1 & 1/2 & 1 \\ 2/5 & 2 & 1 & 2 \\ 1/8 & 1 & 1/2 & 1 \end{bmatrix}:$$

$> AnalyzeComparisonMat(PCM);$

$$[[\ 0.5927 \quad 0.1046 \quad 0.2092 \quad 0.09356\ ], 0.01037]$$

Note: The eigenvector can be quickly approximated by taking row-sums of PCM and dividing by the sum of all elements. This quick technique gives the criteria weights of $[0.60, 0.10, 0.20, 0.10]$ for PCM—very close to the eigenvector weights. Note that quite often different factors will have the same, or very close, weights.

The Entropy Method

In 1947, Shannon and Weaver[11] developed the entropy method for weighting criteria. Entropy, in the context of weightings, is the probabilistic measure of uncertainty in the information. A broad distribution represents more uncertainty in specific values than a sharply peaked distribution. To see this concept, compare the graphs of two normal distributions with mean 0, one having a small standard deviation and the other a large. The entropy method computations can be decomposed into simple steps. Begin with a matrix $\mathbf{M}_{m\times n}$ ranking the alternatives.

STEP 1. Calculate $\mathbf{R} = [r_{ij}]$ which is $\mathbf{M}$ normalized by dividing each column by its column-sum.

STEP 2. Compute the *degree of divergence* vector $e = [e_j]$ for $j = 1..m$ with

$$e_j = \frac{-1}{\ln(n)} \cdot \sum_{i=1}^{n} r_{ij} \ln(r_{ij}).$$

[11]See C. Shannon and W. Weaver, "The Mathematical Theory of Communication," Univ. Illinois Press, 1949.

STEP 3. The normalized weights $w = [w_j]$ are then given by

$$w_j = \frac{1 - e_j}{\sum_{k=1}^{m}(1 - e_k)}.$$

The degree of divergence vector $d = [d_j] = 1 - e$ indicates the average amount of information contained by each attribute with higher being more.

The book's *PSMv2* package contains *EntropyWeights* to calculate weights from a matrix via the entropy method and *Normalize* to normalize a matrix via column-sums. As an example, use the *AltM* preference matrix from the Car Selection example (pg. 377) above.

$$> AltM := \begin{bmatrix} 27.8 & 9.4 & 3 & 7.5 & 44 & 40 & 8.7 \\ 28.5 & 9.6 & 4 & 8.4 & 47 & 47 & 8.1 \\ 38.7 & 9.6 & 3 & 8.2 & 35 & 40 & 6.3 \\ 25.5 & 9.4 & 5 & 7.8 & 43 & 39 & 7.5 \\ 27.5 & 9.6 & 5 & 7.6 & 36 & 40 & 8.3 \\ 36.2 & 9.4 & 3 & 8.1 & 36.2 & 40 & 8.0 \end{bmatrix} :$$

> *EntropyWeights*(*AltM*);

$$\begin{bmatrix} 0.2300 & 0.0010 & 0.4995 & 0.0155 & 0.1227 & 0.0389 & 0.0924 \end{bmatrix}$$

Compare these weights with the other methods' values. These weights can be used in our methods, SAW, AHP, or TOPSIS, in lieu of the pairwise comparison weight method.

Comparison

Table 8.11 shows the weights each method produces for the Agency Project of Example 8.10.

TABLE 8.11: Comparing Weight Generating Methods

Method	Schedule	Project Cost	Agency Control	Competition
ROC	0.52	0.272	0.12	0.06
Ratio	0.417	0.333	0.167	0.083
Pairwise Comp	0.593	0.105	0.209	0.0936
Entropy	0.2492	0.203	0.203	0.345

Exercises

Repeat the exercises from Sections 8.3, 8.4, and 8.5. In each case, compare your new results to the results found using the pairwise comparison method.

1. Use the *rank order centroid method* to generate weights.

2. Use the *ratio method* to generate weights.

3. Use the *entropy method* to generate weights.

Projects

Project 1. Prepare a presentation on "State of Art Surveys of Overviews on MCDM/MADM Methods" by Zavadskas, Turskis, and Kildienė in *Technological And Economic Development Of Economy*, 2014 Vol. 20(1), 165–179.

Project 2. Mary Burton (www.maryburton.com) has discovered that she is being stalked by a deranged editor after switching her successful book, *The Last Move*, to a new publisher. She has decided to quietly relocate to one of the Windward Islands of the Caribbean. Her main criteria are:

- Unemployment rate (as a proxy for crime rate)
- Military expenditures (as a proxy for security)
- Airport capacity
- GDP per capita
- Population growth rate

Choose a preference order for her criteria; explain your choices. Consult the CIA's *World Factbook* "Country Comparisons"[12] to compile data on each island. Select the best island using SAW, then re-select using AHP and TOPSIS; use the different weight selection methods with each technique. Write a report to Ms. Burton describing your preference choices and the islands (with their preference ordering) you recommend. After sending your report and verifying receipt, destroy all of your work so that the unhinged editor cannot use your recommendations to track her.

Project 3. Each year, the *US News and World Report* produces a ranking of the "Best States" in the US.[13] Load the 2019 data from the book's *PSMv2*

[12] Available at https://www.cia.gov/library/publications/resources/the-world-factbook/.
[13] Available at https://www.usnews.com/news/best-states/rankings.

package by executing

```
> with(PSMv2):
  Describe(GetStateRankings);
  TheRankings := GetStateRankings();
```

Examine the criteria used and create your own preference ordering. Use each of the methods of generating weights and each of the MADM methods to rank the 50 states. Prepare a standard conference $4' \times 3'$ poster to describe your results and work.

References and Further Reading

[AlinezhadAmini2011] Alireza Alinezhad and Abbas Amini, *Sensitivity Analysis of TOPSIS Technique: The Results of Change in the Weight of One Attribute on the Final Ranking of Alternatives*, *J. Optimization in Industrial Engineering* **Volume 4** (2011), no. 7, 23–28.

[BarkerZabinsky2011] T. Barker and Z. Zabinsky, *A Multicriteria Decision Making Model for Reverse Logistics Using Analytical Hierarchy Process*, *Omega* **39** (2011), no. 5, 558–573.

[Bauldry2009a] W Bauldry, *Data Envelopment Analysis with Maple in Operations Research and Modeling Courses*, Proc. ICTCM 21, Addison-Wesley, 2009.

[BurdenFaires2005] R. Burden and J. D. Faires, *Numerical Analysis*, 8th ed., Thomson Books, 2005.

[ButlerJiaDyer1997] J. Butler, J. Jia, and J. Dyer, *Simulation Techniques for the Sensitivity Analysis of Multi-Criteria Decision Models*, *European J. Operational Research* **103** (1997), no. 3, 531–546.

[Callen1991] J. Callen, *Data Envelopment Analysis: Partial Survey and Applications for Management Accounting*, *J. Management Accounting Research* **3** (1991), 35–26.

[ORA2011] K. Carley, *Organizational Risk Analyzer (ORA)*, Carnegie Mellon University, 2011.

[CharnesCooperRhodes1978] A. Charnes, W. Cooper, and E. Rhodes, *Measuring the Efficiency of Decision-Making Units, European J. Operational Research,* **2** (1978), 429–444.

[ChenKocaoglu2008] H. Chen and D.F. Kocaoglu, *A Sensitivity Analysis Algorithm for Hierarchical Decision Models*, *European J. Operational Research* **185** (2008), no. 1, 266–288.

[ChurchmanAckoff1954] C. Churchman and R. Ackoff, "An Approximate Measure of Value," *J. Operations Research Soc. Am.*, 2, No. 2, (1954).

[CooperLSTTZ2001] W. Cooper, S. Li, L. Seiford, K. Tone, R. Thrall, and J. Zhu, *Sensitivity and Stability Analysis in DEA: Some Recent Developments, J. Productivity Analysis* **15** (2001), 217–246.

[CooperSeifordTone2007] W. Cooper, L. Seiford, and K. Tone, *Data Envelopment Analysis: A Comprehensive Text with Models, Applications, References and DEA-Solver Software*, Kluwer Academic Publishers, 2007.

[Figueroa2014] S. Figueroa, *Improving Recruiting in the 6th Recruiting Brigade through Statistical Analysis and Efficiency Measures*, Master's thesis, Naval Postgraduate School, Monterey, CA, Dec 2014.

[Fishburn1967] P. Fishburn, "Additive Utilities with Incomplete Product Set: Applications to Priorities and Assignments," *Operations Research Soc. Am (ORSA),* 15 (1967), 537–542.

[Fox2014] W. Fox, *Using Mathematical Models in Decision Making Methodologies to Find Key Nodes in the Noordin Dark Network, Am. J. Operations Research* **4** (2014), 255–267.

[FoxEverton2013] W. Fox and S. Everton, *Mathematical Modeling in Social Network Analysis: Using TOPSIS to Find Node Influences in a Social Network, J. Mathematics and System Science* **3** (2013), 531–541.

[FoxEverton2015] W. Fox and S. Everton, *Using Data Envelopment Analysis and Analytical Hierarchy Process to Find Node Influences in a Social Network, J. Defense Modeling and Simulation* **12** (2015), no. 2, 157–165.

[GiordanoFoxHorton2014] F. Giordano, W. Fox, and S. Horton, *A First Course in Mathematical Modeling*, 5th ed., Nelson Education, 2014.

[Hartwich1999] F. Hartwich, *Weighting of Agricultural Research Results: Strength and Limitations of the Analytic Hierarchy Process (AHP)*, 1999.

[Hurley2001] W. Hurley, *The Analytic Hierarchy Process: A Note on an Approach to Sensitivity which Preserves Rank Order, Comput. Operations Research* **28** (2001), no. 2, 185–188.

[HwangLL1993] C.-L. Hwang, Y.-J. Lai, and T.-Y. Liu, *A New Approach for Multiple Objective Decision Making, Computers & Operations Research* **20** (1993), no. 8, 889–899.

[HwangYoon1981] C.-L. Hwang and K. Yoon, *Multiple Attribute Decision Making*, Lecture Notes in Economics and Mathematical Systems, vol. 186, Springer-Verlag, 1981.

[Krackhardt1990] D. Krackhardt, *Assessing the Political Landscape: Structure, Cognition, and Power in Organizations, Administrative Science Quarterly* **35** (1990), no. 2, 342–369.

[Leonelli2012] R. Leonelli, *Enhancing a Decision Support Tool with Sensitivity Analysis*, Master's thesis, University of Manchester, 2012.

[Neralic1998] L. Neralić, *Sensitivity Analysis in Models of Data Envelopment Analysis, Mathematical Communications* **3** (1998), 41–59.

[Saaty1980] T. Saaty, *The Analytic Hierarchy Process*, McGraw-Hill, 1980.

[Temesi2006] J. Temesi, "Consistency of the decision maker in Pair-Wise Comparisons," *Int. J. Management and Decision Making,* 7, No. 2/3, (2006), pg. 267–274.

[Thanassoulis2001] E. Thanassoulis, *Introduction to the Theory and Application of Data Envelopment Analysis*, Springer Science+Business Media, 2001.

[Trick2014] M. Trick, *Data Envelopment Analysis.* URL: http://mat.gsia.cmu.edu/classes/QUANT/NOTES/chap12.pdf. Accessed June, 2019.

[Trick1996] M. Trick, *Multiple Criteria Decision Making for Consultants.* URL: http://mat.gsia.cmu.edu/classes/mstc/multiple/multiple.html. Accessed June, 2019.

[Winston2002] W. Winston, *Introduction to Mathematical Programming Applications and Algorithms*, 4th ed., Duxbury Press, 2002.

[Yoon1987] K. Yoon, *A Reconciliation Among Discrete Compromise Situations, J. Operational Research Society* **38** (1987), 277–286.

[Zhenhua2009] G. Zhenhua, *The Application of DEA/AHP Method to Supplier Selection, Proc. 2009 International Conference on Information Management, Innovation Management and Industrial Engineering*, vol. 2, IEEE Computer Society, 2009, pp. 449–451.

Index